Stochastic Modelling for Systems Biology

Third Edition

Chapman & Hall/CRC Mathematical and Computational Biology

About the Series

This series aims to capture new developments and summarize what is known over the entire spectrum of mathematical and computational biology and medicine. It seeks to encourage the integration of mathematical, statistical, and computational methods into biology by publishing a broad range of textbooks, reference works, and handbooks. The titles included in the series are meant to appeal to students, researchers, and professionals in the mathematical, statistical, and computational sciences and fundamental biology and bioengineering, as well as interdisciplinary researchers involved in the field. The inclusion of concrete examples and applications and programming techniques and examples is highly encouraged.

Series Editors

Xihong Lin
Mona Singh
N. F. Britton
Anna Tramontano
Maria Victoria Schneider
Nicola Mulder

Chromatin: Structure, Dynamics, Regulation
Ralf Blossey

Mathematical Models of Plant-Herbivore Interactions
Zhilan Feng, Donald DeAngelis

Computational Exome and Genome Analysis
Peter N. Robinson, Rosario Michael Piro, Marten Jager

Gene Expression Studies Using Affymetrix Microarrays
Hinrich Gohlmann, Willem Talloen

Big Data in Omics and Imaging
Association Analysis
Momiao Xiong

Introduction to Proteins
Structure, Function, and Motion, Second Edition
Amit Kessel, Nir Ben-Tal

Big Data in Omics and Imaging
Integrated Analysis and Causal Inference
Momiao Xiong

Computational Blood Cell Mechanics
Road Towards Models and Biomedical Applications
Ivan Cimrak, Iveta Jancigova

For more information about this series please visit: https://www.crcpress.com/ Chapman--*HallCRC*-Mathematical-and-Computational-Biology/book-series/ CHMTHCOMBIO

Stochastic Modelling
for Systems Biology

Third Edition

Darren J. Wilkinson

CRC Press
Taylor & Francis Group
Boca Raton London New York

CRC Press is an imprint of the
Taylor & Francis Group, an **informa** business

A CHAPMAN & HALL BOOK

CRC Press
Taylor & Francis Group
6000 Broken Sound Parkway NW, Suite 300
Boca Raton, FL 33487-2742

First issued in paperback 2020

ISBN-13: 978-1-138-54928-9 (hbk)
ISBN-13: 978-0-367-65693-5 (pbk)

Library of Congress Cataloging-in-Publication Data

Names: Wilkinson, Darren James, author.
Title: Stochastic modelling for systems biology / by Darren J. Wilkinson.
Description: Third edition. | Boca Raton, Florida : CRC Press, [2019] |
Series: Chapman & Hall/CRC mathematical and computational biology |
Includes bibliographical references and index.
Identifiers: LCCN 2018040390| ISBN 9781138549289 (hardback : alk. paper) |
ISBN 9781351000918 (e-book)
Subjects: LCSH: Biological systems--Mathematical models. | Systems biology.
Classification: LCC QH323.5 .W54 2019 | DDC 570--dc23
LC record available at https://lccn.loc.gov/2018040390

Visit the Taylor & Francis Web site at
http://www.taylorandfrancis.com

and the CRC Press Web site at
http://www.crcpress.com

Contents

Author

Darren J. Wilkinson is professor of stochastic modelling at Newcastle University in the United Kingdom. He was educated at the nearby University of Durham, where he took his first degree in mathematics followed by a PhD in Bayesian statistics which he completed in 1995. He moved to a lectureship in statistics at Newcastle University in 1996, where he has remained since, being promoted to his current post in 2007. Professor Wilkinson is interested in computational statistics and Bayesian inference and in the application of modern statistical technology to problems in statistical bioinformatics and systems biology. He is also interested in some of the 'big data' challenges that arise in bioscience and more generally. He currently serves on Biotechnology and Biological Sciences Research Council's Strategy Advisory Panel for Exploiting New Ways of Working, and is co-director of Newcastle's Engineering and Physical Sciences Research Council Centre for Doctoral Training in Cloud Computing for Big Data. He is also a fellow of the Alan Turing Institute for data science and artificial intelligence.

Acknowledgments

I would like to acknowledge the support of everyone at Newcastle University who is involved with systems biology research. Unfortunately, there are far too many people to mention by name, but particular thanks are due to everyone involved in the BBRSC CISBAN project, without whom the first edition of this book would never have been written. Particular thanks are also due to all of the students who have been involved in the MSc in bioinformatics and computational systems biology programme at Newcastle, and especially those who took my course on stochastic systems biology, as it was the teaching of that course which persuaded me that it was necessary to write this book.

The production of the second edition of this book was greatly facilitated by funding from the Biotechnology and Biological Sciences Research Council, both through their funding of CISBAN (BBSRC grant number BBC0082001) and the award to me of a BBSRC Research Development Fellowship (grant number BBF0235451). In addition, a considerable amount of work on this second edition was carried out during a visit I made to the Statistical and Applied Mathematical Sciences Institute in North Carolina during the spring of 2011, as part of their research programme on the Analysis of Object-Oriented Data.

The development of the third edition was facilitated by two visits to the The Isaac Newton Institute for Mathematical Sciences, especially the programme stochastic dynamical systems in biology (SDB) during 2016 (grant numbers EP/K032208/1 and EP/R014604/1).

Last, but by no means least, I would like to thank my family for supporting me in everything that I do.

Preface to the third edition

I was fairly pleased with the way that the second edition of this book, published in 2011, turned out. In particular, the incorporation of the R package, smfsb, proved to be a very popular development, improving the accessibility and reproducibility of the algorithms described. With this third edition I have not substantially re-written the existing text, but instead concentrated on incorporating new material which covers some important topics missing from the second edition, together with appropriately extending the software and accompanying documentation associated with the book.

The first development relates to associated software, which has little impact on the text, but has potentially significant impact on the way that the book is used for both teaching and research. For example, I've produced a small software library for reading SBML models into R for subsequent analysis and simulation using the smfsb R package. So it is now possible to directly use SBML models (designed with discrete stochastic simulation in mind) as the basis for simulation experiments. This was a fairly obvious omission from the second edition of this book, but at the time of producing the second edition the software libraries required to facilitate this (libSBML R bindings) were too immature to make this practical. There are other improvements and developments relating to software associated with the book, and these are now all discussed in a new Appendix. Further information is provided about tools for working with the 'SBML shorthand' notation I use in the text, and there is a description of a new software library associated with the text. This library, written in a fast, efficient, compiled language, Scala, replicates most of the functionality associated with the R packages, and therefore provides a convenient route for people wanting to move beyond working in (slow) dynamically typed languages like R to better, more efficient, general–purpose languages. This will probably be more relevant to researchers than students, but should prove to be very useful for people needing to make that transition.

The second significant addition is the inclusion of a new chapter on spatially extended systems (stochastic reaction–diffusion models). Again, this was a fairly obvious omission from the second edition, but for that edition I really wanted to concentrate on ensuring that there was adequate coverage of the well-mixed case. Spatial stochastic modelling is a huge topic, and this new chapter barely scratches the surface, but some of the most important topics are now covered, and code is included that will allow readers to easily take a well–mixed stochastic biochemical reaction network model and investigate its behaviour as a reaction–diffusion system in 1 and 2 dimensions.

The third major addition concerns the chapter on inference for stochastic kinetic models. An important addition for the second edition was discussion of likelihood–

free methods and particle MCMC, including some example code. These sections have been significantly updated for the third edition, and new sections have been added covering the use of approximate Bayesian computation (ABC) methods for parameter inference, including sequential ABC methods, and associated code. This reflects the increasing use of ABC methods in systems biology in recent years.

There are also numerous other minor updates throughout. Hopefully these significant new additions to the text and the associated software ecosystem improve the range of courses for which this text is applicable.

Darren J. Wilkinson
Newcastle upon Tyne

Preface to the second edition

I was keen to write a second edition of this book even before the first edition was published in the spring of 2006. The first edition was written during the latter half of 2004 and the first half of 2005 when the use of stochastic modelling within computational systems biology was still very much in its infancy. Based on an inter-disciplinary masters course I was teaching, I saw an urgent need for an introductory textbook in this area, and tried in the first edition to lay down all of the key ingredients needed to get started. I think that I largely succeeded, but the emphasis there was very much on the 'bare essentials' and accessibility to non-mathematical readers, and my goal was to get the book published in a timely fashion, in order to help advance the field. I would like to think that the first edition of this text has played a small role in helping to make stochastic modelling a much more mainstream part of computational systems biology today. But naturally there were many limitations of the first edition. There were several places where I would have liked to have elaborated further, providing additional details likely to be of interest to the more mathematically or statistically inclined reader. Also, the latter chapters on inference from data were rather limited and lacking in concrete examples. This was partly due to the fact that the whole area of inference for stochastic kinetic models was just developing, and so it wasn't possible to give a coherent overview of the problem from an introductory viewpoint. Since publishing the first edition there have been many interesting developments in the use of 'likelihood-free' methods of Bayesian inference for complex stochastic models, and so the latter chapters have now been re-written to reflect this more modern perspective, including a detailed case study accompanied by working code examples.

Of course the whole field has moved on considerably since 2005, and so the second edition is also an opportunity to revise and update, and to change the emphasis of the text slightly. The Systems Biology Markup Language (SBML) has continued to evolve, and SBML Level 3 is now finalised. Consequently, I have updated all of the examples to Level 3, which is likely to remain the standard encoding for dynamic biological models for the foreseeable future. I have also taken the opportunity to revise and update the R code examples associated with the book, and to bundle them all together as an R package (`smfsb`). This should make it much easier for people to try out the examples given in the book. I have also re-written and re-structured all of the code relating to simulation, analysis, and inference for stochastic kinetic models. The code is now structured in a more modular way (using a functional programming style), making it easy to 'bolt together' different models, simulation algorithms, and analysis tools. I've created a new website specific to this second edition, where I will keep links, resources, an errata, and up-to-date information on installation and use of the associated R package.

The new edition contains more background material on the theory of Markov processes and stochastic differential equations, providing more substance for mathematically inclined readers. This allows discussion of some of the more advanced concepts relating to stochastic kinetic models, such as random time-change representations, Kolmogorov equations, Fokker–Planck equations and the linear noise approximation. It also enables simple modelling of 'extrinsic' in addition to 'intrinsic' noise. This should make the text suitable for use in a greater range of courses. Naturally, in keeping with the spirit of the first edition, all of the new theory is presented in a very informal and intuitive way, in order to keep the text accessible to the widest possible readership. This is not a rigorous text on the theory of Markov processes (there are plenty of other good texts in that vein) — the book is still intended for use in courses for students with a life sciences background.

I've also updated the references, and provided new pointers to recent publications in the literature where this is especially pertinent. However, it should be emphasised that the book is not intended to provide a comprehensive survey of the stochastic systems biology literature — I don't think that is necessary (or even helpful) for an introductory textbook, and I hope that people working in this area accept this if I fail to cite their work.

So here it is, the second edition, completed at last. I hope that this text continues to serve as an effective introduction to the area of stochastic modelling in computational systems biology, and that this new edition adds additional mathematical detail and computational methods which will provide a stronger foundation for the development of more advanced courses in stochastic biological modelling.

Darren J. Wilkinson
Newcastle upon Tyne, June 2011

Preface to the first edition

Stochastic models for chemical and biochemical reactions have been around for a long time. The standard algorithm for simulating the dynamics of such processes on a computer (the 'Gillespie algorithm') was published nearly 30 years ago (and most of the relevant theory was sorted out long before that). In the meantime there have been dozens of papers published on stochastic kinetics, and several books on stochastic processes in physics, chemistry, and biology. Biological modelling and biochemical kinetic modelling have been around even longer. These distinct subjects have started to merge in recent years as technology has begun to give real insight into intra-cellular processes. Improvements in experimental technology are enabling quantitative real-time imaging of expression at the single-cell level, and improvement in computing technology is allowing modelling and stochastic simulation of such systems at levels of detail previously impossible. The message that keeps being repeated is that the kinetics of biological processes at the intra-cellular level are stochastic, and that cellular function cannot be properly understood without building that stochasticity into *in silico* models. It was this message that first interested me in systems biology and in the many challenging statistical problems that follow naturally from this observation.

It was only when I came to try and teach this interesting view of computational systems biology to graduate students that I realised there was no satisfactory text on which to base the course. The papers assumed far too much background knowledge, the standard biological texts didn't cover stochastic modelling, and the stochastic processes texts were too far removed from the practical applications of systems biology, paying little attention to stochastic biochemical kinetics. Where stochastic models do crop up in the mainstream systems biology literature, they tend to be treated as an add-on or after-thought, in a slightly superficial way. As a statistician I see this as problematic. The stochastic processes formalism provides a beautiful, elegant, and coherent foundation for chemical kinetics, and there is a wealth of associated theory every bit as powerful and elegant as that for conventional continuous deterministic models. Given the increasing importance of stochastic models in systems biology, I thought it would be particularly appropriate to write an introductory text in this area from this perspective.

This book assumes a basic familiarity with what might be termed high school mathematics. That is, a basic familiarity with algebra and calculus. It is also helpful to have had some exposure to linear algebra and matrix theory, but not vital. Since the teaching of probability and statistics at school is fairly patchy, essentially nothing will be assumed, though obviously a good background in this area will be very helpful. Starting from here, the book covers everything that is necessary for a good appreciation of stochastic kinetic modelling of biological networks in the systems biology context. There is an emphasis on the necessary probabilistic and stochastic methods, but the theory is rooted in the intended application, and no time is wasted covering interesting theory that is not necessary for stochastic kinetic modelling. On the other hand, more-or-less everything that is necessary is covered, and the text (at least up to Chapter 8) is intended to be self-contained. The final chapters are necessarily a little more technical in nature, as they concern the difficult problem of inference for stochastic kinetic models from experimental data. This is still an active research area, and so the main aim here is to give pointers to the existing literature and provide enough background information to render that literature more accessible to the non-specialist.

The decision to make the book practically oriented necessitated some technological choices that will not suit everyone. The two key technologies chosen for illustrating the theory in this book are SBML and R. I hope that the choice of the Systems Biology Markup Language (SBML) for model representation is not too controversial. It is the closest thing to a standard that exists in the systems biology area, and there are dozens of software tools that support it. Of course, most people using SBML are using it to encode continuous deterministic models. However, SBML Level 2 and beyond are perfectly capable of encoding discrete stochastic models, and so one of the reasons for using it in this text is to provide some working examples of SBML models constructed with discrete stochastic simulation in mind.

The other technological choice was the use of the statistical programming language R. This is likely to be more controversial, as there are plenty of other languages that could have been used. It seemed to me that in the context of a textbook, using a very high-level language was most appropriate. In the context of stochastic modelling, a language with good built-in mathematical and statistical support also seemed highly desirable. R stood out as being the best choice in terms of built-in language support for stochastic simulation and statistical analysis. It also has the great advantage over some of the other possible choices of being completely free open-source software, and therefore available to anyone reading the book. In addition, R is being used increasingly in bioinformatics and other areas of systems biology, so hopefully for this reason too it will be regarded as a positive choice.

This book is intended for a variety of audiences (advanced undergraduates, graduate students, postdocs, and academics from a variety of backgrounds), and exactly how it is read will depend on the context. The book is certainly suitable for a variety of graduate programs in computational biology. I will be using it as a text for a second-semester course on a masters in bioinformatics programme that will cover much of Chapters 1, 2, 4, 5, 6, and 7 (most of the material from Chapter 3 will be covered in a first-semester course). However, the book has also been written with self-study in mind, and here it is intended that the entire book be read in sequence, with some chapters skipped depending on background knowledge. It is intended to be suitable for computational systems biologists from a continuous deterministic background who would like to know more about the stochastic approach, as well as for statisticians who are interested in learning more about systems biology (though statisticians will probably want to skip most of Chapters 3, 4, and 5). It is worth pointing out that Chapters 9 and 10 will be easier to read with a reasonable background in probability and statistics. Though it should be possible to read these chapters without such a background and still appreciate the key concepts and ideas, some of the technical details may be difficult to understand fully from an elementary viewpoint.

Writing this book has been more effort than I anticipated, and there were one or two moments when I doubted the wisdom of taking the project on. On the whole, however, it has been an interesting and rewarding experience for me, and I am pleased with the result. I know it is a book that *I* will find useful and will refer to often, as it integrates a fairly diverse literature into a single convenient and notationally consistent source. I can only hope that others share the same view, and that the book will help make a stochastic approach to computational systems biology more widely appreciated.

Darren J. Wilkinson
Newcastle upon Tyne, October 2005

## PART I

Modelling and networks

CHAPTER 1

Introduction to biological modelling

1.1 What is modelling?

Modelling is an attempt to describe, in a precise way, an understanding of the elements of a system of interest, their states, and their interactions with other elements. The model should be sufficiently detailed and precise so that it can in principle be used to simulate the behaviour of the system on a computer. In the context of molecular cell biology, a model may describe (some of) the mechanisms involved in transcription, translation, gene regulation, cellular signalling, DNA damage and repair processes, homeostatic processes, the cell cycle, or apoptosis. Indeed any biochemical mechanism of interest can, in principle, be modelled. At a higher level, modelling may be used to describe the functioning of a tissue, organ, or even an entire organism. At still higher levels, models can be used to describe the behaviour and time evolution of populations of individual organisms.

The first issue to confront when embarking on a modelling project is to decide on exactly which features to include in the model, and in particular, the *level* of detail the model is intended to capture. So, a model of an entire organism is unlikely to describe the detailed functioning of every individual cell, but a model of a cell is likely to include a variety of very detailed descriptions of key cellular processes. Even then, however, a model of a cell is unlikely to contain details of every single gene and protein.

Fortunately, biologists are used to thinking about processes at different scales and different levels of detail. Consider, for example, the process of photosynthesis. When studying photosynthesis for the first time at school, it is typically summarised by a single chemical reaction mixing water with carbon dioxide to get glucose and oxygen (catalysed by sunlight). This could be written very simply as

$$\text{Water} + \text{Carbon Dioxide} \overset{\text{Sunlight}}{\longrightarrow} \text{Glucose} + \text{Oxygen},$$

or more formally by replacing the molecules by their chemical formulae and balancing to get

$$6H_2O + 6CO_2 \longrightarrow C_6H_{12}O_6 + 6O_2.$$

Of course, further study reveals that photosynthesis consists of many reactions, and that the single reaction was simply a summary of the overall effect of the process. However, it is important to understand that the above equation is not really *wrong*, it just represents the overall process at a higher level than the more detailed description that biologists often prefer to work with. Whether a single overall equation or a full breakdown into component reactions is necessary depends on whether intermediaries such as ADP and ATP are elements of interest to the modeller. Indeed, really accurate modelling of the process would require a model far more detailed and complex than most biologists would be comfortable with, using molecular dynamic simulations that explicitly manage the position and momentum of every molecule in the system.

The 'art' of building a good model is to capture the essential features of the biology without burdening the model with non-essential details. Every model is to some extent a simplification of the biology, but models are valuable because they take ideas that might have been expressed verbally or diagrammatically and make them more explicit, so that they can begin to be understood in a *quantitative* rather than purely *qualitative* way.

1.2 Aims of modelling

The features of a model depend very much on the aims of the modelling exercise. We therefore need to consider why people model and what they hope to achieve by so doing. Often the most basic aim is to make clear the current state of knowledge regarding a particular system, by attempting to be precise about the elements involved and the interactions between them. Doing this can be a particularly effective way of highlighting gaps in understanding. In addition, having a detailed model of a system allows people to *test* that their understanding of a system is correct, by seeing if the implications of their models are consistent with observed experimental data. In practice, this model validation stage is central to the systems biology approach. However, this work will often represent only the initial stage of the modelling process. Once people have a model they are happy with, they often want to use their models *predictively*, by conducting 'virtual experiments' that might be difficult, time-consuming, or impossible to do in the lab. Such experiments may uncover important indirect relationships between model components that would be hard to predict otherwise. An additional goal of modern biological modelling is to pool a number of small models of well-understood mechanisms into a large model in order to investigate the effect of interactions between the model components. Models can also be extremely useful for informing the design and analysis of complex biological experiments.

In summary, modelling and computer simulation are becoming increasingly important in post-genomic biology for integrating knowledge and experimental data and making testable predictions about the behaviour of complex biological systems.

1.3 Why is stochastic modelling necessary?

Ignoring quantum mechanical effects, current scientific wisdom views biological systems as essentially deterministic in character, with dynamics entirely predictable given sufficient knowledge of the state of the system (together with complete knowledge of the physics and chemistry of interacting biomolecules). At first this perhaps suggests that a deterministic approach to the modelling of biological systems is likely to be successful. However, despite the rapid advancements in computing technology, we are still a very long way away from a situation where we might expect to be able to model biological systems of realistic size and complexity over interesting time scales using such a molecular dynamic approach. We must therefore use models that leave out many details of the 'state' of a system (such as the position, orientation, and momentum of every single molecule under consideration), in favour of a higher-level view. Viewed at this higher level, the dynamics of the system are not deterministic, but intrinsically stochastic, and consideration of statistical physics is necessary to uncover the precise nature of the stochastic processes governing the system dynamics. A more detailed discussion of this issue will have to be deferred until much later in the book, once the appropriate concepts and terminology have been established. In the meantime, it is helpful to highlight the issues using a very simple example that illustrates the importance of stochastic modelling, both for simulation and inference.

The example we will consider is known as the *linear birth–death process*. In the first instance, it is perhaps helpful to view this as a model for the number of bacteria in a bacterial colony. It is assumed that each bacterium in the colony gives rise to new individuals at rate λ (that is, on average, each bacterium will produce λ offspring per unit time). Similarly, each bacterium dies at rate μ (that is, on average, the proportion of bacteria that die per unit time is μ). These definitions are not quite right, but we will define such things much more precisely later. Let the number of bacteria in the colony at time t be denoted $X(t)$. Assume that the number of bacteria in the colony at time zero is known to be x_0. Viewed in a continuous deterministic manner, this description of the system leads directly to the ordinary differential equation

$$\frac{dX(t)}{dt} = (\lambda - \mu)X(t),$$

which can be solved analytically to give the complete dynamics of the system as

$$X(t) = x_0 \exp\{(\lambda - \mu)t\}.$$

So, predictably, in the case $\lambda > \mu$ the population size will increase exponentially as $t \longrightarrow \infty$, and will decrease in size exponentially if $\lambda < \mu$. Similarly, it will remain at constant size x_0 if $\lambda = \mu$. Five such solutions are given in Figure 1.1. There are other things worth noting about this solution. In particular, the solution clearly only depends on $\lambda - \mu$ and not on the particular values that λ and μ take (so, for example, $\lambda = 0.5, \mu = 0$ will lead to exactly the same solution as $\lambda = 1, \mu = 0.5$). In some sense, therefore, $\lambda - \mu$ (together with x_0) is a 'sufficient' description of the system dynamics. At first this might sound like a good thing, but it is clear that there is a flip-side: namely that studying *experimental data* on bacteria numbers can only provide information about $\lambda - \mu$, and not on the particular values of λ and μ separately (as the data can only provide information about the 'shape' of the curve, and the shape of the curve is determined by $\lambda - \mu$). Of course, this is not a problem if the continuous deterministic model is really appropriate, as then $\lambda - \mu$ is the only thing one needs to know and the precise values of λ and μ are not important for predicting system behaviour. Note, however, that the lack of identifiability of λ and μ has implications for *network inference*, as well as inference for rate constants. It is clear that in this model we cannot know from experimental data if we have a pure birth or death process, or a process involving both births and deaths, as it is not possible to know if λ or μ is zero.*

---

* This also illustrates another point that is not widely appreciated — the fact that reliable network in-

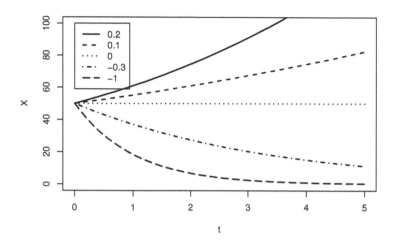

Figure 1.1 *Five deterministic solutions of the linear birth–death process for values of $\lambda - \mu$ given in the legend ($x_0 = 50$).*

The problem, of course, is that bacteria don't vary in number continuously and deterministically. They vary discretely and stochastically. Using the techniques that will be developed later in this book, it is straightforward to understand the stochastic process associated with this model as a *Markov jump process*, and to simulate it on a computer. By their very nature, such stochastic processes are random, and each time they are simulated they will look different. In order to understand the behaviour of such processes it is therefore necessary (in general) to study many realisations of the process. Five realisations are given in Figure 1.2, together with the corresponding deterministic solution.

It is immediately clear that the stochastic realisations exhibit much more interesting behaviour and match much better with the kind of experimental data one is likely to encounter. They also allow one to ask questions and get answers to issues that can't be addressed using a continuous deterministic model. For example, according to the deterministic model, the population size at time $t = 2$ is given by $X(2) = 50/e^2 \simeq 6.77$. Even leaving aside the fact that this is not an integer, we see from the stochastic realisations that there is considerable uncertainty for the value of $X(2)$, and stochastic simulation allows us to construct, *inter alia*, a likely range of values for $X(2)$. Another quantity of considerable practical interest is the 'time to extinction' (the time, t, at which $X(t)$ first becomes zero). Under the deterministic model, $X(t)$ *never* reaches zero, but simply tends to zero as $t \longrightarrow \infty$. We see from the stochastic realisations that these do go extinct, and that there is considerable ran-

---

ference is necessarily more difficult than rate-parameter inference, as determining the existence of a reaction is equivalent to deciding whether the rate of that reaction is zero.

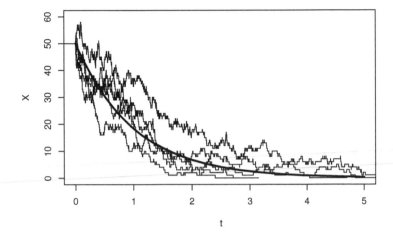

Figure 1.2 *Five realisations of a stochastic linear birth–death process together with the continuous deterministic solution* $(x_0 = 50, \lambda = 3, \mu = 4)$.

domness associated with the time that this occurs. Again, stochastic simulation will allow us to understand the *distribution* associated with the time to extinction, something that simply isn't possible using a deterministic framework.

Another particularly noteworthy feature of the stochastic process representation is that it depends explicitly on both λ and μ, and not just on $\lambda - \mu$. This is illustrated in Figure 1.3. It is clear that although $\lambda - \mu$ controls the essential 'shape' of the process, $\lambda + \mu$ controls the degree of 'noise' or 'volatility' in the system. This is a critically important point to understand — it tells us that if stochastic effects are present in the system, we cannot properly understand the system dynamics unless we know both λ and μ. Consequently, *we cannot simply fit a deterministic model to available experimental data and then use the inferred rate constants in a stochastic simulation*, as it is not possible to infer the stochastic rate constants using a deterministic model.

This has important implications for the use of stochastic models for inference from experimental data. It suggests that given some data on the variation in colony size over time, it ought to be possible to get information about $\lambda - \mu$ from the overall shape of the data, and information about $\lambda + \mu$ from the volatility of the data. If we know both $\lambda - \mu$ and $\lambda + \mu$, we can easily determine both λ and μ separately. Once we know both λ and μ, we can accurately simulate the dynamics of the system we are interested in (as well as inferring network structure, as we could also test to see if either λ or μ is zero). However, it is only possible to make satisfactory inferences for both λ and μ if the stochastic nature of the system is taken into account at the inference stage of the process.

Although we have here considered a trivial example, the implications are broad. In particular, they apply to the genetic and biochemical network models that much of

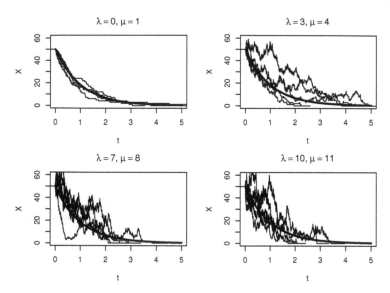

Figure 1.3 *Five realisations of a stochastic linear birth–death process together with the continuous deterministic solution for four different (λ, μ) combinations, each with $\lambda - \mu = -1$ and $x_0 = 50$.*

this book will be concerned with. This is because genetic and biochemical networks involve the interaction of integer numbers of molecules that react when they collide after random times, driven by Brownian motion. Although it is now becoming increasingly accepted that stochastic mechanisms are important in many (if not most) genetic and biochemical networks, routine use of stochastic simulation in order to understand system dynamics is still not as ubiquitous as it might be. This could be because inference methods regularly used in practice work by fitting continuous deterministic models to experimental data. We have just seen that such methods cannot in general give us reliable information about all of the parameters important for determining the stochastic dynamics of a system, and so stochastic simulation cannot be done reliably until we have good methods of inference for stochastic models. It turns out that it is possible to formalise the problem of inference for stochastic kinetic models from time-course experimental data, and this is the subject matter of the latter chapters. However, it should be pointed out at the outset that inference for stochastic models is even more difficult than inference for deterministic models (in terms of the mathematics required, algorithmic complexity, and computation time), and is still the subject of a great deal of ongoing research.

1.4 Chemical reactions

There are a number of ways one could represent a model of a biological system. Biologists have traditionally favoured diagrammatic schemas coupled with verbal explanations in order to convey qualitative information regarding mechanisms. At the other extreme, applied mathematicians traditionally prefer to work with systems of ordinary or partial differential equations (ODEs or PDEs). These have the advantage of being more precise and fully quantitative, but also have a number of disadvantages. In some sense differential equation models are too low level a description, as they not only encode the essential features of the model, but also a wealth of accompanying baggage associated with a particular interpretation of chemical kinetics that is not always well suited to application in the molecular biology context. Somewhere between these two extremes, the biochemist will tend to view systems as networks of coupled chemical reactions, and it appears that most of the best ways of representing biochemical mechanisms exist at this level of detail, though there are many ways of representing networks of this type. Networks of coupled chemical reactions are sufficiently general that they can be simulated in different ways using different algorithms depending on assumptions made about the underlying kinetics. On the other hand, they are sufficiently detailed and precise that once the kinetics have been specified, they can be used directly to construct full dynamic simulations of the system behaviour on a computer.

A general chemical reaction takes the form

$$m_1 R_1 + m_2 R_2 + \cdots + m_r R_r \longrightarrow n_1 P_1 + n_2 P_2 + \cdots + n_p P_p,$$

where r is the number of reactants and p is the number of products. R_i represents the ith reactant molecule and P_j is the jth product molecule. m_i is the number of molecules of R_i consumed in a single reaction step, and n_j is the number of molecules of P_j produced in a single reaction step. The coefficients m_i and n_j are known as *stoichiometries*. The stoichiometries are usually (though not always) assumed to be integers, and in this case it is assumed that there is no common factor of the stoichiometries. That is, it is assumed that there is no integer greater than one which exactly divides each stoichiometry on both the left and right sides. There is no assumption that the R_i and P_j are distinct, and it is perfectly reasonable for a given molecule to be both consumed and produced by a single reaction.[†] The reaction equation describes precisely which chemical species[‡] react together, and in what proportions, along with what is produced.

In order to make things more concrete, consider the dimerisation of a protein P. This is normally written

$$2P \longrightarrow P_2,$$

as two molecules of P react together to produce a single molecule of P_2. Here P has a stoichiometry of 2 and P_2 has a stoichiometry of 1. Stoichiometries of 1 are not usually written explicitly. Similarly, the reaction for the dissociation of the dimer would be written

$$P_2 \longrightarrow 2P.$$

[†] Note that a chemical species that occurs on both the left- and right-hand sides with the same stoichiometry is somewhat special, and is sometimes referred to as a *modifier*. Clearly the reaction will have no effect on the amount of this species. Such a species is usually included in the reaction because the *rate* at which the reaction proceeds depends on the level of this species.

[‡] The use of the term 'species' to refer to a particular type of molecule will be explained later in the chapter.

A reaction that can happen in both directions is known as *reversible*. Reversible reactions are quite common in biology and tend not to be written as two separate reactions. They can be written with a double-headed arrow such as

$$2P \rightleftharpoons P_2 \quad \text{or} \quad 2P \longleftrightarrow P_2 \quad \text{or} \quad 2P \Longleftrightarrow P_2.$$

If one direction predominates over the other, this is sometimes emphasised in the notation. So if the above protein prefers the dimerised state, this may be written something like

$$2P \underset{\leftarrow}{\rightharpoonup} P_2 \quad \text{or} \quad 2P \underset{\leftarrow}{\longrightarrow} P_2.$$

It is important to remember that the notation for a reversible reaction is simply a convenient shorthand for the two separate reaction processes taking place. In the context of the discrete stochastic models to be studied in this book, *it will not usually be acceptable to replace the two separate reactions by a single reaction proceeding at some kind of overall combined rate.*

1.5 Modelling genetic and biochemical networks

Before moving on to look at different ways of representing and working with systems of coupled chemical reactions in the next chapter, it will be helpful to end this chapter by looking in detail at some basic biochemical mechanisms and how their essential features can be captured with fairly simple systems of coupled chemical reactions. Although biological modelling can be applied to biological systems at a variety of different scales, it turns out that stochastic effects are particularly important and prevalent at the scale of genetic and biochemical networks, and these will therefore provide the main body of examples for this book.

1.5.1 Transcription (prokaryotes)

Transcription is a key cellular process, and control of transcription is a fundamental regulation mechanism. As a result, virtually any model of genetic regulation is likely to require some modelling of the transcription process. This process is much simpler in prokaryotic organisms, so it will be helpful to consider this in the first instance. Here, typically, a promoter region exists just upstream of the gene of interest. RNA-polymerase (RNAP) is able to bind to this promoter region and initiate the transcription process, which ultimately results in the production of an mRNA transcript and the release of RNAP back into the cell. The transcription process itself is complex, but whether it will be necessary to model this explicitly will depend very much on the modelling goals. If the modeller is primarily interested in control and the downstream effects of the transcription process, it may not be necessary to model transcription itself in detail.

The process is illustrated diagrammatically in Figure 1.4. Here, g is the gene of interest, p is the upstream promoter region, and r is the mRNA transcript of g. A very simple representation of this process as a system of coupled chemical reactions can

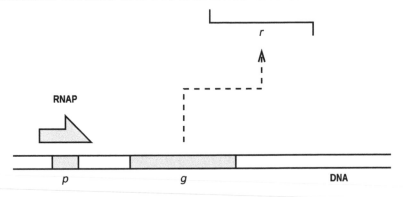

Figure 1.4 *Transcription of a single prokaryotic gene.*

be written as follows:

$$p + \text{RNAP} \longrightarrow p \cdot \text{RNAP}$$
$$p \cdot \text{RNAP} \longrightarrow p + \text{RNAP} + r.$$

As discussed, the second reaction is really the end result of a very large number of reactions. It is also worth emphasising that the reactions do not represent a closed system, as r appears to be produced out of nothing. In reality, it is created from other chemical species within the cell, but we have chosen here not to model at such a fine level of detail. One detail not included here that may be worth considering is the reversible nature of the binding of RNAP to the promoter region. It is also worth noting that these two reactions form a simple linear chain, whereby the product of the first reaction is the reactant for the second. Indeed, we could write the pair of reactions as

$$p + \text{RNAP} \longrightarrow p \cdot \text{RNAP} \longrightarrow p + \text{RNAP} + r.$$

It is therefore tempting to summarise this chain of reactions by the single reaction

$$p + \text{RNAP} \longrightarrow p + \text{RNAP} + r,$$

and this is indeed possible, but is likely to be inadequate for any model of regulation or control where the intermediary compound $p \cdot \text{RNAP}$ is important, such as any model for competitive binding of RNAP and a repressor in the promoter region.

If modelling the production of the entire RNA molecule in a single step is felt to be an oversimplification, it is relatively straightforward to model the explicit elongation of the molecule. As a first attempt, consider the following model for the transcription of an RNA molecule consisting of n nucleotides.

$$p + \text{RNAP} \longrightarrow p \cdot \text{RNAP}$$
$$p \cdot \text{RNAP} \longrightarrow p \cdot \text{RNAP} \cdot r_1$$
$$p \cdot \text{RNAP} \cdot r_1 \longrightarrow p \cdot \text{RNAP} \cdot r_2$$
$$\vdots \quad \longrightarrow \quad \vdots$$
$$p \cdot \text{RNAP} \cdot r_{n-1} \longrightarrow p \cdot \text{RNAP} \cdot r_n$$
$$p \cdot \text{RNAP} \cdot r_n \longrightarrow p + \text{RNAP} + r.$$

This still does not model the termination process in detail; see Arkin, Ross, & McAdams (1998) for details of how this could be achieved. One problem with the above model is that the gene is blocked in the first reaction and is not free for additional RNAP binding until it is released again after the last reaction. This prevents concurrent transcription from occurring. An alternative would be to model the process as follows:

$$p + \text{RNAP} \longrightarrow p \cdot \text{RNAP}$$
$$p \cdot \text{RNAP} \longrightarrow p + \text{RNAP} \cdot r_1$$
$$\text{RNAP} \cdot r_1 \longrightarrow \text{RNAP} \cdot r_2$$
$$\vdots \quad \longrightarrow \quad \vdots$$
$$\text{RNAP} \cdot r_{n-1} \longrightarrow \text{RNAP} \cdot r_n$$
$$\text{RNAP} \cdot r_n \longrightarrow \text{RNAP} + r.$$

This model frees the gene for further transcription as soon as the transcription process starts. In fact, it is probably more realistic to free the gene once a certain number of nucleotides have been transcribed, and this is easily incorporated into the above model. Another slightly undesirable feature is that it does not prevent one RNAP from 'overtaking' another during concurrent transcription (but this is not usually particularly important, nor is it very easy to fix).

1.5.2 Eukaryotic transcription (a very simple case)

The transcription process in eukaryotic cells is rather more complex than in prokaryotes. This book is not an appropriate place to explore the many and varied mechanisms for control and regulation of eukaryotic transcription, so we will focus on a simple illustrative example, shown in Figure 1.5. In this model, there are two transcription factor (TF) binding sites upstream of a gene, g. Transcription factor TF1 reversibly binds to site tf1, and TF2 reversibly binds to tf2, but is only able to bind if TF1 is already in place. Also, TF1 cannot dissociate if TF2 is in place. The transcription process cannot initiate (starting with RNAP binding) unless both TFs are in place.

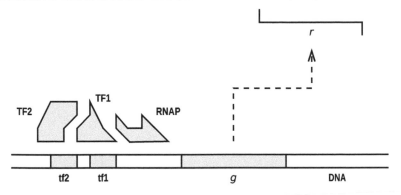

Figure 1.5 *A simple illustrative model of the transcription process in eukaryotic cells.*

We can model this as follows:

$$g + \text{TF1} \rightleftharpoons \text{TF1} \cdot g$$
$$\text{TF1} \cdot g + \text{TF2} \rightleftharpoons \text{TF2} \cdot \text{TF1} \cdot g$$
$$\text{RNAP} + \text{TF2} \cdot \text{TF1} \cdot g \rightleftharpoons \text{RNAP} \cdot \text{TF2} \cdot \text{TF1} \cdot g$$
$$\text{RNAP} \cdot \text{TF2} \cdot \text{TF1} \cdot g \longrightarrow \text{TF2} \cdot \text{TF1} \cdot g + \text{RNAP} + r.$$

Note that we have not explicitly included tf1, tf2, and g separately in the model, as they are all linked on a DNA strand and hence are a single entity from a modelling perspective. Instead we use g to represent the gene of interest together with its regulatory region (including tf1 and tf2). Note that this system, like the previous one, also forms a linear progression of ordered reactions and does not involve a 'feedback' loop of any sort.

1.5.3 Gene regulation (prokaryotes)

Regulation and control are fundamental to biological systems. These necessarily involve feedback and a move away from a simple ordered set of reactions (hence the term biochemical *network*). Sometimes such systems are large and complex, but feedback, and its associated non-lincarity, can be found in small and apparently simple systems. We will look here at a simple control mechanism (repression of a prokaryotic gene) and see how this can be embedded into a regulatory feedback system in a later example.

Figure 1.6 illustrates a model where a repressor protein R can bind to regulatory site q, downstream of the RNAP binding site p but upstream of the gene g, thus preventing transcription of g. We can formulate a set of reactions for this process in the following way:

$$g + R \rightleftharpoons g \cdot R$$
$$g + \text{RNAP} \rightleftharpoons g \cdot \text{RNAP}$$
$$g \cdot \text{RNAP} \longrightarrow g + \text{RNAP} + r.$$

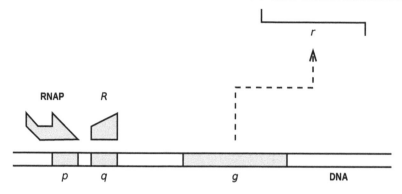

Figure 1.6 *A simple prokaryotic transcription repression mechanism.*

This set of equations no longer has a natural ordering and hence cannot be read from top to bottom to go from reactants to products in an obvious way. Each reaction represents a possible direction for the system to move in. Also note that there are actually five reactions represented here, as two of the three listed are reversible. The crucial thing to appreciate here is that from a modelling perspective, $g \cdot R$ is a different species from g, and so the fact that RNAP can bind to g does not suggest that RNAP can bind to $g \cdot R$. Thus, this set of reactions precisely captures the mechanism of interest, namely that RNAP can bind to g when it is free but not when it is repressed by R.

At this point it is worth mentioning another critically important genetic regulation mechanism often used in prokaryotes such as bacteria. *Sigma factors* (often written *σ-factors*) are special proteins which bind to RNAP and allow the RNAP to recognise the (sigma-factor-specific) promoter region upstream of the gene. Different sigma factors have different promoter sequences, and hence the regulation of sigma factors is a key control mechanism. Typically the number of distinct sigma factor proteins is very small relative to the total number of genes, and therefore sigma factor regulation is used to turn on and off large numbers of genes which share the same sigma factor. As a concrete example, the sigma factor *SigD* in the Gram positive bacterium *Bacillus subtilis* is used to turn on a large number of genes relating to cell motility. A simple model for transcription which includes a sigma factor can be constructed as follows:

$$\text{SigF} + \text{RNAP} \rightleftharpoons \text{SigF} \cdot \text{RNAP}$$
$$g + \text{SigF} \cdot \text{RNAP} \rightleftharpoons g \cdot \text{SigF} \cdot \text{RNAP}$$
$$g \cdot \text{SigF} \cdot \text{RNAP} \longrightarrow g + \text{SigF} \cdot \text{RNAP} + r.$$

It is straightforward to combine such a promoter model with a model for repression to obtain a combined model of prokaryotic regulation.

1.5.4 Translation

Translation (like transcription) is a complex process involving several hundred reactions to produce a single protein from a single mRNA transcript. Again, however, it will not always be necessary to model every aspect of the translation process — just those features pertinent to system features of interest. The really key stages of the translation process are the binding of a ribosome (Rib) to the mRNA, the translation of the mRNA, and the folding of the resulting polypeptide chain into a functional protein. These stages are easily coded as a set of reactions:

$$r + \text{Rib} \rightleftharpoons r \cdot \text{Rib}$$
$$r \cdot \text{Rib} \longrightarrow r + \text{Rib} + P_u$$
$$P_u \longrightarrow P.$$

Here, P_u denotes unfolded protein and P denotes the folded protein. In some situations, it will also be necessary to model various post-translational modifications such as *phosphorylation*. Clearly the second and third reactions are gross simplifications of the full translation process. Elongation can be modelled in more detail using an approach similar to that adopted for transcription; see Arkin et al. (1998) for further details. Folding could also be modelled similarly if necessary.

1.5.5 Degradation

The simplest model for degradation is just

$$r \longrightarrow \emptyset,$$

where $\emptyset$ is the 'empty set' symbol, meaning that r is transformed to nothing (as far as the model is concerned). A more appropriate model for RNA degradation would be

$$r + \text{RNase} \longrightarrow r \cdot \text{RNase}$$
$$r \cdot \text{RNase} \longrightarrow \text{RNase},$$

where RNase denotes ribonuclease (an RNA-degrading enzyme). Modelling in this way is probably only important if there is limited RNase availability, but is interesting in conjunction with a translation model involving Rib, as it will then capture the competitive binding of Rib and RNase to r.

Models for protein degradation can be handled similarly. Here one would typically model the tagging of the protein with a cell signalling molecule (such as *ubiquitin*), t, and then subsequent degradation in a separate reaction. A minimal model would therefore look like the following:

$$P + t \longrightarrow P \cdot t$$
$$P \cdot t \longrightarrow t.$$

In fact, cell protein degradation machinery is rather complex; see Proctor et al. (2005) for a more detailed treatment of this problem.

1.5.6 Transport

In eukaryotes, mRNA is transported out of the cell nucleus before the translation process can begin. Often, the modelling of this process will be unnecessary, but could be important if the transportation process itself is of interest, if the delay associated with transportation is relevant, or if the number of available transportation points is limited. A model for this could be as simple as

$$r_n \longrightarrow r_c,$$

where r_n denotes the mRNA pool within the nucleus, and r_c the corresponding pool in the cytoplasm. However, this would not take into account the limited number of transport points. A more realistic model would therefore be

$$r_n + N \longrightarrow r_n \cdot N$$
$$r_n \cdot N \longrightarrow r_c + N,$$

where N denotes the set of available mRNA transport points embedded in the outer shell of the cell nucleus. In fact, this system is very closely related to the Michaelis–Menten enzyme kinetic system that will be examined in more detail later in the book. Here, the transport points behave like an enzyme whose abundance limits the flow from r_n to r_c.

1.5.7 Prokaryotic auto-regulation

Now that we have seen how to generate very simple models of key processes involved in gene expression and regulation, we can put them together in the form of a simple prokaryotic auto-regulatory network.

   Figure 1.7 illustrates a simple gene expression auto-regulation mechanism often present in prokaryotic gene networks. Here dimers of the protein P coded by the gene g repress their own transcription by binding to a (repressive) regulatory region upstream of g,

$$
\begin{array}{ll}
g + P_2 \rightleftharpoons g \cdot P_2 & \text{Repression} \\
g \longrightarrow g + r & \text{Transcription} \\
r \longrightarrow r + P & \text{Translation} \\
2P \rightleftharpoons P_2 & \text{Dimerisation} \\
r \longrightarrow \emptyset & \text{mRNA degradation} \\
P \longrightarrow \emptyset & \text{Protein degradation.}
\end{array}
$$

Notice that this model is minimal in terms of the level of detail included. In particular, the transcription part ignores sigma factor and RNAP binding, the translation/mRNA degradation parts ignore Rib/RNase competitive binding, and so on. However, as we will see later, this model contains many of the interesting features of an auto-regulatory feedback network. See Bundschuh et al. (2003) for discussion of this dimer-autoregulation model in the context of the λ repressor protein cI of phage-λ in *E. coli*.

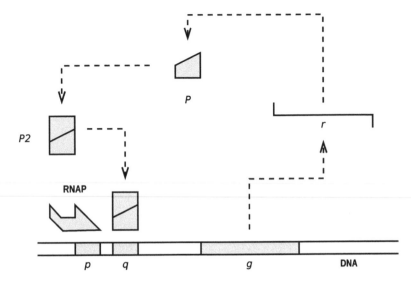

Figure 1.7 *A very simple model of a prokaryotic auto-regulatory gene network. Here dimers of a protein P coded for by a gene g repress their own transcription by binding to a regulatory region q upstream of g and downstream of the promoter p.*

1.5.8 lac operon

We will finish this section by looking briefly at a classic example of prokaryotic gene regulation — probably the first well-understood genetic regulatory network. The genes in the operon code for enzymes required for the respiration of lactose (Figure 1.8). The enzymes convert lactose to glucose, which is then used as the 'fuel' for respiration in the usual way. These enzymes are only required if there is a shortage of glucose and an abundance of lactose, and so there is a transcription control mechanism regulating their production. Upstream of the *lac* operon there is a gene coding for a protein which represses transcription of the operon by binding to the DNA just downstream of the RNAP binding site. Under normal conditions (absence of lactose), transcription of the *lac* operon is turned off. However, in the presence of lactose, the inhibitor protein preferentially binds to lactose, and in the bound state can no longer bind to the DNA. Consequently, the repression of transcription is removed, and production of the required enzymes can take place. We can represent this with the following simple set of reactions:

$$i \longrightarrow i + r_I$$
$$r_I \longrightarrow r_I + I$$
$$I + \text{Lactose} \rightleftharpoons I \cdot \text{Lactose}$$
$$I + o \rightleftharpoons I \cdot o$$

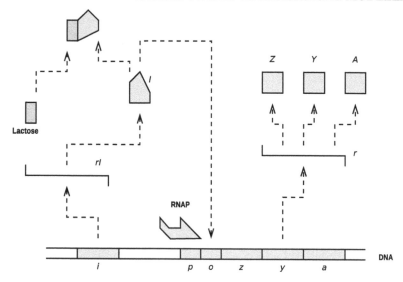

Figure 1.8 *Key mechanisms involving the lac operon. Here an inhibitor protein I can repress transcription of the lac operon by binding to the operator o. However, in the presence of lactose, the inhibitor preferentially binds to it, and in the bound state can no longer bind to the operator, thereby allowing transcription to proceed.*

$$o + \text{RNAP} \rightleftharpoons \text{RNAP} \cdot o$$
$$\text{RNAP} \cdot o \longrightarrow o + \text{RNAP} + r$$
$$r \longrightarrow r + A + Y + Z$$
$$\text{Lactose} + Z \longrightarrow Z.$$

Here i represents the gene for the inhibitor protein, r_I the associated mRNA, and I the inhibitor protein itself. The *lac* operon is denoted o and is treated as a single entity from a modelling viewpoint. The mRNA transcript from the operon is denoted by r, and this codes for all three *lac* proteins. The final reaction represents the transformation of lactose to something not directly relevant to the regulation mechanism.

Again this system is fairly minimal in terms of the detail included, and all of the degradation reactions have been omitted, along with what happens to lactose once it has been acted on by β-galactosidase (Z). In fact, there is also another mechanism we have not considered here that ensures that transcription of the operon will only occur when there is a shortage of glucose (as respiration of glucose is always preferred).

1.6 Modelling higher-level systems

We have concentrated so far on fairly low-level biochemical models where the concept of modelling with 'chemical reactions' is perhaps most natural. However, it is important to recognise that we use the notation of chemical reactions simply to describe things that combine and the things that they produce, and that this framework can be used to model higher-level phenomena in a similar way. In Section 1.3 the linear birth–death process was introduced as a model for the number of bacteria present in a colony. We can use our chemical reaction notation to capture the qualitative structure of this model:

$$X \longrightarrow 2X$$
$$X \longrightarrow \emptyset.$$

The first equation represents 'birth' of new bacteria and the second 'death'. There are many possible extensions of this simple model, including the introduction of immigration of new bacteria from another source and emigration of bacteria to another source.

The model represents a 'population' of individuals (here the individuals are bacteria), and it is possible to extend such models to populations involving more than one 'species'. Consider the Lotka–Volterra predator prey model for two interacting species:

$$Y_1 \longrightarrow 2Y_1$$
$$Y_1 + Y_2 \longrightarrow 2Y_2$$
$$Y_2 \longrightarrow \emptyset.$$

Again this is not a real reaction system in the strictest sense, but it is interesting and useful, as it is the simplest model exhibiting the kind of non-linear auto-regulatory feedback behaviour considered earlier. Also, as it only involves two species and three reactions, it is relatively easy to work with without getting lost in detail. Here, Y_1 represents a 'prey' species (such as rabbits) and Y_2 represents a 'predator' species (such as foxes).[§] The first reaction is a simple representation of prey reproduction. The second reaction is an attempt to capture predator–prey interaction (consumption of prey by predator, in turn influencing predator reproduction rate). The third reaction represents death of predators due to natural causes. We will revisit this model in greater detail in later chapters.

Another widely studied individual level model is the so-called SIR model for disease epidemiology, where the initials stand for *susceptible*, *infectious*, and *recovered*. The idea is that individuals are initially susceptible to catching a disease from an infectious person. Should they contract the disease, they will make the transition from susceptible to infectious, where they will have the possibility of infecting susceptibles. Eventually the infectious individual will make the transition to the 'recovered' category, when they will no longer be able to infect susceptibles, but will have immunity to the disease, and hence will not be themselves any longer susceptible to infection. Of course for some diseases, this 'recovered' category will include individuals who are in fact dead! In this case, the 'R' category is sometimes used to stand for *Removed*. The simplest variant of this model can be summarised with just

---

[§] Note that the use of reactions to model the interaction of 'species' in a population dynamics context explains the use of the term 'species' to refer to a particular type of chemical molecule in a set of coupled chemical reactions.

two reactions as

$$S \longrightarrow I \longrightarrow R,$$

although, since the rate of the first reaction depends on the number of infectives, it is clearer to write this pair as

$$S + I \longrightarrow 2I$$

$$I \longrightarrow R.$$

There are obviously many variants on this basic model. For example, some individuals may develop immunity without ever becoming infectious ($S \longrightarrow R$) and some recovered individuals may lose their immunity ($R \longrightarrow S$), etc. Another commonly studied variant is the SEIR model which introduces an additional category, *exposed*, representing individuals who have been infected with the disease but are not yet themselves infectious:

$$S \longrightarrow E \longrightarrow I \longrightarrow R.$$

1.7 Exercises

1. Write out a more detailed and realistic model for the simple auto-regulatory network considered in Section 1.5.7. Include a sigma factor, RNAP binding, Rib/RNase competitive binding, and so on.
2. Consider the *lac* operon model from Section 1.5.8.

   (a) First add more detail to the model, as in the previous exercise.
   (b) Look up the β-galactosidase pathway and add detail from this to the model.
   (c) Find details of the additional regulation mechanism mentioned, which ensures lactose is only respired in an absence of glucose, and try to incorporate that into the model.

1.8 Further reading

See Wilkinson (2009) for a review of stochastic modelling approaches with applications to systems biology, which provides a more comprehensive survey of the recent literature than this text. See Bower and Bolouri (2000) for more detailed information on modelling, and the different possible approaches to modelling genetic and biochemical networks. Kitano (2001) gives a more general overview of biological modelling and systems biology. McAdams and Arkin (1997) and Arkin et al. (1998) explore biological modelling in the context of the discrete stochastic models we will consider later. The *lac* operon is discussed in many biochemistry texts, including Stryer (1988), and is modelled in Santillán et al. (2007). Elowitz et al. (2002) and Swain et al. (2002) were key papers investigating the sources of stochasticity in gene expression at the single-cell level. Latchman (2002) is the classic text on eukaryotic gene regulation. The original references for the Lotka–Volterra predator-prey models are Lotka (1925) and Volterra (1926).

The website associated with this (edition of the) book* contains a range of links

---

* URL: `https://github.com/darrenjw/smfsb`

to on-line information of relevance to the various chapters of the book. I will also include an errata for any typos which are notified to me. Now would probably be a good time to have a quick look at it and 'bookmark' it for future reference. The links for this chapter contain pointers to various pertinent Wikipedia pages for further reading.

## CHAPTER 2

Representation of biochemical networks

2.1 Coupled chemical reactions

As was illustrated in the first chapter, a powerful and flexible way to specify a model is to simply write down a list of reactions corresponding to the system of interest. Note, however, that the reactions themselves specify only the qualitative structure of a model and must be augmented with additional information before they can be used to carry out a dynamic simulation on a computer. The model is completed by specifying the *rate* of every reaction, together with initial amounts of each reacting species.

Reconsider the auto-regulation example from Section 1.5.7:

$$g + P_2 \rightleftharpoons g \cdot P_2$$
$$g \longrightarrow g + r$$
$$r \longrightarrow r + P$$
$$2P \rightleftharpoons P_2$$
$$r \longrightarrow \emptyset, \qquad P \longrightarrow \emptyset.$$

Although only six reactions are listed, there are actually eight, as two are reversible. Each of those eight reactions must have a rate law associated with it. We will defer a complete discussion of rate laws until Chapter 6. For now, it is sufficient to know that the rate laws quantify the propensity of particular reactions to take place and are likely to depend on the *current* amounts of available reactants. In addition there must be an *initial* amount for each of the five chemical species involved: $g \cdot P_2$, g, r, P, and P_2. Given the reactions, the rate laws, and the initial amounts (together with some assumptions regarding the underlying kinetics, which are generally not regarded as part of the model), the model is specified and can in principle be simulated dynamically on a computer.

The problem is that even this short list of reactions is hard to understand on its own, whereas the simple biologist's diagram (Figure 1.7) is not sufficiently detailed and explicit to completely define the model. What is needed is something between the biologist's diagram and the list of reactions.

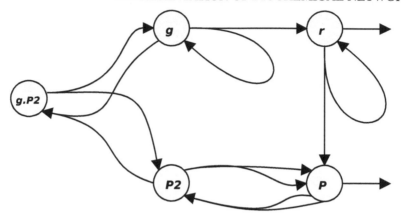

Figure 2.1 *A simple graph of the auto-regulatory reaction network.*

2.2 Graphical representations

2.2.1 Introduction

One way to begin to understand a reaction network is to display it as a pathway diagram of some description. The diagram in Figure 2.1 is similar to that used by some biological model-building tools such as CellDesigner.*

Such a diagram is easier to understand than the reaction list, yet it contains the same amount of information and hence could be used to generate a reaction list. Note that auto-regulation by its very nature implies a 'loop' in the reaction network, which is very obvious and explicit in the associated diagram. One possible problem with diagrams such as this, however, is the fact that it can sometimes be hard to distinguish which are the reactants and which are the products in any given reaction (particularly in large complex networks), and this can make it difficult to understand the flow of species through the network. Also, the presence of 'branching' and 'looping' arcs makes them slightly unnatural to work with directly in a mathematical sense.

Such problems are easily overcome by formalising the notion of pathway diagrams using the concept of a *graph* (here we mean the mathematical notion of a graph, not the idea of the graph of a function), where each *node* represents either a chemical species or a reaction, and arcs are used to indicate reaction pathways. In order to make this explicit, some elementary graph theoretic notation is helpful.

2.2.2 Graph theory

Definition 2.1 *A directed graph or digraph,* $\mathcal{G}$ *is a tuple* (V, E)*, where* $V = \{v_1, \ldots, v_n\}$ *is a set of* nodes *(or* vertices*) and* $E = \{(v_i, v_j)|v_i, v_j \in V,\ v_i \to v_j\}$ *is a set of directed* edges *(or* arcs*), where we use the notation* $v_i \to v_j$ *if and only if there is a directed edge from node* v_i *to* v_j*.*

---

* Software web links tend to go out of date rather quickly, so a regularly updated list is available from the book's web page. See the links for Chapter 2.

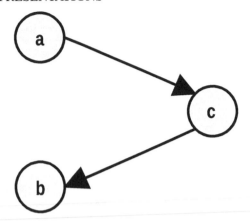

Figure 2.2 *A simple digraph.*

So the graph shown in Figure 2.2 has mathematical representation

$$\mathcal{G} = (\{a, b, c\}, \{(a, c), (c, b)\}).$$

Definition 2.2 *A graph is described as* simple *if there do not exist edges of the form* (v_i, v_i) *and there are no repeated edges. A* bipartite *graph is a simple graph where the nodes are partitioned into two distinct subsets V_1 and V_2 (so that $V = V_1 \cup V_2$ and $V_1 \cap V_2 = \emptyset$) such that there are no arcs joining nodes from the same subset.*

Referring back to the previous example, the partition $V_1 = \{a, b\}$, $V_2 = \{c\}$ gives a bipartite graph ($\mathcal{G}$ is said to be bipartite over the partition), and the partition $V_1 = \{a\}$, $V_2 = \{b, c\}$ does not (as the edge (c, b) would then be forbidden). A *weighted* graph is a graph which has (typically positive) numerical values associated with each edge.

2.2.3 Reaction graphs

It turns out that it is very natural to represent sets of coupled chemical reactions using weighted bipartite digraphs where the nodes are partitioned into two sets representing the species and reactions. An arc from a species node to a reaction node indicates that the species is a reactant for that reaction, and an arc from a reaction node to a species node indicates that the species is a product of the reaction. The weights associated with the arcs represent the stoichiometries associated with the reactants and products. There is a very strong correspondence between reaction graphs modelled this way, and the theory of Petri nets, which are used extensively in computing science for a range of modelling problems. A particular advantage of Petri net theory is that it is especially well suited to the discrete-event stochastic simulation models this book is mainly concerned with. It is therefore helpful to have a basic familiarity with Petri nets and their application to biological modelling.

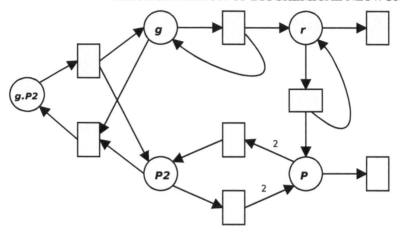

Figure 2.3 *A Petri net for the auto-regulatory reaction network.*

2.3 Petri nets

2.3.1 Introduction

Petri nets are a mathematical framework for systems modelling together with an associated graphical representation. Goss and Peccoud (1998) were among the first to use stochastic Petri nets for biological modelling. More recent reviews of the use of Petri nets for biological modelling include Pinney et al. (2003), Hardy and Robillard (2004), and Chaouiya (2007). A brief introduction to key elements of Petri net theory will be outlined here — see Reisig (1985) and Murata (1989) for further details.

By way of an informal introduction, the Petri net corresponding to the network shown in Figure 2.1 is shown in Figure 2.3. The rectangular boxes in Figure 2.3 represent individual reactions. The arcs into each box denote reactants, and the arcs out of each box denote products. Numbers on arcs denote the *weight* of the arc (un-numbered arcs are assumed to have a weight of 1). The weights represent reaction stoichiometries. The Petri net graph is only a slight refinement of the basic graph considered earlier, but it is easier to comprehend visually and more convenient to deal with mathematically (it is a bipartite graph). It is easy to see how to work through the graph and enumerate the full list of chemical reactions.

Traditionally each place (species) node of a place/transition (P/T) Petri net has an integer number of 'tokens' associated with it, representing the abundance of that 'species'. This fits in particularly well with the discrete stochastic molecular kinetics models we will consider in more detail later. Here, the number of tokens at a given node may be interpreted as the number of molecules of that species in the model at a given time (Figure 2.4). The collection of all token numbers at any given point in time is known as the current *marking* of the net (which corresponds here to the *state* of the reaction system). The Petri net shows what happens when particular transitions 'fire' (reactions occur). For example, in the above state, if two reactions occur, one a repression binding and the other a translation, the new Petri net will be as given in

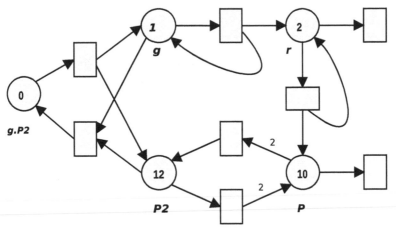

Figure 2.4 *A Petri net labelled with tokens.*

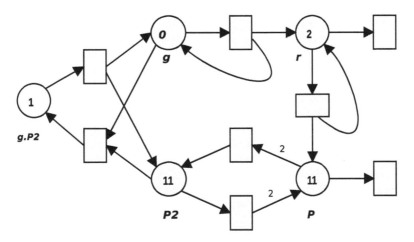

Figure 2.5 *A Petri net with new numbers of tokens after reactions have taken place.*

Figure 2.5. This is because the repression binding has the effect of increasing $g \cdot P_2$ by 1, and decreasing g and P_2 by 1, and the translation has the effect of increasing P by 1 (as r is both increased and decreased by 1, the net effect is that it remains unchanged). So the old and new Petri net markings can be written as

Species	No. tokens
$g \cdot P_2$	0
g	1
r	2
P	10
P_2	12

and

Species	No. tokens
$g \cdot P_2$	1
g	0
r	2
P	11
P_2	11

Table 2.1 *The auto-regulatory system displayed in tabular (matrix) form (zero stoichiometries omitted for clarity)*

	Reactants (*Pre*)					Products (*Post*)				
Species	$g \cdot P_2$	g	r	P	P_2	$g \cdot P_2$	g	r	P	P_2
Repression		1			1	1				
Reverse repression	1						1			1
Transcription		1					1	1		
Translation			1					1	1	
Dimerisation				2						1
Dissociation					1				2	
mRNA degradation			1							
Protein degradation				1						

respectively. A transition (reaction) can only fire (take place) if there are sufficiently many tokens (molecules) associated with each input place (reactant species). We are now in a position to consider Petri nets more formally.

2.3.2 Petri net formalism and matrix representations

Definition 2.3 *A Petri net, N, is an n-tuple $(P, T, Pre, Post, M)$, where $P = \{p_1, \ldots, p_u\}$, $(u > 0)$ is a finite set of places, $T = \{t_1, \ldots, t_v\}$, $(v > 0)$ is a finite set of transitions, and $P \cap T = \emptyset$. Pre is a $v \times u$ integer matrix containing the weights of the arcs going from places to transitions (the (i, j)th element of this matrix is the weight of the arc going from place j to transition i), and Post is a $v \times u$ integer matrix containing the weights of arcs from transitions to places (the (i, j)th element of this matrix is the weight of the arc going from transition i to place j).[†] Note that Pre and Post will both typically be* sparse *matrices.[‡] M is a u-dimensional integer vector representing the current marking of the net (i.e. the current state of the system).*

The initial marking of the net is typically denoted M_0. Note that the form of Pre and $Post$ ensure that arcs only exist between nodes of different types, so the resulting network is a bipartite graph. A particular transition, t_i can only fire if $M_j \geq Pre_{ij}$, $j = 1, \ldots, u$.

In order to make this concrete, let us now write out the reaction list for Figure 2.3 in the form of a table, shown in Table 2.1. We can then use this to give a formal Petri net specification of the system.

[†] Non-existent arcs have a weight of zero.

[‡] A *sparse* matrix is a matrix consisting mainly of zeros. There are special algorithms for working with sparse matrices that are much more efficient than working with the full matrix directly. It is hard to be precise about how sparse a matrix has to be before it is worth treating as a sparse matrix, but for an $n \times n$ matrix, if the number of non-zero elements is closer to order n than order n^2, it is likely to be worthwhile using sparse matrix algorithms.

Table 2.2 *Table representing the overall effect of each transition (reaction) on the marking (state) of the network*

Species	$g \cdot P_2$	g	r	P	P_2
Repression	1	−1	0	0	−1
Reverse repression	−1	1	0	0	1
Transcription	0	0	1	0	0
Translation	0	0	0	1	0
Dimerisation	0	0	0	−2	1
Dissociation	0	0	0	2	−1
mRNA degradation	0	0	−1	0	0
Protein degradation	0	0	0	−1	0

$$N = (P, T, Pre, Post, M), \quad P = \begin{pmatrix} g \cdot P_2 \\ g \\ r \\ P \\ P_2 \end{pmatrix}, \quad T = \begin{pmatrix} \text{Repression} \\ \text{Reverse repression} \\ \text{Transcription} \\ \text{Translation} \\ \text{Dimerisation} \\ \text{Dissociation} \\ \text{mRNA degradation} \\ \text{Protein degradation} \end{pmatrix}$$

$$Pre = \begin{pmatrix} 0 & 1 & 0 & 0 & 1 \\ 1 & 0 & 0 & 0 & 0 \\ 0 & 1 & 0 & 0 & 0 \\ 0 & 0 & 1 & 0 & 0 \\ 0 & 0 & 0 & 2 & 0 \\ 0 & 0 & 0 & 0 & 1 \\ 0 & 0 & 1 & 0 & 0 \\ 0 & 0 & 0 & 1 & 0 \end{pmatrix}, \quad Post = \begin{pmatrix} 1 & 0 & 0 & 0 & 0 \\ 0 & 1 & 0 & 0 & 1 \\ 0 & 1 & 1 & 0 & 0 \\ 0 & 0 & 1 & 1 & 0 \\ 0 & 0 & 0 & 0 & 1 \\ 0 & 0 & 0 & 2 & 0 \\ 0 & 0 & 0 & 0 & 0 \\ 0 & 0 & 0 & 0 & 0 \end{pmatrix}, \quad M = \begin{pmatrix} 0 \\ 1 \\ 2 \\ 10 \\ 12 \end{pmatrix}.$$

Now when a particular transition (reaction) occurs (given by a particular row of Table 2.1), the numbers of tokens associated with each place (species) will decrease according to the numbers on the LHS (Pre) and increase according to the numbers on the RHS ($Post$). So, it is the *difference* between the RHS and the LHS that is important for calculating the change in state associated with a given transition (or reaction). We can write this matrix out in the form of a table, shown in Table 2.2, or more formally as a matrix

$$A = Post - Pre = \begin{pmatrix} 1 & -1 & 0 & 0 & -1 \\ -1 & 1 & 0 & 0 & 1 \\ 0 & 0 & 1 & 0 & 0 \\ 0 & 0 & 0 & 1 & 0 \\ 0 & 0 & 0 & -2 & 1 \\ 0 & 0 & 0 & 2 & -1 \\ 0 & 0 & -1 & 0 & 0 \\ 0 & 0 & 0 & -1 & 0 \end{pmatrix}.$$

This 'net effect' matrix, A, is of fundamental importance in the theory and application of Petri nets and chemical reaction networks. Unfortunately there is no agreed standard notation for this matrix within either the Petri net or biochemical network literature. Within the Petri net literature, it is usually referred to as the *incidence matrix* and is often denoted by the letter A, C, or I.[§] Within the biochemical network field, it is usually referred to as the *reaction* or *stoichiometry matrix* and often denoted by the letter A or S. As if this were not confusing enough, the matrix is often (but by no means always) defined to be the *transpose* of the matrix we have called A,[¶] as this is often more convenient to work with. Suffice to say that care needs to be taken when exploring and interpreting the wider literature in this area. For clarity and convenience throughout this book, A (the reaction matrix) will represent the matrix as we have already defined it, and S (the stoichiometry matrix), will be used to denote its transpose.

Definition 2.4 *The* reaction matrix

$$A = Post - Pre$$

is the $v \times u$-dimensional matrix whose rows represent the effect of individual transitions (reactions) on the marking (state) of the network. Similarly, the stoichiometry matrix

$$S = A^T$$

is the $u \times v$-dimensional matrix whose columns represent the effect of individual transitions on the marking of the network.[||]

However, it must be emphasised that this is not a universally adopted notation.

Now suppose that we have some reaction events. For example, suppose we have one repression binding reaction and one translation reaction. We could write this list of reactions in a table as

---

[§] I is a particularly bad choice, as this is typically used in linear algebra to denote the *identity matrix*, which has '1's along the diagonal and '0's elsewhere.

[¶] The transpose of a matrix is the matrix obtained by interchanging the rows and columns of a matrix. So in this case it would be a matrix where the rows represented places (species) and the columns represented transitions (reactions).

[||] Here and elsewhere, the notation ^{T} is used to denote the transpose of a vector or matrix.

Reaction	No. transitions
Repression	1
Reverse repression	0
Transcription	0
Translation	1
Dimerisation	0
Dissociation	0
mRNA degradation	0
Protein degradation	0

or more neatly as a vector

$$r = \begin{pmatrix} 1 \\ 0 \\ 0 \\ 1 \\ 0 \\ 0 \\ 0 \\ 0 \end{pmatrix}.$$

Note that in order to save space, column vectors such as r are sometimes written as the transpose of row vectors, e.g., $r = (1, 0, 0, 1, 0, 0, 0, 0)^{\mathsf{T}}$. We can now use matrix algebra to update the marking (state) of the network.

Proposition 2.1 *If r represents the transitions that have taken place subsequent to the marking M, the new marking $\tilde{M}$ is related to the old marking via the matrix equation*[**]

$$\tilde{M} = M + Sr. \tag{2.1}$$

Note that this equation (2.1) is of fundamental importance both for the mathematical analysis of Petri nets and biochemical networks and also for the development of simulation and inference algorithms. We will use this equation extensively in a variety of different ways throughout the book. Before we justify it, it is probably helpful to see a simple and direct application of it in practice.

In the context of our example, we can compute the new marking from the old

---

[**] Note that in matrix equations, addition is defined element-wise, but multiplication is defined in a special way. The product of the $n \times m$ matrix A and the $m \times p$ matrix B is the $n \times p$ matrix whose (i, j)th element is $\sum_{k=1}^{m} a_{ik} b_{kj}$. A vector is treated as a matrix with one column.

marking as

$$\tilde{M} = M + Sr$$
$$= M + A^{\mathsf{T}}r$$

$$= \begin{pmatrix} 0 \\ 1 \\ 2 \\ 10 \\ 12 \end{pmatrix} + \begin{pmatrix} 1 & -1 & 0 & 0 & -1 \\ -1 & 1 & 0 & 0 & 1 \\ 0 & 0 & 1 & 0 & 0 \\ 0 & 0 & 0 & 1 & 0 \\ 0 & 0 & 0 & -2 & 1 \\ 0 & 0 & 0 & 2 & -1 \\ 0 & 0 & -1 & 0 & 0 \\ 0 & 0 & 0 & -1 & 0 \end{pmatrix}^{\mathsf{T}} \begin{pmatrix} 1 \\ 0 \\ 0 \\ 1 \\ 0 \\ 0 \\ 0 \\ 0 \end{pmatrix}$$

$$= \begin{pmatrix} 0 \\ 1 \\ 2 \\ 10 \\ 12 \end{pmatrix} + \begin{pmatrix} 1 & -1 & 0 & 0 & 0 & 0 & 0 & 0 \\ -1 & 1 & 0 & 0 & 0 & 0 & 0 & 0 \\ 0 & 0 & 1 & 0 & 0 & 0 & -1 & 0 \\ 0 & 0 & 0 & 1 & -2 & 2 & 0 & -1 \\ -1 & 1 & 0 & 0 & 1 & -1 & 0 & 0 \end{pmatrix} \begin{pmatrix} 1 \\ 0 \\ 0 \\ 1 \\ 0 \\ 0 \\ 0 \\ 0 \end{pmatrix}$$

$$= \begin{pmatrix} 0 \\ 1 \\ 2 \\ 10 \\ 12 \end{pmatrix} + \begin{pmatrix} 1 \\ -1 \\ 0 \\ 1 \\ -1 \end{pmatrix}$$

$$= \begin{pmatrix} 1 \\ 0 \\ 2 \\ 11 \\ 11 \end{pmatrix}.$$

We can see that (2.1) appears to have worked in this particular example, but we need to establish why it is true in general.

Proof. The ith row of A represents the effect on the marking of the ith reaction. The same is true of the ith column of S, which we denote s^i. Clearly r_i is the number of type i reactions that take place, so the change in marking, $\tilde{M} - M$, is given by

$$\tilde{M} - M = s^1 r_1 + \cdots + s^v r_v$$
$$= \sum_{i=1}^{v} s^i r_i$$

$$= \sum_{i=1}^{v} S e_i r_i$$

$$= S \sum_{i=1}^{v} e_i r_i$$

$$= S r,$$

where e_i is the ith unit v-vector.[††]

$\square$

2.3.3 Network invariants and conservation laws

There are two Petri net concepts that are of particular relevance to biochemical networks: P- and T-invariants.

Definition 2.5 *A P-invariant (sometimes referred to in the literature as an S-invariant) is a non-zero u-vector y that is a solution to the matrix equation $Ay = 0$. That is, y is any non-zero vector in the null-space of A.*[‡‡]

The null-space of A therefore characterises the set of P-invariants. These P-invariants are interesting because they correspond to *conservation laws* of the network.

In the example we have been studying, it is clear that the vector $y = (1, 1, 0, 0, 0)^{\mathsf{T}}$ is a P-invariant (as $Ay = 0$). This vector corresponds to the fairly obvious conservation law

$$g \cdot P_2 + g = \text{Constant}.$$

That is, the total number of copies of the gene does not change. It is true in general that if y is a P-invariant then the linear combination of states, $y^{\mathsf{T}} M$, is conserved by the reaction network. To see why this works, we can evaluate the current linear combination by computing $y^{\mathsf{T}} M$, where M is the current marking. Similarly, the value of the linear combination when the marking is $\tilde{M}$ is $y^{\mathsf{T}} \tilde{M}$. So the change in the linear combination is

$$y^{\mathsf{T}} \tilde{M} - y^{\mathsf{T}} M = y^{\mathsf{T}} (\tilde{M} - M)$$
$$= y^{\mathsf{T}} S r$$
$$= (S^{\mathsf{T}} y)^{\mathsf{T}} r$$
$$= (Ay)^{\mathsf{T}} r$$
$$= 0,$$

where the second line follows from (2.1) and the last line follows from the fact that y is a P-invariant.[§§]

[††] The ith unit vector, e_i, is the vector with a 1 in the ith position and zeros elsewhere. Multiplying a matrix by e_i has the effect of picking out the ith column.

[‡‡] The null-space of a matrix (sometimes known as the kernel) is defined to be the set of all vectors that get mapped to zero by the matrix.

[§§] We also used the fact that for arbitrary (conformable) matrices A and B, we have $(AB)^{\mathsf{T}} = B^{\mathsf{T}} A^{\mathsf{T}}$.

Definition 2.6 *A T-invariant is a non-zero, non-negative (integer-valued) v-vector x that is a solution to the matrix equation $Sx = 0$. That is, x is in the null-space of S.*

These invariants are of interest because they correspond to sequences of transitions (reactions) that return the system to its original marking (state). This is clear immediately from (2.1). If the primary concern is continuous deterministic modelling, then any non-negative solution is of interest. However, if the main interest is in discrete stochastic models of biochemical networks, only non-negative integer vector solutions correspond to sequences of transitions (reactions) that can actually take place. Hence, we will typically want to restrict our attention to these.

In the example we have been considering, it is easily seen that the vectors $x = (1, 1, 0, 0, 0, 0, 0, 0)^\mathsf{T}$ and $\tilde{x} = (0, 0, 0, 0, 1, 1, 0, 0)^\mathsf{T}$ are both T-invariants of our network. The first corresponds to a repression binding and its reverse reaction, and the second corresponds to a dimerisation and corresponding dissociation. However, not all T-invariants are associated with reversible reactions.

Although it is trivial to verify whether a given vector is a P- or T-invariant, it is perhaps less obvious how to systematically find such invariants and classify the set of all such invariants for a given Petri net. In fact, the singular value decomposition (SVD) is a classic matrix algorithm (Golub and Van Loan, 1996) that completely characterises the null-space of a matrix and its transpose and hence helps considerably in this task. However, if we restrict our attention to positive integer solutions, there is still more work to be done even once we have the SVD. We will revisit this issue in Chapter 11 when the need to find invariants becomes more pressing.

Before leaving the topic of invariants, it is worth exploring the relationship between the number of (linearly independent) P- and T-invariants. The *column-rank* of a matrix is the dimension of the image-space of the matrix (the space spanned by the columns of the matrix). The *row-rank* is the column-rank of the transpose of the matrix. It is a well-known result from linear algebra that the row and column ranks are the same, and so we can refer unambiguously to the *rank* of a matrix. By the *rank-nullity theorem*, the dimension of the image-space and null-space must sum to the dimension of the space being operated on, which is the number of columns of the matrix. So, if we fix on S, which has dimension $u \times v$, suppose the rank of the matrix is k. Let the dimension of the null-space of S be t and the dimension of the null-space of $A(= S^\mathsf{T})$ be p. Then we have $k + p = u$ and $k + t = v$. This leads immediately to the following result.

Proposition 2.2 *The number of linearly independent P-invariants, p, and the number of linearly independent T-invariants, t, are related by*

$$t - p = v - u. \tag{2.2}$$

In the context of our example, we have $u = 5$ and $v = 8$. As we have found a P-invariant, we know that $p \geq 1$. Now using (2.2) we can deduce that $t \geq 4$. So there are at least four linearly independent T-invariants for this network, and we have so far found only two.

2.3.4 Reachability

Another Petri net concept of considerable relevance is that of *reachability*.

Definition 2.7 *A marking* $\tilde{M}$ *is* reachable *from marking* M *if there exists a finite sequence of transitions leading from* M *to* $\tilde{M}$.

If such a sequence of transitions is summarised in the v-vector r, it is clear that r will be a non-negative integer solution to (2.1). However, it is important to note that the converse does not follow: the existence of a non-negative integer solution to (2.1) does not guarantee the reachability of $\tilde{M}$ from M. This is because the markings have to be non-negative. A transition cannot occur unless the number of tokens at each place is at least that required for the transition to take place. It can happen that there exists a set of non-negative integer transitions between two valid markings M and $\tilde{M}$, but all possible sequences corresponding to this set are impossible.

This issue is best illustrated with an example. Suppose we have the reaction network

$$2X \longrightarrow X_2$$
$$X_2 \longrightarrow 2X$$
$$X_2 \longrightarrow X_2 + Y.$$

So X can dimerise and dissociate, and dimers of X somehow catalyse the production of Y. If we formulate this as a Petri net with $P = (X, X_2, Y)$, and the transitions in the above order, we get the stoichiometry matrix

$$S = \begin{pmatrix} -2 & 2 & 0 \\ 1 & -1 & 0 \\ 0 & 0 & 1 \end{pmatrix}.$$

If we begin by thinking about getting from initial marking $M = (2,0,0)^\mathsf{T}$ to marking $\tilde{M} = (2,0,1)^\mathsf{T}$, we see that it is possible and can be achieved with the sequence of transitions t_1, t_3, and t_2, giving reaction vector $r = (1,1,1)^\mathsf{T}$. We can now ask if marking $\tilde{M} = (1,0,1)^\mathsf{T}$ is reachable from $M = (1,0,0)^\mathsf{T}$. If we look for a non-negative integer solution to (2.1), we again find that $r = (1,1,1)^\mathsf{T}$ is appropriate, as $\tilde{M} - M$ is the same in both scenarios. However, this does not correspond to any legal sequence of transitions, as *no* transitions are legal from the initial marking $M = (1,0,0)^\mathsf{T}$. In fact, $r = (0,0,1)^\mathsf{T}$ is another solution, since $(1,1,0)^\mathsf{T}$ is a T-invariant. This solution is forbidden despite the fact that firing of t_3 will not cause the marking to go negative.

This is a useful warning that although the matrix representation of Petri net (and biochemical network) theory is powerful and attractive, it is not a complete characterisation — discrete-event systems analysis is a delicate matter, and there is much that systems biologists interested in discrete stochastic models can learn from the Petri net literature.

2.4 Stochastic process algebras

Stochastic process algebras are another way of representing reactive systems as collections of concurrent and independent 'agents' or 'processes' which interact with other processes via 'reaction channels'. There are many variations on this approach; here we consider briefly the two methods that are most commonly used in the context of modelling biochemical networks.

2.4.1 Stochastic π-calculus

The usual approach to modelling biochemical networks with π-calculus is to model each molecule in the system as an independent 'agent' or 'process' which interacts with other processes via reaction channels. The stochastic π-calculus is a simple description of how processes are transformed, and in interaction with what other processes. This is most easily explained in the context of a very simple example. We will use the Lotka–Volterra example from Section 1.6:

$$Y_1 \longrightarrow 2Y_1$$
$$Y_1 + Y_2 \longrightarrow 2Y_2$$
$$Y_2 \longrightarrow \emptyset.$$

In its most basic form, the π-calculus requires all processes to pairwise interact, which corresponds to chemical reactions involving exactly two species. Here this can be accomplished by adding in dummy species D_1 and D_2 whose levels are unaffected by the reactions as follows:

$$D_1 + Y_1 \xrightarrow{r_1} D_1 + 2Y_1$$
$$Y_1 + Y_2 \xrightarrow{r_2} 2Y_2$$
$$D_2 + Y_2 \xrightarrow{r_3} D_2.$$

This is now in the form of a reaction system involving four species (processes) and three reaction channels (labelled r_1, r_2, r_3) which can be converted directly into the π-calculus. For each of the four species in the reaction system, we consider the reactions it can participate in, and write out a corresponding process rule. There are several variations on the precise notation used, but typically the resulting system will be presented in something like the following way:

$$D_1 = ?r_1.D_1$$
$$Y_1 = !r_1.(Y_1|Y_1) + ?r_2.\emptyset$$
$$Y_2 = !r_2.(Y_2|Y_2) + !r_3.\emptyset$$
$$D_2 = ?r_3.D_2.$$

The first line says that D_1 can only participate in reaction r_1, and that when it does, it is transformed to itself. The ? means that it is an 'input' for channel r_1, and as such can only interact with processes that 'output' on r_1. Note that the terms 'input' and 'output' are *not* synonymous with 'reactants' and 'products'. They are just terms to identify the respective ends of the pairwise interaction between (reacting) processes.

The second line says that Y_1 can participate in either r_1 (as an output) or r_2 (as an input). If it participates in r_1, the result will be two copies of Y_1, and if it participates in r_2, it will simply disappear. The third and fourth lines are similar. Together the rules represent a description of the reaction system equivalent to the list of biochemical reactions.

In the context of the small networks considered in this text, it is difficult to see the advantage of this representation over the basic reaction list or corresponding Petri net representation, but advocates claim that the π-calculus scales much better than methods based on lists of reactions in the context of larger, more complex models. See Phillips and Cardelli (2007) and Blossey et al. (2006) for further details of this modelling approach.

2.4.2 Bio-PEPA

PEPA is a stochastic process algebra closely related to the π-calculus, aimed at analysis and evaluation of concurrent systems. Bio-PEPA is an extension specifically targeted at the modelling of biochemical reaction networks. In many ways, it is an attempt to formalise reaction models as described in SBML (discussed in the next section). Bio-PEPA adopts the 'species as processes' abstraction model, similarly to the Petri-net approach. In Bio-PEPA, the Lotka–Volterra model previously discussed could be represented as follows:

$$Y_1 \stackrel{\text{def}}{=} (r_1, 1){\downarrow}Y_1 + (r_1, 2){\uparrow}Y_1 + (r_2, 1){\downarrow}Y_1$$
$$Y_2 \stackrel{\text{def}}{=} (r_2, 1){\downarrow}Y_2 + (r_2, 2){\uparrow}Y_2 + (r_3, 1){\downarrow}Y_2$$

$$Y_1(y_{1,0}) \underset{\{r_2\}}{\bowtie} Y_2(y_{2,0}).$$

The first line defines the transformations affecting species Y_1. The first term: $(r_1, 1){\downarrow}$ Y_1 means that Y_1 participates as a reactant (hence $\downarrow$) in reaction r_1, where it has stoichiometry 1. In the second term, $\uparrow$ is used to flag a product (which has stoichiometry 2 in reaction r_1). Just as for π-calculus, $+$ is used to delimit choices. The second line describes the transformations affecting Y_2. Line 3 completes the definition of the model, and describes the synchronisation requirements. Here species Y_1 and Y_2 must be synchronised for a reaction of type r_2. The values $y_{1,0}$ and $y_{2,0}$ represent the initial amounts (or concentrations) of Y_1 and Y_2, respectively, and hence can be used to initialise the state of the model.

Again, in the context of simple biochemical network models, it is difficult to see any great advantage of this approach over the stochastic Petri net representation. The advantages of Bio-PEPA over the stochastic π-calculus is that it more closely matches the structure of biochemical network models. For example, it explicitly encodes stoichiometry, allows synchronisations involving more than two reactants, and also allows the use of general rate laws (not discussed here). For further information about this approach to network modelling, see Ciocchetta and Hillston (2009) and references therein.

2.5 Systems Biology Markup Language (SBML)

Different representations of biochemical networks are useful for different purposes. Graphical representations (including Petri nets) are useful both for visualisation and analysis, and matrix representations (and other formal representations, such as process algebras) are useful for mathematical and computational analysis. The Systems Biology Markup Language (SBML), described in Hucka et al. (2003), is a way of representing biochemical networks that is intended to be convenient for computer software to generate and parse, thereby enabling communication of biochemical network models between disparate modelling and simulation tools. It is essentially an eXtensible Markup Language (XML) encoding (DuCharme, 1999) of the reaction list, together with the additional information required for quantitative modelling and simulation. It is intended to be independent of particular kinetic theories and should be as appropriate for discrete stochastic models as for continuous deterministic ones. The existence of a widely adopted exchange format for systems biology models makes it possible to develop a web-based community resource for archiving and sharing of models. The BioModels database, described in Le Novère et al. (2006), is just such a resource, and contains quantitative models covering an range of organisms and cellular processes, mainly encoded using SBML.

We will concentrate here on the version of SBML known as Level 3 (version 1, core), as this is the most popular variant of the recent specifications (at the time of writing) and contains sufficient features for the biochemical network models considered in this book. Further details regarding SBML, including the specification and XML Schema, can be obtained from the SBML.org website. Note that SBML should perhaps not be regarded as an alternative to other representations, but simply as an electronic format which could in principle be used in conjunction with any of the representations we have considered (though it corresponds most closely to the Petri net representation). Also note that it is not intended that SBML models should be generated and manipulated 'by hand' using a text editor, but rather by software tools which present to the user a more human-oriented representation. It is also worth bearing in mind that SBML continues to evolve. At the time of writing, SBML Level 3 (version 2, core) is the current specification, and various packages are available which extend the functionality and feature coverage of SBML in various ways. SBML (Levels 2 and 3) encodes all mathematical formulae using MathML (an XML encoding for mathematical notation) rather than as strings containing algebraic expressions. SBML Level 2 can also be used for encoding the discrete stochastic models covered in this text; this was discussed in detail in the first edition of this book (Wilkinson, 2006). SBML Level 1 is not sufficiently expressive to unambiguously encode discrete stochastic models, and should now be considered obsolete.

2.5.1 Basic document structure

An SBML (Level 3) model consists of lists of *functions*, *units*, *compartments*, *species*, *parameters*, *initial assignments*, *rules*, *constraints*, *reactions*, and *events*. Each of these lists is optional. We will concentrate here on units, compartments, species, parameters, and reactions, as these are sufficient for adequately describing most simple discrete stochastic models. This basic structure is encoded in SBML as follows.

```
<?xml version="1.0" encoding="UTF-8"?>
<sbml xmlns="http://www.sbml.org/sbml/level3/version1/core"
    level="3" version="1">
  <model id="MyBiochemicalNetwork" name="My␣biochemical␣
      network" substanceUnits="item" timeUnits="second"
      volumeUnits="litre" extentUnits="item">
    <listOfUnitDefinitions>
      ...
    </listOfUnitDefinitions>
    <listOfCompartments>
      ...
    </listOfCompartments>
    <listOfSpecies>
      ...
    </listOfSpecies>
    <listOfParameters>
      ...
    </listOfParameters>
    <listOfReactions>
      ...
    </listOfReactions>
  </model>
</sbml>
```

Note that as part of the model declaration it is possible to declare default units for the model. This is a new feature in Level 3. The declaration given is likely to be appropriate for many discrete stochastic models, and means that in many cases a separate `<listOfUnitDefinitions>` section of the model will not be required. However, this is only the case if the model uses built-in SBML units, otherwise a units section will still be necessary. The `extentUnits` are used to define the units of the kinetic laws (to be discussed later), which have units of `extentUnits` per `timeUnits`.

2.5.2 Units

The (optional) units list allows definition and redefinition of the units used by the model. Discrete stochastic models encoded in SBML Level 2 often contained the following units declaration.

```
<listOfUnitDefinitions>
  <unitDefinition id="substance">
    <listOfUnits>
      <unit kind="item"/>
    </listOfUnits>
  </unitDefinition>
</listOfUnitDefinitions>
```

This declaration had the effect of changing the default substance units from the default value (*mole*) to *item*. The effect of this is that subsequent (unqualified) specifications of (and references to) amounts will be assumed to be in the unit of *item*. That is, amounts are interpreted as numbers of molecules rather than the default of numbers of moles. There are no default units in SBML Level 3, but this construct can still be used in an SBML Level 3 document if the model declaration includes `substanceUnits="substance"` rather than `substanceUnits="item"`. Although this is no longer necessary, it is likely to be a commonly encountered declaration in models that have been converted to SBML Level 3 from SBML Level 2. Units turn out to be a rather delicate issue. We will examine units in some detail later in the book when we look at kinetics and rate laws. For further details on using units with SBML, see the specification document which includes an example of a model encoded with discrete stochastic simulation in mind. Note that many simulators in current use ignore the units section of the SBML document. In practice, this means that many deterministic simulators will assume units of mole, and most stochastic simulators will assume units of item, irrespective of the content of this section. However, it is important to ensure that models are encoded correctly so that they are not misinterpreted in the future.

2.5.3 Compartments

The compartment list simply states the compartments in the model. So for a model with two compartments, the declaration might be as follows.

```
<listOfCompartments>
  <compartment id="Cell" spatialDimensions="3" size="1"/>
  <compartment id="Nucleus" spatialDimensions="3" size="1"/>
</listOfCompartments>
```

A simulatable model must have at least one compartment, and each compartment should be given an *id*. You may also specify the number of spatial dimensions the compartment has (typically 3), and a size (or volume) in the current model size units appropriate for the number of spatial dimensions (`volumeUnits` for compartments with 3 spatial dimensions).

2.5.4 Species

The species list simply states all species in the model. So, for the auto-regulatory network model we have been considering throughout this chapter, these could be declared using

```
<listOfSpecies>
  <species id="Gene" compartment="Cell" initialAmount="10"
      hasOnlySubstanceUnits="true"/>
  <species id="P2Gene" name="P2.Gene" compartment="Cell"
      initialAmount="0" hasOnlySubstanceUnits="true"/>
  <species id="Rna" compartment="Cell" initialAmount="0"
      hasOnlySubstanceUnits="true"/>
```

```
      <species id="P" compartment="Cell" initialAmount="0"
          hasOnlySubstanceUnits="true"/>
      <species id="P2" compartment="Cell" initialAmount="0"
          hasOnlySubstanceUnits="true"/>
  </listOfSpecies>
```

There are several things worth pointing out about this declaration. First, the initial amounts are assumed to be in the model substance units, unless explicit units are specified (see the specification for how to do this). Also note that each species is declared with the attribute `hasOnlySubstanceUnits`. This has the effect of ensuring that wherever the species is referred to elsewhere in the model (for example, in rate laws), it will be interpreted as an amount (in the appropriate substance units), and not a concentration (substance per size). Most stochastic simulators will make this assumption anyway, but it is important to encode models correctly so that they will not be misinterpreted by tools which correctly interpret the specification. Each species also declares the compartment in which it resides. For species using substance units, this is largely a model annotation, but for species with concentration units, it provides a link to the size information needed to calculate the concentration.

2.5.5 Parameters

The parameters section can be used to declare names for numeric values to be used in algebraic formulae. They are most often used to declare rate constants for the kinetic laws of biochemical reactions, but can be used for other variables as well. An example parameter section might be as follows.

```
<listOfParameters>
  <parameter id="k1" value="0.01"/>
  <parameter id="k2" value="0.1"/>
</listOfParameters>
```

Parameters defined here are 'global' to the whole model. In contrast, any parameters defined in the context of a particular reaction will be local to the kinetic law for that reaction only.

2.5.6 Reactions

The reaction list consists of a list of reactions. A reaction in turn consists of a list of reactants, a list of products, and a rate law. A reaction may also declare 'modifier' species — species that are not created or destroyed by the reaction but figure in the rate law. For example, consider the following encoding of a dimerisation reaction.

```
      <reaction id="Dimerisation" reversible="false">
        <listOfReactants>
          <speciesReference species="P" stoichiometry="2"/>
        </listOfReactants>
        <listOfProducts>
          <speciesReference species="P2" stoichiometry="1"/>
        </listOfProducts>
```

```
<kineticLaw>
  <math xmlns="http://www.w3.org/1998/Math/MathML">
    <apply>
      <times/>
      <ci> k4 </ci>
      <cn> 0.5 </cn>
      <ci> P </ci>
      <apply>
        <minus/>
        <ci> P </ci>
        <cn type="integer"> 1 </cn>
      </apply>
    </apply>
  </math>
  <listOfLocalParameters>
    <localParameter id="k4" value="1"/>
  </listOfLocalParameters>
</kineticLaw>
</reaction>
```

There are several things to note about this declaration. One thing that perhaps appears strange is that the reaction is declared to be not reversible when we know that dimerisation is typically a reversible process. However, reactions should only be flagged as reversible when the associated kinetic law represents the combined effects of both forward and backward reactions. It turns out that while this is fine for continuous deterministic models, there is no satisfactory way to do this for discrete stochastic models. As a result, when developing a model for discrete stochastic simulation, the forward and backward reactions must be specified separately (and declared not reversible) along with their separate kinetic laws. We will examine the meaning and specification of reaction rates and kinetic laws later in this book. The specification of reactants and products and their associated stoichiometries is fairly self-explanatory. The kinetic law itself is a MathML (W3C, 2000) encoding of the simple algebraic formula `k4*0.5*P*(P-1)`. Rate laws will be discussed in more detail later, but it is important to know that the units of this law are of the form *substance / time*, using the model substance and time units. The kinetic law given above uses a local parameter `k4`. This constant will always be used in the formula of the kinetic law, masking any global parameter of the same name. To use a global parameter called `k4`, the entire section

```
<listOfParameters>
  <parameter name="k4" value="1"/>
</listOfParameters>
```

should be removed from the kinetic law. Kinetic laws can use a mixture of global and local parameters. Any reference to a compartment will be replaced by the size (volume) of the compartment. A list of reactions should be included in the SBML file between `<listOfReactions>` and `</listOfReactions>` tags.

2.5.7 The full SBML model

The various model components are embedded into the basic model structure. For completeness, Appendix A.1 lists a full SBML model for the simple auto-regulatory network we have been considering. Note that as this model uses locally specified parameters rather than globally defined parameters, there is no <listOfParameters> section in the model definition. This model can also be downloaded from the book's website.

2.6 SBML-shorthand

2.6.1 Introduction

SBML has become the *lingua franca* for electronic representation and exchange of models of interest in systems biology. Dozens of different software tools provide SBML support to varying degrees, many providing both SBML import and export provisions, and some using SBML as their native format. However, while SBML is a good format for computers to parse and generate, its verbosity, pedantic syntax, and low signal-to-noise ratio make it rather inconvenient for humans to read and write. I have found it helpful to develop a shorthand notation for SBML that is much easier for humans to read and write, and can easily be 'compiled' into SBML for subsequent import into other SBML-aware software tools. The notation can be used as a partial substitute for the numerous GUI-based model-building tools that are widely used for systems biology model development. An additional advantage of the notation is that it is much more suitable than raw SBML for presentation in a book such as this (because it is more concise and readable). Many of the examples discussed in subsequent chapters will be presented using the shorthand notation, so it is worth presenting the essential details here. Here we describe a particular version of the shorthand notation, known as 3.1.1. A compiler for translating the shorthand notation into full SBML is freely available; see Appendix B.2 for details.

2.6.2 Basic structure

The description format is plain ASCII text. The suggested file extension is .mod, but this is not required. All whitespace other than carriage returns is insignificant (unless it is contained within a quoted 'name' element). Carriage returns are significant. The description is case-sensitive. Blank lines are ignored. The comment character is # — all text from a # to the end of the line is ignored.

The model description must begin with the characters **@model**:3.1.1= (the 3.1.1 corresponds to the version number of the specification). The text following the = on the first line is the model identification string (ID). An optional model name may also be specified, following the ID, enclosed in double quotes. This model declaration line may be followed by an optional additional line declaring model units. For example, the shorthand text

```
@model:3.1.1=AutoRegulatoryNetwork "Auto-regulatory network"
 s=item, t=second, v=litre, e=item
```

will be translated to

```
<?xml version="1.0" encoding="UTF-8"?>
<sbml xmlns="http://www.sbml.org/sbml/level3/version1/core"
    level="3" version="1">
  <model id="AutoRegulatoryNetwork" name="Auto-regulatory␣
      network" substanceUnits="item" timeUnits="second"
      volumeUnits="litre" extentUnits="item">
```

The model is completed with the specification of up to seven additional sections: **@units**, **@compartments**, **@species**, **@parameters**, **@rules**, **@reactions**, and **@events**, each corresponding to an analogous SBML section. Here we concentrate on the five sections, **@units**, **@compartments**, **@species**, **@parameters**, and **@reactions**, corresponding to the SBML sections, <listOfUnitDefinitions >, <listOfCompartments>, <listOfSpecies>, <listOfParameters>, and < listOfReactions>, respectively. The sections must occur in the stated order. Sections are optional, but if present, may not be empty. These are the only sections covered by this specification.

2.6.3 Units

The format of the individual sections will be explained mainly by example. The following SBML-shorthand

@units
```
 substance=item
 mmls=mole:s=-3; litre:e=-1; second:e=-1
```

would be translated to

```
    <listOfUnitDefinitions>
      <unitDefinition id="substance">
        <listOfUnits>
          <unit kind="item" exponent="1" scale="0" multiplier=
            "1"/>
        </listOfUnits>
      </unitDefinition>
      <unitDefinition id="mmls">
        <listOfUnits>
          <unit kind="mole" exponent="1" scale="-3" multiplier
            ="1"/>
          <unit kind="litre" exponent="-1" scale="0"
            multiplier="1"/>
          <unit kind="second" exponent="-1" scale="0"
            multiplier="1"/>
        </listOfUnits>
      </unitDefinition>
    </listOfUnitDefinitions>
```

The unit attributes exponent, multiplier, and scale are denoted by the letters e, m, and s, respectively. Note that because there is no way to refer to units elsewhere in

SBML-shorthand, the only function for this section is to redefine model-wide units such as `substance` and `size` (named in the model declaration).

2.6.4 Compartments

The following SBML-shorthand

```
@compartments
 Cell=1
 Nucleus=0.1
 Cyto=0.8 "Cytoplasm"
 Mito "Mitochondria"
```

would be translated to

```
    <listOfCompartments>
      <compartment id="Cell" size="1"/>
      <compartment id="Nucleus" size="0.1"/>
      <compartment id="Cyto" name="Cytoplasm" size="0.8"/>
      <compartment id="Mito" name="Mitochondria"/>
    </listOfCompartments>
```

Note that if a `name` attribute is to be specified, it should be specified at the end of the line in double quotes. This is true for other SBML elements too.

2.6.5 Species

The following shorthand

```
@species
 cell:Gene = 10b "The Gene"
 cell:P2=0
 cell:S1=100 s
 cell:[S2]=20 sc
 cell:[S3]=1000 bc
 mito:S4=0 b
```

would be translated to

```
    <listOfSpecies>
      <species id="Gene" name="The Gene" compartment="cell"
              initialAmount="10" boundaryCondition="true"/>
      <species id="P2" compartment="cell" initialAmount="0"/>
      <species id="S1" compartment="cell" initialAmount="100"
              hasOnlySubstanceUnits="true"/>
      <species id="S2" compartment="cell"
              initialConcentration="20"
              hasOnlySubstanceUnits="true" constant="true"/>
      <species id="S3" compartment="cell"
              initialConcentration="1000"
              boundaryCondition="true" constant="true"/>
```

```
        <species id="S4" compartment="mito" initialAmount="0"
            boundaryCondition="true"/>
    </listOfSpecies>
```

Compartments are compulsory. An `initialConcentration` (as opposed to an `initialAmount`) is flagged by enclosing the species `id` in brackets. The boolean attributes `hasOnlySubstanceUnits`, `boundaryCondition`, and `constant` can be set to `true` by appending the letters s, b, and c, respectively. The order of the flags is not important.

2.6.6 Parameters

The section

```
@parameters
 k1=1
 k2=10
```

would be translated to

```
  <listOfParameters>
   <parameter name="k1" value="1"/>
   <parameter name="k2" value="10"/>
  </listOfParameters>
```

2.6.7 Reactions

Each reaction is specified by exactly two or three lines of text. The first line declares the reaction name and whether the reaction is reversible (`@rr=` for reversible and `@r=` otherwise). The second line specifies the reaction itself using a fairly standard notation. The (optional) third line specifies the full rate law for the kinetics. If local parameters are used, they should be declared on the same line in a comma-separated list (separated from the rate law using a :).

So for example,

```
@reactions
@r=RepressionBinding "Repression Binding"
 Gene + 2P -> P2Gene
 k2*Gene
@rr=Reverse
 P2Gene -> Gene+2P
 k1r*P2Gene : k1r=1,k2=3
@r=NoKL
 Harry->Jim
@r=Test
 Fred -> Fred2
 k4*Fred : k4=1
```

would translate to

```
<listOfReactions>
  <reaction id="RepressionBinding" name="Repression␣
    Binding" reversible="false">
    <listOfReactants>
      <speciesReference species="Gene" stoichiometry="1"/>
      <speciesReference species="P" stoichiometry="2"/>
    </listOfReactants>
    <listOfProducts>
      <speciesReference species="P2Gene" stoichiometry="1"
        />
    </listOfProducts>
    <kineticLaw>
      <math xmlns="http://www.w3.org/1998/Math/MathML">
        <apply>
          <times/>
          <ci> k2 </ci>
          <ci> Gene </ci>
        </apply>
      </math>
    </kineticLaw>
  </reaction>
  <reaction id="Reverse" reversible="true">
    <listOfReactants>
      <speciesReference species="P2Gene" stoichiometry="1"
        />
    </listOfReactants>
    <listOfProducts>
      <speciesReference species="Gene" stoichiometry="1"/>
      <speciesReference species="P" stoichiometry="2"/>
    </listOfProducts>
    <kineticLaw>
      <math xmlns="http://www.w3.org/1998/Math/MathML">
        <apply>
          <times/>
          <ci> k1r </ci>
          <ci> P2Gene </ci>
        </apply>
      </math>
      <listOfLocalParameters>
        <localParameter id="k1r" value="1"/>
        <localParameter id="k2" value="3"/>
      </listOfLocalParameters>
    </kineticLaw>
  </reaction>
  <reaction id="NoKL" reversible="false">
    <listOfReactants>
      <speciesReference species="Harry" stoichiometry="1"/
        >
    </listOfReactants>
```

```
        <listOfProducts>
          <speciesReference species="Jim" stoichiometry="1"/>
        </listOfProducts>
      </reaction>
      <reaction id="Test" reversible="false">
        <listOfReactants>
          <speciesReference species="Fred" stoichiometry="1"/>
        </listOfReactants>
        <listOfProducts>
          <speciesReference species="Fred2" stoichiometry="1"/
            >
        </listOfProducts>
        <kineticLaw>
          <math xmlns="http://www.w3.org/1998/Math/MathML">
            <apply>
              <times/>
              <ci> k4 </ci>
              <ci> Fred </ci>
            </apply>
          </math>
          <listOfLocalParameters>
            <localParameter id="k4" value="1"/>
          </listOfLocalParameters>
        </kineticLaw>
      </reaction>
  </listOfReactions>
```

For information on the other SBML-shorthand sections, see the specification document on the SBML-shorthand website.

2.6.8 Example

The auto-regulatory network whose SBML is given in Appendix A.1 can be represented in SBML-shorthand in the following way.

```
@model:3.1.1=AutoRegulatoryNetwork "Auto-regulatory network"
 s=item, t=second, v=litre, e=item
@compartments
 Cell
@species
 Cell:Gene=10 s
 Cell:P2Gene=0 s "P2.Gene"
 Cell:Rna=0 s
 Cell:P=0 s
 Cell:P2=0 s
@reactions
@r=RepressionBinding "Repression binding"
 Gene+P2 -> P2Gene
 k1*Gene*P2 : k1=1
@r=ReverseRepressionBinding "Reverse repression binding"
```

```
P2Gene -> Gene+P2
k1r*P2Gene : k1r=10
@r=Transcription
 Gene -> Gene+Rna
 k2*Gene : k2=0.01
@r=Translation
 Rna -> Rna+P
 k3*Rna : k3=10
@r=Dimerisation
 2P -> P2
 k4*0.5*P*(P-1) : k4=1
@r=Dissociation
 P2 -> 2P
 k4r*P2 : k4r=1
@r=RnaDegradation "RNA Degradation"
 Rna ->
 k5*Rna : k5=0.1
@r=ProteinDegradation "Protein degradation"
 P ->
 k6*P : k6=0.01
```

2.7 Exercises

1. Go to the book's website* and follow the Chapter 2 links to explore the world of SBML and SBML-aware software tools. In particular, download and read the SBML Level 3 specification document.

2. Consider this simple model of Michaelis–Menten enzyme kinetics

$$S + E \longrightarrow SE$$
$$SE \longrightarrow S + E$$
$$SE \longrightarrow P + E.$$

(a) Represent this reaction network graphically using a Petri net–style diagram.

(b) Represent it mathematically as a Petri net, $N = (P, T, Pre, Post, M)$, assuming that there are currently 100 molecules of substrate S, 20 molecules of the enzyme E, and no molecules of the substrate-enzyme complex SE or the product P.

(c) Calculate the reaction and stoichiometry matrices A and S.

(d) If the first reaction occurs 20 times, the second 10, and the last 5, what will be the new state of the system?

(e) Can you find a different set of transitions that will lead to the same state?

(f) Can you identify any P- or T-invariants for this system?

(g) Write the model using SBML-shorthand (do not attempt to specify any kinetic laws yet).

* URL: `https://github.com/darrenjw/smfsb`

(h) Hand-translate the SBML-shorthand into SBML, then validate it using the on-line validation tool at the the SBML.org website.

(i) Download and install the SBML-shorthand compiler and use it to translate SBML-shorthand into SBML.

(j) Download and install some SBML-aware model-building tools. Try loading your valid SBML model into the tools, and also try building it from scratch (at the time of writing, COPASI and CellDesigner are popular tools — there should be links to these from the SBML.org website).

2.8 Further reading

Murata (1989) provides a good tutorial introduction to the general theory of P/T Petri nets and Haas (2002) is the definitive guide to stochastic Petri nets. Pinney et al. (2003) and Hardy and Robillard (2004) give introductions to the use of Petri nets in systems biology, and Goss and Peccoud (1998) explore the use of stochastic Petri nets (SPNs) for discrete-event simulation of stochastic kinetic models. Also see Chaouiya (2007) for a more recent review, and Koch et al. (2010) for a recent comprehensive guide to the use of Petri nets in systems biology modelling.

Information on SBML and all things related can be found at the SBML.org website. In particular, background papers on SBML, the various SBML specifications, SBML models, tools for model validation and visualisation, and other software supporting SBML can all be found there.

See Phillips and Cardelli (2007) and Blossey et al. (2006) for further information regarding stochastic π-calculus, and Ciocchetta and Hillston (2009) for an overview of the Bio-PEPA stochastic process algebra.

## PART II

Stochastic processes and simulation

CHAPTER 3

Probability models

3.1 Probability

3.1.1 Sample spaces, events, and sets

The models and representations considered in the previous chapter provide a framework for thinking about the state of a biochemical network, the reactions that can take place, and the change in state that occurs as a result of particular chemical reactions. As yet, however, little has been said about *which* reactions are likely to occur or *when*. The state of a biochemical network evolves continuously through time with discrete changes in state occurring as the result of reaction events. These reaction events are *random*, governed by *probabilistic laws*. It is therefore necessary to have a fairly good background in probability theory in order to properly understand these processes. In a short text such as this, it is impossible to provide complete coverage of all of the necessary material. However, this chapter is meant to provide a quick summary of the essential concepts in a form that should be accessible to anyone with a high school mathematics education who has ever studied some basic probability and statistics. Readers with a strong background in probability will want to skip through this chapter. Note, however, that particular emphasis is placed on the properties of the exponential distribution, as these turn out to be central to understanding the various stochastic simulation algorithms that will be examined in detail in later chapters.

Any readers finding this chapter difficult should go back to a classic introductory text such as Ross (2009). The material in this chapter should provide sufficient background for the next few chapters (concerned with stochastic processes and simulation of biochemical networks). However, it does not cover sufficient statistical theory for the later chapters concerned with inference from data. Suitable additional reading matter for those chapters will be discussed at an appropriate point in the text.

Probability theory is used as a model for situations for which the outcomes occur randomly. Generically such situations are called *experiments*, and the set of all possible outcomes of the experiment is known as the *sample space* corresponding to an experiment. The sample space is usually denoted by S, and a generic element of the sample space (a possible outcome) is denoted by s. The sample space is chosen so that exactly one outcome will occur. The size of the sample space is *finite, countably infinite*, or *uncountably infinite*.

Definition 3.1 *A subset of the sample space (a collection of possible outcomes) is known as an* event. *We write $E \subset S$ if E is a subset of S. Events may be classified into four types:*

the null event *is the empty subset of the sample space;*

an atomic event *is a subset consisting of a single element of the sample space;*

a compound event *is a subset consisting of more than one element of the sample space;*

the sample space *itself is also an event (the* certain *event).*

Definition 3.2

The union *of two events E and F (written $E \cup F$) is the event that at least one of E or F occurs. The union of the events can be obtained by forming the union of the sets.*

The intersection *of two events E and F (written $E \cap F$) is the event that both E and F occur. The intersection of two events can be obtained by forming the intersection of the sets.*

The complement *of an event, A, denoted A^c or $\bar{A}$, is the event that A does* not *occur, and hence consists of all those elements of the sample space that are not in A.*

Two events A and B are disjoint *or* mutually exclusive *if they cannot both occur. That is, their intersection is empty*

$$A \cap B = \emptyset.$$

Note that for any event A, the events A and A^c are disjoint, and their union is the whole of the sample space:

$$A \cap A^c = \emptyset \quad and \quad A \cup A^c = S.$$

The event A is true *if the outcome of the experiment, s, is contained in the event A; that is, if $s \in A$. If $s \notin A$, A is* false. *We say that the event A* implies *the event B, and write $A \Rightarrow B$, if the truth of B automatically follows from the truth of A. If A is a subset of B, then occurrence of A necessarily implies occurrence of the event B. That is*

$$(A \subseteq B) \iff (A \cap B = A) \iff (A \Rightarrow B).$$

In order to carry out more sophisticated manipulations of events, it is helpful to know some basic rules of set theory.

Proposition 3.1

Commutative laws:

$$A \cup B = B \cup A$$
$$A \cap B = B \cap A$$

Associative laws:

$$(A \cup B) \cup C = A \cup (B \cup C)$$
$$(A \cap B) \cap C = A \cap (B \cap C)$$

Distributive laws:

$$(A \cup B) \cap C = (A \cap C) \cup (B \cap C)$$
$$(A \cap B) \cup C = (A \cup C) \cap (B \cup C)$$

DeMorgan's laws:

$$(A \cup B)^c = A^c \cap B^c$$
$$(A \cap B)^c = A^c \cup B^c$$

Disjoint union:

$$A \cup B = (A \cap B^c) \cup (A^c \cap B) \cup (A \cap B)$$

and $A \cap B^c$, $A^c \cap B$, and $A \cap B$ are disjoint.

3.1.2 Probability axioms

Once a suitable mathematical framework for understanding events in terms of sets has been established, it is possible to construct a corresponding framework for understanding probabilities of events in terms of sets.

Definition 3.3 *The real valued function* $P(\cdot)$ *is a* probability measure *if it acts on subsets of S and obeys the following axioms:*

I. $P(S) = 1$.

II. If $A \subseteq S$ *then* $P(A) \geq 0$.

III. If A *and* B *are* disjoint $(A \cap B = \emptyset)$ *then*

$$P(A \cup B) = P(A) + P(B).$$

Repeated use of Axiom III gives the more general result that if $A_1, A_2, \ldots, A_n$ *are mutually disjoint, then*

$$P\left(\bigcup_{i=1}^{n} A_i\right) = \sum_{i=1}^{n} P(A_i).$$

Indeed, we will assume further that the above result holds even if we have a countably *infinite collection of disjoint events* $(n = \infty)$.

These axioms seem to fit well with most people's intuitive understanding of probability, but there are a few additional comments worth making.

1. Axiom I says that one of the possible outcomes must occur. A probability of 1 is assigned to the event 'something occurs'. This fits in exactly with the definition of sample space. Note, however, that the implication does not go the other way! When dealing with infinite sample spaces, there are often events of probability 1, which are not the sample space, and events of probability zero, which are not the empty set.

2. Axiom II simply states that we wish to work only with positive probabilities, because in some sense, probability measures the *size* of the set (event).

3. Axiom III says that probabilities 'add up' in a natural way. Allowing this result to hold for countably infinite unions is slightly controversial, but it makes the mathematics much easier, so it will be assumed throughout this text. This assumption is known as the *countable additivity* assumption. Note that this additivity property will not hold (in general) for an *uncountable* collection of events.

These axioms are (almost) all that is needed in order to develop a theory of probability, but there are a collection of commonly used properties which follow directly from these axioms and are used extensively when carrying out probability calculations.

Proposition 3.2

1. $\mathrm{P}(A^c) = 1 - \mathrm{P}(A)$
2. $\mathrm{P}(\emptyset) = 0$
3. If $A \Rightarrow B$ *(that is, $A \subseteq B$), then* $\mathrm{P}(A) \le \mathrm{P}(B)$
4. *(Addition Law)* $\mathrm{P}(A \cup B) = \mathrm{P}(A) + \mathrm{P}(B) - \mathrm{P}(A \cap B)$

Proof. For 1, since $A \cap A^c = \emptyset$, and $S = A \cup A^c$, Axiom III tells us that $\mathrm{P}(S) = \mathrm{P}(A \cup A^c) = \mathrm{P}(A) + \mathrm{P}(A^c)$. From axiom I we know that $\mathrm{P}(S) = 1$. Re-arranging gives the result. Property 2 follows from property 1 as $\emptyset = S^c$. It simply says that the probability of no outcome is zero, which again fits in with our definition of a sample space. For property 3 write B as

$$B = A \cup (B \cap A^c),$$

where A and $B \cap A^c$ are disjoint. Then, from the third axiom,

$$\mathrm{P}(B) = \mathrm{P}(A) + \mathrm{P}(B \cap A^c)$$

so that

$$\mathrm{P}(A) = \mathrm{P}(B) - \mathrm{P}(B \cap A^c) \le \mathrm{P}(B).$$

Property 4 is one of the exercises at the end of the chapter.
□

3.1.3 Interpretations of probability

Most people have an intuitive feel for the notion of probability, and the axioms seem to capture its essence in a mathematical form. However, for probability theory to be anything other than an interesting piece of abstract pure mathematics, it must have an interpretation that in some way connects it to reality. If you wish only to study probability as a mathematical theory, there is no need to have an interpretation. However, if you are to use probability theory as your foundation for a theory which makes probabilistic statements about the world around us, then there must be an interpretation of probability which makes some connection between the mathematical theory and reality.

While there is (almost) unanimous agreement about the mathematics of probability, the axioms, and their consequences, there is considerable disagreement about the interpretation of probability. The three most common interpretations are given below.

Classical interpretation

The classical interpretation of probability is based on the assumption of underlying equally likely events. That is, for any events under consideration, there is always a sample space which can be considered where all atomic events are equally likely. If this sample space is given, then the probability axioms may be deduced from set-theoretic considerations.

This interpretation is fine when it is obvious how to partition the sample space into equally likely events and is in fact entirely compatible with the other two interpretations to be described in that case. The problem with this interpretation is that for many situations it is not at all obvious what the partition into equally likely events is. For example, consider the probability that it will rain in a particular location tomorrow. This is clearly a reasonable event to consider, but it is not at all clear what sample space we should construct with equally likely outcomes. Consequently, the classical interpretation falls short of being a good interpretation for real-world problems. However, it provides a good starting point for a mathematical treatment of probability theory and is the interpretation adopted by many mathematicians and theoreticians.

Frequentist interpretation

An interpretation of probability widely adopted by statisticians is the relative frequency interpretation. This interpretation makes a much stronger connection with reality than the previous one and fits in well with traditional statistical methodology. Here probability only has meaning for events from experiments which could in principle be repeated arbitrarily many times under essentially identical conditions. Here, the probability of an event is simply the 'long-run proportion' of times that the event occurs under many repetitions of the experiment. It is reasonable to suppose that this proportion will settle down to some limiting value eventually, which is the probability of the event. In such a situation, it is possible to derive the axioms of probability from consideration of the long-run frequencies of various events. The probability p, of an event E, is defined by

$$p = \lim_{n \to \infty} \frac{r}{n},$$

where r is the number of times E occurred in n repetitions of the experiment.

Unfortunately it is hard to know precisely why such a limiting frequency should exist. A bigger problem, however, is that the interpretation only applies to outcomes of repeatable experiments, and there are many 'one-off' events, such as 'rain here tomorrow,' on which we would like to be able to attach probabilities.

Subjective interpretation

This final common interpretation of probability is somewhat controversial, but does not suffer from the problems the other interpretations do. It suggests that the association of probabilities to events is a personal (subjective) process, relating to your *degree of belief* in the likelihood of the event occurring. It is controversial because it accepts that *different* people will assign *different* probabilities to the *same event*.

While in some sense it gives up on an objective notion of probability, it is in no sense arbitrary. It can be defined in a precise way, from which the axioms of probability may be derived as requirements of self-consistency.

A simple way to define *your* subjective probability that some event E will occur is as follows. Your probability is the number p such that you consider £p to be a *fair price* for a gamble which will pay you £1 if E occurs and nothing otherwise.

So, if you consider 40 pence to be a fair price for a gamble which pays you £1 if it rains tomorrow, then 0.4 is your subjective probability for the event.* The subjective interpretation is sometimes known as the *degree of belief interpretation*, which is the interpretation of probability underlying the theory of *Bayesian statistics* — a powerful theory of statistical inference named after Thomas Bayes, the 18th-century Presbyterian minister who first proposed it. Consequently, this interpretation of probability is sometimes also known as the *Bayesian interpretation*.

Summary

While the interpretation of probability is philosophically very important, all interpretations lead to the same set of axioms, from which the rest of probability theory is deduced. Consequently, for much of this text, it will be sufficient to adopt a fairly classical approach, taking the axioms as given and investigating their consequences independently of the precise interpretation adopted. However, the inferential theory considered in the later chapters is distinctly Bayesian in nature, and Bayesian inference is most naturally associated with a subjective interpretation of probability.

3.1.4 Classical probability

Classical probability theory is concerned with carrying out probability calculations based on *equally likely outcomes*. That is, it is assumed that the sample space has been constructed in such a way that every subset of the sample space consisting of a single element has the same probability. If the sample space contains n possible outcomes ($\#S = n$), we must have for all $s \in S$,

$$P(\{s\}) = \frac{1}{n}$$

and hence for all $E \subseteq S$

$$P(E) = \frac{\#E}{n}.$$

More informally, we have

$$P(E) = \frac{\text{number of ways } E \text{ can occur}}{\text{total number of outcomes}}.$$

_____

* For non-British readers, replace £ with $ and pence with cents.

Example

Suppose that a fair coin is thrown twice, and the results are recorded. The sample space is

$$S = \{HH, HT, TH, TT\}.$$

Let us assume that each outcome is equally likely — that is, each outcome has a probability of $1/4$. Let A denote the event *head on the first toss*, and B denote the event *head on the second toss*. In terms of sets

$$A = \{HH, HT\}, \; B = \{HH, TH\}.$$

So

$$P(A) = \frac{\#A}{n} = \frac{2}{4} = \frac{1}{2},$$

and similarly $P(B) = 1/2$. If we are interested in the event $C = A \cup B$ we can work out its probability from the set definition as

$$P(C) = \frac{\#C}{4} = \frac{\#(A \cup B)}{4} = \frac{\#\{HH, HT, TH\}}{4} = \frac{3}{4}$$

or by using the addition formula

$$P(C) = P(A \cup B) = P(A) + P(B) - P(A \cap B) = \frac{1}{2} + \frac{1}{2} - P(A \cap B).$$

Now $A \cap B = \{HH\}$, which has probability 1/4, so

$$P(C) = \frac{1}{2} + \frac{1}{2} - \frac{1}{4} = \frac{3}{4}.$$

In this simple example, it seems easier to work directly with the definition. However, in more complex problems, it is usually much easier to work out how many elements there are in an intersection than in a union, making the addition law very useful.

The multiplication principle

In the above example we saw that there were two distinct experiments — *first throw* and *second throw*. There were two equally likely outcomes for the first throw and two equally likely outcomes for the second throw. This leads to a combined experiment with $2 \times 2 = 4$ possible outcomes. This is an example of the *multiplication principle*.

Proposition 3.3 *If there are p experiments and the first has n_1 equally likely outcomes, the second has n_2 equally likely outcomes, and so on until the pth experiment has n_p equally likely outcomes, then there are*

$$n_1 \times n_2 \times \cdots \times n_p = \prod_{i=1}^{p} n_i$$

equally likely possible outcomes for the p experiments.

3.1.5 Conditional probability and the multiplication rule

We now have a way of understanding the probabilities of events, but so far we have no way of *modifying* those probabilities when certain events occur. For this, we need an extra axiom which can be justified under any of the interpretations of probability.

Definition 3.4 *The* conditional probability of A given B, *written* $P(A|B)$, *is defined by*

$$P(A|B) = \frac{P(A \cap B)}{P(B)}, \quad for\ P(B) > 0.$$

Note that we can only condition on events with strictly positive probability.

Under the classical interpretation of probability, we can see that if we are told that B has occurred, then all outcomes in B are equally likely, and all outcomes not in B have zero probability — so B is the new sample space. The number of ways that A can occur is now just the number of ways $A \cap B$ can occur, and these are all equally likely. Consequently we have

$$P(A|B) = \frac{\#(A \cap B)}{\#B} = \frac{\#(A \cap B)/\#S}{\#B/\#S} = \frac{P(A \cap B)}{P(B)}.$$

Because conditional probabilities really just correspond to a new probability measure defined on a smaller sample space, they obey all of the properties of 'ordinary' probabilities. For example, we have

$$P(B|B) = 1$$
$$P(\emptyset|B) = 0$$
$$P(A \cup C|B) = P(A|B) + P(C|B), \quad \text{for } A \cap C = \emptyset$$

and so on.

The definition of conditional probability simplifies when one event is a special case of the other. If $A \Rightarrow B$, then $A \cap B = A$, and so

$$P(A|B) = \frac{P(A)}{P(B)}, \quad \text{for } A \subseteq B.$$

Example

A dice is rolled, and the number showing recorded. Given that the number rolled was even, what is the probability that it was a six?

Let E denote the event 'even' and F denote the event 'a six'. Clearly $F \Rightarrow E$, so

$$P(F|E) = \frac{P(F)}{P(E)} = \frac{1/6}{1/2} = \frac{1}{3}.$$

The formula for conditional probability is useful when we want to calculate $P(A|B)$ from $P(A \cap B)$ and $P(B)$. However, more commonly we want to know $P(A \cap B)$ and we know $P(A|B)$ and $P(B)$. A simple rearrangement gives us the multiplication rule.

Proposition 3.4 (multiplication rule)

$$P(A \cap B) = P(B) \times P(A|B)$$

Example

Two cards are dealt from a deck of 52 cards. What is the probability that they are both aces?

Let A_1 be the event 'first card an ace' and A_2 be the event 'second card an ace'. $P(A_2|A_1)$ is the probability of a second ace. Given that the first card has been drawn and was an ace, there are 51 cards left, 3 of which are aces, so $P(A_2|A_1) = 3/51$. So,

$$P(A_1 \cap A_2) = P(A_1) \times P(A_2|A_1)$$
$$= \frac{4}{52} \times \frac{3}{51}$$
$$= \frac{1}{221}.$$

The multiplication rule generalises to more than two events. For example, for three events we have

$$P(A_1 \cap A_2 \cap A_3) = P(A_1) P(A_2|A_1) P(A_3|A_1 \cap A_2).$$

3.1.6 Independent events, partitions, and Bayes' theorem

Recall the multiplication rule (Proposition 3.4):

$$P(A \cap B) = P(B) P(A|B).$$

For some events A and B, knowing that B has occurred will not alter the probability of A, so that $P(A|B) = P(A)$. When this is so, the multiplication rule becomes

$$P(A \cap B) = P(A) P(B),$$

and the events A and B are said to be *independent events*. Independence is a very important concept in probability theory, and it is often used to build up complex events from simple ones. Do not confuse the independence of A and B with the exclusivity of A and B — they are entirely different concepts. If A and B both have positive probability, then they cannot be both independent and exclusive (this is an end-of-chapter exercise).

When it is clear that the occurrence of B can have no influence on A, we will *assume* independence in order to calculate $P(A \cap B)$. However, if we can calculate $P(A \cap B)$ directly, we can check the independence of A and B by seeing if it is true that

$$P(A \cap B) = P(A) P(B).$$

We can generalise independence to collections of events as follows.

Definition 3.5 *The set of events $A = \{A_1, A_2, \ldots, A_n\}$ is* mutually independent *if for any subset, $B \subseteq A$, $B = \{B_1, B_2, \ldots, B_r\}$, $r \leq n$ we have*

$$P(B_1 \cap \cdots \cap B_r) = P(B_1) \times \cdots \times P(B_r).$$

Note that mutual independence is much stronger than *pair-wise* independence, where

we only require independence of subsets of size $r = 2$. That is, pair-wise independence *does not* imply mutual independence.

Definition 3.6 *A* partition *of a sample space is simply the decomposition of the sample space into a collection of mutually* exclusive *events with positive probability. That is, $\{B_1, \ldots, B_n\}$ forms a* partition *of S if*

- $S = B_1 \cup B_2 \cup \cdots \cup B_n = \bigcup_{i=1}^{n} B_i,$

- $B_i \cap B_j = \emptyset, \ \forall i \neq j,$

- $P(B_i) > 0, \ \forall i.$

Theorem 3.1 (theorem of total probability) *Suppose that we have a partition $\{B_1, \ldots, B_n\}$ of a sample space, S. Suppose further that we have an event A. Then*

$$P(A) = \sum_{i=1}^{n} P(A|B_i) P(B_i).$$

Proof. A can be written as the disjoint union

$$A = (A \cap B_1) \cup \cdots \cup (A \cap B_n),$$

and so the probability of A is given by

$$
\begin{aligned}
P(A) &= P((A \cap B_1) \cup \cdots \cup (A \cap B_n)) \\
&= P(A \cap B_1) + \cdots + P(A \cap B_n) && \text{(by Axiom III)} \\
&= P(A|B_1) P(B_1) + \cdots + P(A|B_n) P(B_n) && \text{(by Proposition 3.4)} \\
&= \sum_{i=1}^{n} P(A|B_i) P(B_i).
\end{aligned}
$$

□

Theorem 3.2 (Bayes' theorem) *For all events A, B such that $P(B) > 0$ we have*

$$P(A|B) = \frac{P(B|A) P(A)}{P(B)}.$$

This is a very important result as it tells us how to 'turn conditional probabilities around' — that is, it tells us how to work out $P(A|B)$ from $P(B|A)$, and this is often very useful.

Proof. By Definition 3.4 we have

$$
\begin{aligned}
P(A|B) &= \frac{P(A \cap B)}{P(B)} \\
&= \frac{P(A) P(B|A)}{P(B)}. && \text{(by Proposition 3.4)}
\end{aligned}
$$

□

Example

A clinic offers you a free test for a very rare, but hideous disease. The test they offer is very reliable. If you have the disease it has a 98% chance of giving a positive result, and if you do not have the disease, it has only a 1% chance of giving a positive result. Despite having no *a priori* reason to suppose that you have the disease, you nevertheless decide to take the test and find that you test positive — what is the probability that you have the disease?

Let P be the event 'test positive' and D be the event 'you have the disease'. We know that

$$P(P|D) = 0.98 \text{ and that } P(P|D^c) = 0.01.$$

We want to know $P(D|P)$, so we use Bayes' theorem.

$$
\begin{aligned}
P(D|P) &= \frac{P(P|D)\,P(D)}{P(P)} \\
&= \frac{P(P|D)\,P(D)}{P(P|D)\,P(D) + P(P|D^c)\,P(D^c)} \qquad \text{(using Theorem 3.1)} \\
&= \frac{0.98\,P(D)}{0.98\,P(D) + 0.01(1 - P(D))}.
\end{aligned}
$$

So we see that the probability you have the disease given the test result depends on the probability that you had the disease in the first place. This is a rare disease, affecting only 1 in 10,000 people, so that $P(D) = 0.0001$. Substituting this in gives

$$P(D|P) = \frac{0.98 \times 0.0001}{0.98 \times 0.0001 + 0.01 \times 0.9999} \simeq 0.01.$$

So, your probability of having the disease has increased from 1 in 10,000 to 1 in 100, but still is not that much to get worried about! Note the *crucial* difference between $P(P|D)$ and $P(D|P)$.

Another important thing to notice about the above example is the use of the theorem of total probability in order to expand the bottom line of Bayes' theorem. In fact, this is done so often that Bayes' theorem is often stated in this form.

Corollary 3.1 *Suppose that we have a partition* $\{B_1, \ldots, B_n\}$ *of a sample space S. Suppose further that we have an event A, with* $P(A) > 0$. *Then, for each* B_j, *the probability of* B_j *given A is*

$$
\begin{aligned}
P(B_j|A) &= \frac{P(A|B_j)\,P(B_j)}{P(A)} \\
&= \frac{P(A|B_j)\,P(B_j)}{P(A|B_1)\,P(B_1) + \cdots + P(A|B_n)\,P(B_n)} \\
&= \frac{P(A|B_j)\,P(B_j)}{\displaystyle\sum_{i=1}^{n} P(A|B_i)\,P(B_i)}.
\end{aligned}
$$

In particular, if the partition is simply $\{B, B^c\}$, then this simplifies to

$$P(B|A) = \frac{P(A|B)\,P(B)}{P(A|B)\,P(B) + P(A|B^c)\,P(B^c)}.$$

3.2 Discrete probability models

3.2.1 Introduction, mass functions, and distribution functions

We have seen how to relate events to sets and how to calculate probabilities for events by working with the sets that represent them. So far, however, we have not developed any special techniques for thinking about *random quantities*. *Discrete probability models* provide a framework for thinking about *discrete random quantities*, and *continuous probability models* (to be considered in the next section) form a framework for thinking about *continuous* random quantities.

Example

Consider the sample space for tossing a fair coin twice:

$$S = \{HH, HT, TH, TT\}.$$

These outcomes are equally likely. There are several random quantities we could associate with this experiment. For example, we could count the number of heads or the number of tails.

Definition 3.7 *A* random quantity *is a real valued function which acts on* elements *of the sample space (outcomes). That is, to each outcome, the random variable assigns a real number.*

Random quantities (sometimes known as *random variables*) are always denoted by upper case letters.

In our example, if we let X be the number of heads, we have

$$X(HH) = 2,$$
$$X(HT) = 1,$$
$$X(TH) = 1,$$
$$X(TT) = 0.$$

The observed value of a random quantity is the number corresponding to the actual outcome. That is, if the outcome of an experiment is $s \in S$, then $X(s) \in \mathbb{R}$ is the observed value. This observed value is always denoted with a lowercase letter — here x. Thus $X = x$ means that the observed value of the random quantity X is the number x. The set of possible observed values for X is

$$S_X = \{X(s)|s \in S\}.$$

For the above example we have

$$S_X = \{0, 1, 2\}.$$

Clearly here the values are not all equally likely.

Example

Roll one dice and call the random number which is uppermost Y. The sample space for the *random quantity Y* is

$$S_Y = \{1, 2, 3, 4, 5, 6\},$$

and these outcomes are all equally likely. Now roll two dice and call their sum Z. The sample space for Z is

$$S_Z = \{2, 3, 4, 5, 6, 7, 8, 9, 10, 11, 12\},$$

and these outcomes are *not* equally likely. However, we know the probabilities of the events corresponding to each of these outcomes, and we could display them in a table as follows.

Outcome	2	3	4	5	6	7	8	9	10	11	12
Probability	1/36	2/36	3/36	4/36	5/36	6/36	5/36	4/36	3/36	2/36	1/36

This is essentially a tabulation of the *probability mass function* for the random quantity Z.

Definition 3.8 (probability mass function) *For any discrete random variable X, we define the* probability mass function *(PMF) to be the function which gives the probability of each $x \in S_X$. Clearly we have*

$$P(X = x) = \sum_{\{s \in S | X(s) = x\}} P(\{s\}).$$

That is, the probability of getting a particular number is the sum of the probabilities of all those outcomes which have that number associated with them. Also $P(X = x) \geq 0$ for each $x \in S_X$, and $P(X = x) = 0$ otherwise.

Definition 3.9 *The set of all pairs $\{(x, P(X = x)) | x \in S_X\}$ is known as the* probability distribution *of X.*

Example

For the example above concerning the sum of two dice, the probability distribution is

$$\{(2, 1/36), (3, 2/36), (4, 3/36), (5, 4/36), (6, 5/36), (7, 6/36),$$
$$(8, 5/36), (9, 4/36), (10, 3/36), (11, 2/36), (12, 1/36)\}$$

and the probability mass function can be tabulated as follows.

x	2	3	4	5	6	7	8	9	10	11	12
$P(X = x)$	1/36	2/36	3/36	4/36	5/36	6/36	5/36	4/36	3/36	2/36	1/36

For any discrete random quantity, X, we clearly have

$$\sum_{x \in S_X} P(X = x) = 1$$

as every outcome has some number associated with it. It can often be useful to know the probability that your random number is no greater than some particular value.

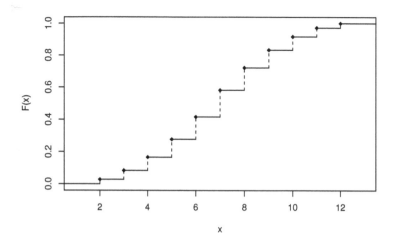

Figure 3.1 *CDF for the sum of a pair of fair dice.*

Definition 3.10 (cumulative distribution function) *The cumulative distribution function (CDF), is defined by*

$$F_X(x) = P(X \le x) = \sum_{\{y \in S_X | y \le x\}} P(X = y).$$

Example

For the sum of two dice, the CDF can be tabulated for the outcomes as

x	2	3	4	5	6	7	8	9	10	11	12
$F_X(x)$	1/36	3/36	6/36	10/36	15/36	21/36	26/36	30/36	33/36	35/36	36/36

but it is important to note that the CDF is defined *for all real numbers* — not just the possible values. In our example we have

$$F_X(-3) = P(X \le -3) = 0,$$
$$F_X(4.5) = P(X \le 4.5) = P(X \le 4) = 6/36,$$
$$F_X(25) = P(X \le 25) = 1.$$

We may plot the CDF for our example as shown in Figure 3.1.

It is clear that for any random variable X, for all $x \in \mathbb{R}$, $F_X(x) \in [0, 1]$ and that $F_X(x) \to 0$ as $x \to -\infty$ and $F_X(x) \to 1$ as $x \to +\infty$.

3.2.2 Expectation and variance for discrete random quantities

It is useful to be able to summarise the distribution of random quantities. A *location measure* often used to summarise random quantities is the *expectation* of the random quantity. It is the 'centre of mass' of the probability distribution.

Definition 3.11 *The expectation of a discrete random quantity X, written $\mathrm{E}(X)$, is defined by*

$$\mathrm{E}(X) = \sum_{x \in S_X} x\, \mathrm{P}(X = x).$$

The expectation is often denoted by μ_X or even just μ. Note that the expectation is a known function of the probability distribution. It is *not* a random quantity, and it is *not* the sample mean of a set of data (random or otherwise). In some sense (to be made precise in Proposition 3.14), it represents the value of the random variable that you expect to get on average.

Example

For the sum of two dice, X, we have

$$\mathrm{E}(X) = 2 \times \frac{1}{36} + 3 \times \frac{2}{36} + 4 \times \frac{3}{36} + \cdots + 12 \times \frac{1}{36} = 7.$$

By looking at the symmetry of the mass function, it is clear that in some sense 7 is the 'central' value of the probability distribution.

As well as a method for summarising the location of a given probability distribution, it is also helpful to have a summary for the *spread*.

Definition 3.12 *For a discrete random quantity X, the* variance *of X is defined by*

$$\mathrm{Var}(X) = \sum_{x \in S_X} \left\{ (x - \mathrm{E}(X))^2 \, \mathrm{P}(X = x) \right\}.$$

The variance is often denoted σ_X^2, or even just σ^2. Again, this is a known function of the probability distribution. It is not random, and it is not the *sample* variance of a set of data. The variance can be rewritten as

$$\mathrm{Var}(X) = \sum_{x_i \in S_X} x_i^2 \, \mathrm{P}(X = x_i) - [\mathrm{E}(X)]^2,$$

and this expression is usually a bit easier to work with. We also define the *standard deviation* of a random quantity by

$$\mathrm{SD}(X) - \sqrt{\mathrm{Var}(X)},$$

and this is usually denoted by σ_X or just σ. The variance represents the average squared distance of X from $\mathrm{E}(X)$. The units of variance are therefore the square of the units of X (and $\mathrm{E}(X)$). Some people prefer to work with the standard deviation because it has the same units as X and $\mathrm{E}(X)$.

Example

For the sum of two dice, X, we have

$$\sum_{x_i \in S_X} x_i^2 \, \mathrm{P}(X = x_i) = 2^2 \times \frac{1}{36} + 3^2 \times \frac{2}{36} + 4^2 \times \frac{3}{36} + \cdots + 12^2 \times \frac{1}{36} = \frac{329}{6}$$

and so

$$\text{Var}(X) = \frac{329}{6} - 7^2 = \frac{35}{6}, \quad \text{and} \quad \text{SD}(X) = \sqrt{\frac{35}{6}}.$$

3.2.3 Properties of expectation and variance

One of the reasons that expectation is widely used as a measure of location for probability distributions is the fact that it has many desirable mathematical properties which make it elegant and convenient to work with. Indeed, many of the nice properties of expectation lead to corresponding nice properties for variance, which is one of the reasons why variance is widely used as a measure of spread.

Suppose that X is a discrete random quantity, and that Y is another random quantity that is a known function of X. That is, $Y = g(X)$ for some function $g(\cdot)$. What is the expectation of Y?

Example

Throw a dice, and let X be the number showing. We have

$$S_X = \{1, 2, 3, 4, 5, 6\}$$

and each value is equally likely. Now suppose that we are actually interested in the square of the number showing. Define a new random quantity $Y = X^2$. Then

$$S_Y = \{1, 4, 9, 16, 25, 36\}$$

and clearly each of these values is equally likely. We therefore have

$$\text{E}(Y) = 1 \times \frac{1}{6} + 4 \times \frac{1}{6} + \cdots + 36 \times \frac{1}{6} = \frac{91}{6}.$$

The above example illustrates the more general result, that for $Y = g(X)$, we have

$$\text{E}(Y) = \sum_{x \in S_X} g(x) \, \text{P}(X = x).$$

Note that in general $\text{E}(g(X)) \neq g(\text{E}(X))$. For the above example, $\text{E}(X^2) = 91/6 \simeq 15.2$, and $\text{E}(X)^2 = 3.5^2 = 12.25$.

We can use this more general notion of expectation in order to redefine variance purely in terms of expectation as follows:

$$\text{Var}(X) = \text{E}([X - \text{E}(X)]^2) = \text{E}(X^2) - [\text{E}(X)]^2.$$

Having said that $\text{E}(g(X)) \neq g(\text{E}(X))$ in general, it does in fact hold in the (very) special but important case where $g(\cdot)$ is a *linear* function.

Lemma 3.1 (expectation of a linear transformation) *If we have a random quantity X, and a linear transformation, $Y = aX + b$, where a and b are known real constants, then we have that*

$$\text{E}(aX + b) = a \, \text{E}(X) + b.$$

Proof.

$$E(aX + b) = \sum_{x \in S_X} (ax + b) P(X = x)$$

$$= \sum_{x \in S_X} ax P(X = x) + \sum_{x \in S_X} b P(X = x)$$

$$= a \sum_{x \in S_X} x P(X = x) + b \sum_{x \in S_X} P(X = x)$$

$$= a E(X) + b.$$

□

Lemma 3.2 (expectation of a sum) *For two random quantities X and Y, the expectation of their sum is given by*

$$E(X + Y) = E(X) + E(Y).$$

Note that this result is true irrespective of whether X and Y are independent. Let us see why.

Proof. First,

$$S_{X+Y} = \{x + y | (x \in S_X) \cap (y \in S_Y)\},$$

and so

$$E(X + Y) = \sum_{(x+y) \in S_{X+Y}} (x + y) P((X = x) \cap (Y = y))$$

$$= \sum_{x \in S_X} \sum_{y \in S_Y} (x + y) P((X = x) \cap (Y = y))$$

$$= \sum_{x \in S_X} \sum_{y \in S_Y} x P((X = x) \cap (Y = y))$$

$$+ \sum_{x \in S_X} \sum_{y \in S_Y} y P((X = x) \cap (Y = y))$$

$$= \sum_{x \in S_X} \sum_{y \in S_Y} x P(X = x) P(Y = y | X = x)$$

$$+ \sum_{y \in S_Y} \sum_{x \in S_X} y P(Y = y) P(X = x | Y = y)$$

$$= \sum_{x \in S_X} x P(X = x) \sum_{y \in S_Y} P(Y = y | X = x)$$

$$+ \sum_{y \in S_Y} y P(Y = y) \sum_{x \in S_X} P(X = x | Y = y)$$

$$= \sum_{x \in S_X} x P(X = x) + \sum_{y \in S_Y} y P(Y = y)$$

$$= E(X) + E(Y).$$

☐

We can now put together the result for the expectation of a sum, and for the expectation of a linear transformation in order to give the following important property, commonly referred to as the *linearity of expectation*.

Proposition 3.5 (expectation of a linear combination) *For any random quantities* $X_1, \ldots, X_n$ *and scalar constants* $a_0, \ldots, a_n$ *we have*

$$E(a_0 + a_1 X_1 + a_2 X_2 + \cdots + a_n X_n) = a_0 + a_1 E(X_1) + a_2 E(X_2) + \cdots$$
$$+ a_n E(X_n).$$

Lemma 3.3 (expectation of an independent product) *If* X *and* Y *are independent random quantities, then*
$$E(XY) = E(X) E(Y).$$

Proof.
$$S_{XY} = \{xy | (x \in S_X) \cap (y \in S_Y)\},$$

and so

$$
\begin{aligned}
E(XY) &= \sum_{xy \in S_{XY}} xy\, P((X = x) \cap (Y = y)) \\
&= \sum_{x \in S_X} \sum_{y \in S_Y} xy\, P(X = x) P(Y = y) \\
&= \sum_{x \in S_X} x\, P(X = x) \sum_{y \in S_Y} y\, P(Y = y) \\
&= E(X) E(Y)
\end{aligned}
$$

☐

Note that here it is vital that X and Y are independent or the result does not necessarily hold. Random quantities possessing this expectation property are said to be *uncorrelated*. The lemma shows that independent random quantities are uncorrelated. Note, however, that the converse of this result is not true; uncorrelated random quantities are not necessarily independent.

Lemma 3.4 (variance of a linear transformation) *If* X *is a random quantity with finite variance* $\mathrm{Var}(X)$, *then*
$$\mathrm{Var}(aX + b) = a^2 \, \mathrm{Var}(X).$$

Proof.

$$
\begin{aligned}
\mathrm{Var}(aX + b) &= E\big([aX + b]^2\big) - [E(aX + b)]^2 \\
&= E\big(a^2 X^2 + 2abX + b^2\big) - [a\, E(X) + b]^2 \\
&= a^2 \Big[E(X^2) - E(X)^2\Big] \\
&= a^2 \, \mathrm{Var}(X).
\end{aligned}
$$

☐

Lemma 3.5 (variance of an independent sum) *If X and Y are* independent *random quantities, then*

$$\mathrm{Var}(X + Y) = \mathrm{Var}(X) + \mathrm{Var}(Y).$$

Proof.

$$\begin{aligned}
\mathrm{Var}(X + Y) &= \mathrm{E}\big([X + Y]^2\big) - [\mathrm{E}(X + Y)]^2 \\
&= \mathrm{E}\big(X^2 + 2XY + Y^2\big) - [\mathrm{E}(X) + \mathrm{E}(Y)]^2 \\
&= \mathrm{E}\big(X^2\big) + 2\,\mathrm{E}(XY) + \mathrm{E}\big(Y^2\big) - \mathrm{E}(X)^2 - 2\,\mathrm{E}(X)\,\mathrm{E}(Y) \\
&\qquad\qquad\qquad\qquad\qquad\qquad\qquad\qquad - \mathrm{E}(Y)^2 \\
&= \mathrm{E}\big(X^2\big) + 2\,\mathrm{E}(X)\,\mathrm{E}(Y) + \mathrm{E}\big(Y^2\big) - \mathrm{E}(X)^2 - 2\,\mathrm{E}(X)\,\mathrm{E}(Y) \\
&\qquad\qquad\qquad\qquad\qquad\qquad\qquad\qquad - \mathrm{E}(Y)^2 \\
&= \mathrm{E}\big(X^2\big) - \mathrm{E}(X)^2 + \mathrm{E}\big(Y^2\big) - \mathrm{E}(Y)^2 \\
&= \mathrm{Var}(X) + \mathrm{Var}(Y)
\end{aligned}$$

☐

Again, it is vital that X and Y are independent or the result does not necessarily hold. Notice that this implies a slightly less attractive result for the standard deviation of the sum of two independent random quantities,

$$\mathrm{SD}(X + Y) = \sqrt{\mathrm{SD}(X)^2 + \mathrm{SD}(Y)^2},$$

which is why it is often more convenient to work with variances.[†]

Putting together previous results, we get the following proposition.

Proposition 3.6 (variance of a linear combination) *For mutually independent X_1, $X_2, \ldots, X_n$ we have*

$$\begin{aligned}
\mathrm{Var}(a_0 + a_1 X_1 + a_2 X_2 + \cdots + a_n X_n) \\
= a_1^2\,\mathrm{Var}(X_1) + a_2^2\,\mathrm{Var}(X_2) + \cdots + a_n^2\,\mathrm{Var}(X_n).
\end{aligned}$$

Before moving on to look at some interesting families of probability distributions, it is worth emphasising the link between expectation and the theorem of total probability.

Proposition 3.7 *If G is an event and X is a discrete random quantity with outcome*

[†] Note the similarity to Pythagoras' theorem for the lengths of the sides of a triangle. Viewed in the correct way, this result is seen to be a special case of Pythagoras' theorem, as the standard deviation of a random quantity can be viewed as a length, and independence leads to orthogonality in an appropriately constructed inner-product space.

space S_X, then by the theorem of total probability (Theorem 3.1) we have

$$P(G) = \sum_{x \in S_X} P(G|X = x) P(X = x)$$
$$= E(P(G|X)),$$

where $P(G|X)$ is the random quantity which takes the value $P(G|X = x)$ when X takes the value x.

3.3 The discrete uniform distribution

The theory of discrete random quantities covered in the previous section provides a generic set of tools for studying distributions and their properties. However, there are several important 'families' of discrete probability models that occur frequently and are therefore worthy of special consideration. The first is the 'discrete uniform distribution,' which corresponds to a generalisation of the number obtained by rolling a single dice. A random quantity X is discrete-uniform on the numbers from 1 to n, written

$$X \sim U\{1 : n\}$$

if each of the integers from 1 to n is equally likely to be the observed value of X. We therefore have outcome space

$$S_X = \{1, 2, \ldots, n\}$$

and PMF

$$P(X = k) = \frac{1}{n}, \quad k \in S_X.$$

The CDF at the points in S_X is therefore clearly given by

$$P(X \leq k) = \frac{k}{n}, \quad k \in S_X.$$

It is straightforward to compute the expectation of X as

$$E(X) = \sum_{k=1}^{n} k \times \frac{1}{n}$$
$$= \frac{1}{n} \sum_{k=1}^{n} k$$
$$= \frac{1}{n} \frac{n(n+1)}{2}$$
$$= \frac{n+1}{2}.$$

The variance can be calculated similarly to be

$$\text{Var}(X) = \frac{n^2 - 1}{12}.$$

In general the discrete uniform distribution can be defined for an arbitrary subset of the integers $S_X \subseteq \mathbb{Z}$, written $U\{S_X\}$, and in this case each $k \in S_X$ has probability

$P(X = k) = 1/\#S_X$. It can be further generalised to any finite set, S_X, where $\#S_X = n < \infty$. Generalisations to infinite sets, such as intervals of the real line, are discussed later.

Although this distributional family seems somewhat contrived, it is especially useful as it typically forms the starting point for the development of any stochastic simulation algorithm.

3.4 The binomial distribution

The binomial distribution is the distribution of the number of 'successes' in a series of n independent 'trials,' each of which results in a 'success' (with probability p) or a 'failure' (with probability $1 - p$). If the number of successes is X, we would write

$$X \sim Bin(n, p)$$

to indicate that X is a binomial random quantity based on n independent trials, each occurring with probability p.

Let us now derive the probability mass function for $X \sim Bin(n, p)$. Clearly X can take on any value from 0 up to n, and no other. Therefore, we simply have to calculate $P(X = k)$ for $k = 0, 1, 2, \ldots, n$. The probability of k successes followed by $n - k$ failures is clearly $p^k(1 - p)^{n-k}$. Indeed, this is the probability of *any* particular sequence involving k successes. There are $\binom{n}{k}$ such sequences, so by the multiplication principle, we have

$$P(X = k) = \binom{n}{k} p^k (1 - p)^{n-k}, \qquad k = 0, 1, 2, \ldots, n.$$

Now, using the binomial theorem, we have

$$\sum_{k=0}^{n} P(X = k) = \sum_{k=0}^{n} \binom{n}{k} p^k (1 - p)^{n-k} = (p + [1 - p])^n = 1^n = 1,$$

and so this does define a valid probability distribution. There is no neat analytic expression for the CDF of a binomial distribution, but it is straightforward to compute and tabulate. A plot of the probability mass function and cumulative distribution function of a particular binomial distribution is shown in Figure 3.2.

The expectation and variance of the binomial distribution can be computed straightforwardly, and are found to be

$$E(X) = np \quad \text{and} \quad \text{Var}(X) = np(1 - p).$$

Note that the binomial distribution has an important additivity property which is clear from the definition. Namely, that if $X_1 \sim Bin(n_1, p)$ and $X_2 \sim Bin(n_2, p)$ are independent, then $Y \equiv X_1 + X_2 \sim Bin(n_1 + n_2, p)$.

3.5 The geometric distribution

The geometric distribution is the distribution of the number of independent binary 'success' or 'fail' trials until the first 'success' is encountered. If X is the number of

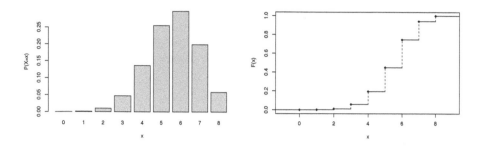

Figure 3.2 *PMF and CDF for a Bin(8, 0.7) distribution.*

trials until a success is encountered, and each independent trial has probability p of being a success, we write

$$X \sim Geom(p).$$

Clearly X can take on any positive integer ($S_X = \{1, 2, \ldots\}$), so to deduce the PMF, we need to calculate $P(X = k)$ for $k = 1, 2, 3, \ldots$. In order to have $X = k$, we must have an ordered sequence of $k - 1$ failures followed by one success. By the multiplication rule, therefore,

$$P(X = k) = (1 - p)^{k-1}p, \qquad k = 1, 2, 3, \ldots.$$

For the geometric distribution, it is possible to calculate an analytic form for the CDF as follows. If $X \sim Geom(p)$, then

$$F_X(k) = P(X \leq k)$$

$$= \sum_{j=1}^{k}(1 - p)^{j-1}p$$

$$= p\sum_{j=1}^{k}(1 - p)^{j-1}$$

$$= p \times \frac{1 - (1 - p)^k}{1 - (1 - p)} \qquad \text{(geometric series)}$$

$$= 1 - (1 - p)^k.$$

Consequently, there is no need to tabulate the CDF of the geometric distribution. Also note that the CDF tends to 1 as k increases. This confirms that the PMF we defined does determine a valid probability distribution.

3.5.1 Useful series in probability

Notice that we used the sum of a geometric series in the derivation of the CDF. There are many other series that crop up in the study of probability. A few of the more commonly encountered series are listed below.

$$\sum_{i=1}^{n} a^{i-1} = \frac{1-a^n}{1-a} \qquad\qquad (a > 0)$$

$$\sum_{i=1}^{\infty} a^{i-1} = \frac{1}{1-a} \qquad\qquad (0 < a < 1)$$

$$\sum_{i=1}^{\infty} i\,a^{i-1} = \frac{1}{(1-a)^2} \qquad\qquad (0 < a < 1)$$

$$\sum_{i=1}^{\infty} i^2 a^{i-1} = \frac{1+a}{(1-a)^3} \qquad\qquad (0 < a < 1)$$

$$\sum_{i=1}^{n} i = \frac{n(n+1)}{2}$$

$$\sum_{i=1}^{n} i^2 = \frac{1}{6}n(n+1)(2n+1).$$

We will use two of these in the derivation of the expectation and variance of the geometric distribution.

3.5.2 Expectation and variance of geometric random quantities

Suppose that $X \sim Geom(p)$. Then

$$E(X) = \sum_{i=1}^{\infty} i\,P(X = i)$$

$$= \sum_{i=1}^{\infty} i(1-p)^{i-1}p$$

$$= p \times \frac{1}{(1-[1-p])^2}$$

$$= \frac{1}{p}.$$

Similarly,

$$E(X^2) = \sum_{i=1}^{\infty} i^2 \, P(X = i)$$

$$= \sum_{i=1}^{\infty} i^2 (1-p)^{i-1} p$$

$$= p \times \frac{1 + [1-p]}{(1 - [1-p])^3}$$

$$= \frac{2-p}{p^2},$$

and so

$$Var(X) = E(X^2) - E(X)^2$$

$$= \frac{2-p}{p^2} - \frac{1}{p^2}$$

$$= \frac{1-p}{p^2}.$$

3.6 The Poisson distribution

The Poisson distribution is a very important discrete probability distribution, which arises in many different contexts in probability and statistics. For example, Poisson random quantities are often used in place of binomial random quantities in situations where n is large, p is small, and the expectation np is stable. The Poisson distribution is particularly important in the context of stochastic modelling of biochemical networks, as the number of reaction events occurring in a short time interval is approximately Poisson.

A Poisson random variable, X with parameter λ is written as

$$X \sim Po(\lambda).$$

3.6.1 Poisson as the limit of a binomial

Let $X \sim Bin(n, p)$. Put $\lambda = E(X) = np$ and let n increase and p decrease so that λ remains constant.

$$P(X = k) = \binom{n}{k} p^k (1-p)^{n-k}, \qquad k = 0, 1, 2, \ldots, n.$$

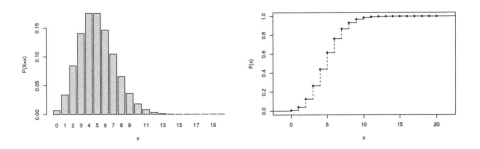

Figure 3.3 *PMF and CDF for a Po(5) distribution.*

Replacing p by λ/n gives

$$
\begin{aligned}
P(X = k) &= \binom{n}{k} \left(\frac{\lambda}{n}\right)^k \left(1 - \frac{\lambda}{n}\right)^{n-k} \\
&= \frac{n!}{k!(n-k)!} \left(\frac{\lambda}{n}\right)^k \left(1 - \frac{\lambda}{n}\right)^{n-k} \\
&= \frac{\lambda^k}{k!} \frac{n!}{(n-k)!n^k} \frac{(1-\lambda/n)^n}{(1-\lambda/n)^k} \\
&= \frac{\lambda^k}{k!} \frac{n}{n} \frac{(n-1)}{n} \frac{(n-2)}{n} \cdots \frac{(n-k+1)}{n} \frac{(1-\lambda/n)^n}{(1-\lambda/n)^k} \\
&\to \frac{\lambda^k}{k!} \times 1 \times 1 \times 1 \times \cdots \times 1 \times \frac{e^{-\lambda}}{1}, \quad \text{as } n \to \infty \\
&= \frac{\lambda^k}{k!} e^{-\lambda}.
\end{aligned}
$$

To see the limit, note that $(1 - \lambda/n)^n \to e^{-\lambda}$ as n increases (compound interest formula).

3.6.2 PMF

If $X \sim Po(\lambda)$, then the PMF of X is

$$
P(X = k) = \frac{\lambda^k}{k!} e^{-\lambda}, \quad k = 0, 1, 2, 3, \ldots.
$$

The PMF and CDF of $X \sim Po(5)$ are given in Figure 3.3. It is easy to verify that the PMF we have adopted for $X \sim Po(\lambda)$ does indeed define a valid probability distribution, as the probabilities sum to 1.

3.6.3 Expectation and variance of Poisson

If $X \sim Po(\lambda)$, the expectation and variance can be computed directly from the PMF, but it is clear from the binomial limit that we must have

$$E(X) = \lambda \quad \text{and} \quad \text{Var}(X) = \lambda.$$

That is, the mean and variance are both λ.

3.6.4 Sum of Poisson random quantities

One of the particularly convenient properties of the Poisson distribution is that the sum of two independent Poisson random quantities is also a Poisson random quantity. If $X \sim Po(\lambda)$ and $Y \sim Po(\mu)$ and X and Y are independent, then $Z = X + Y \sim Po(\lambda + \mu)$. Clearly this result extends to the sum of many independent Poisson random variables. The proof is straightforward, but omitted. Again, the result should be clear anyway, as the limit of the corresponding additivity property of the binomial distribution.

This property of the Poisson distribution is fundamental to its usefulness for modelling 'counts'. Consider, for example, modelling the number of counts recorded by a Geiger counter of relatively stable radioactive isotope (with a very long half-life). If we consider the number of counts in one second, the binomial limit argument suggests that since we are dealing with a huge number of molecules, and each independently has a constant but tiny probability of decaying and triggering a count, then the number of counts in that one-second interval should be Poisson. Suppose that the rate of decay is such that the count is $Po(\lambda)$, for some λ. Then in each one-second interval the count will have the same distribution, independently of all other intervals. So, using the above additive property, the number of counts in a two-second interval will be $Po(2\lambda)$, and in a three-second interval it will be $Po(3\lambda)$, etc. So in an interval of length t seconds, the number of counts will be $Po(\lambda t)$.[‡] This is the Poisson process.

3.6.5 The Poisson process

A sequence of timed observations is said to follow a *Poisson process* with *rate* λ if the number of observations, X, in any interval of length t is such that

$$X \sim Po(\lambda t),$$

and the numbers of events in disjoint intervals are independent of one another. We see that this is a simple model for discrete events occurring continuously in time. It turns out that the Poisson process is too simple to realistically model the time-evolution of biochemical networks, but a detailed understanding of this simple process is key

---

[‡] Of course we are ignoring the fact that once a molecule has decayed, it cannot decay again. But we are assuming that the number of available molecules is huge, and the number of decay events over the time scale of interest is so small that this effect can be ignored. For an isotope with a much shorter half-life, a *pure death process* would provide a much more suitable model. However, a formal discussion of processes of this nature will have to wait until Chapter 5.

to understanding more sophisticated stochastic processes. It is interesting to wonder about the time between successive events of the Poisson process, but it is clear that such random times do not have a discrete distribution. This, therefore, motivates the study of continuous random quantities.

3.7 Continuous probability models

3.7.1 Introduction, PDF and CDF

We now turn to techniques for handling continuous random quantities. These are random quantities with a sample space which is neither finite nor countably infinite. The sample space is usually taken to be the real line, or a part thereof. Continuous probability models are appropriate if the result of an experiment is a continuous *measurement*, rather than a *count* of a discrete set. In the context of systems biology, such measurements often (but not always) correspond to measurements of time.

If X is a continuous random quantity with sample space S_X, then for any particular $a \in S_X$, we generally have that

$$P(X = a) = 0.$$

This is because the sample space is so 'large' and every possible outcome so 'small' that the probability of any *particular* value is vanishingly small. Therefore, the probability mass function we defined for discrete random quantities is inappropriate for understanding continuous random quantities. In order to understand continuous random quantities, a little calculus is required.

Definition 3.13 (probability density function) *If X is a continuous random quantity, then there exists a function $f_X(x)$, called the probability density function (PDF), which satisfies the following:*

1. $f_X(x) \geq 0, \quad \forall x ;$ *(the symbol '$\forall$' means 'for all')*

2. $\int_{-\infty}^{\infty} f_X(x)\, dx = 1;$

3. $P(a \leq X \leq b) = \int_a^b f_X(x)\, dx$ *for any $a \leq b$.*

Consequently we have

$$P(x \leq X \leq x + \delta x) = \int_x^{x+\delta x} f_X(y)\, dy$$
$$\simeq f_X(x)\delta x, \qquad \text{(for small } \delta x)$$
$$\Rightarrow f_X(x) \simeq \frac{P(x \leq X \leq x + \delta x)}{\delta x},$$

and so we may interpret the PDF as

$$f_X(x) = \lim_{\delta x \to 0} \frac{P(x \leq X \leq x + \delta x)}{\delta x}.$$

Note that PDFs are *not* probabilities. For example, the density can take values greater than 1 in some regions as long as it still integrates to 1. Also note that because $P(X = a) = 0$, we have $P(X \leq k) = P(X < k)$ for continuous random quantities.

Definition 3.14 (cumulative distribution function) *Earlier in this chapter we defined the* cumulative distribution function *of a random variable X to be*

$$F_X(x) = P(X \leq x), \quad \forall x.$$

This definition works just as well for continuous random quantities, and is one of the many reasons why the distribution function is so useful. For a discrete random quantity we had

$$F_X(x) = P(X \leq x) = \sum_{\{y \in S_X | y \leq x\}} P(X = y),$$

but for a continuous random quantity we have the continuous analogue

$$\begin{aligned} F_X(x) &= P(X \leq x) \\ &= P(-\infty \leq X \leq x) \\ &= \int_{-\infty}^{x} f_X(z) \, dz. \end{aligned}$$

Just as in the discrete case, the distribution function is defined for all $x \in \mathbb{R}$, even if the sample space S_X is not the whole of the real line.

Proposition 3.8

1. *Since it represents a probability, $F_X(x) \in [0, 1]$.*
2. *$F_X(-\infty) = 0$ and $F_X(\infty) = 1$.*
3. *If $a < b$, then $F_X(a) \leq F_X(b)$. i.e., $F_X(\cdot)$ is a non-decreasing function.*
4. *When X is continuous, $F_X(x)$ is continuous. Also, by the fundamental theorem of calculus, we have*

$$\frac{d}{dx} F_X(x) = f_X(x),$$

and so the slope *of the CDF $F_X(x)$ is the PDF $f_X(x)$.*

Proposition 3.9 *The* median *of a random quantity is the value m which is the 'middle' of the distribution. That is, it is the value m such that*

$$P(X \leq m) = \frac{1}{2}.$$

Equivalently, it is the value m such that

$$F_X(m) = 0.5.$$

Similarly, the lower quartile *of a random quantity is the value l such that*

$$F_X(l) = 0.25,$$

and the upper quartile *is the value u such that*

$$F_X(u) = 0.75.$$

3.7.2 Properties of continuous random quantities

Proposition 3.10 *The* expectation *or* mean *of a continuous random quantity X is given by*

$$E(X) = \int_{-\infty}^{\infty} x\, f_X(x)\, dx,$$

which is just the continuous analogue of the corresponding formula for discrete random quantities. Similarly, the variance *is given by*

$$Var(X) = \int_{-\infty}^{\infty} [x - E(X)]^2 f_X(x)\, dx$$
$$= \int_{-\infty}^{\infty} x^2\, f_X(x)\, dx - [E(X)]^2.$$

Note that the expectation of $g(X)$ is given by

$$E(g(X)) = \int_{-\infty}^{\infty} g(x) f_X(x)\, dx$$

and so the variance is just

$$Var(X) = E([X - E(X)]^2) = E(X^2) - [E(X)]^2$$

as in the discrete case.

Note that all of the properties of expectation and variance derived for discrete random quantities also hold true in the continuous case. It is also worth explicitly noting the continuous version of Proposition 3.7.

Proposition 3.11 *If G is an event and X is a continuous random quantity, then*

$$P(G) = \int_{-\infty}^{\infty} P(G|X = x)\, f(x)\, dx$$
$$= E(P(G|X)),$$

where $P(G|X)$ is the random quantity which takes the value $P(G|X = x)$ when X takes the value x.

It should be noted that there is a subtlety here, as we are effectively conditioning on an event with probability zero, which is something that we explicitly ruled out when conditional probability was introduced. It turns out that the above formula is fine in most cases, though care has to be taken with the construction of the conditional probability.

Proposition 3.12 (PDF of a linear transformation) *Let X be a continuous random quantity with PDF $f_X(x)$ and CDF $F_X(x)$, and let $Y = aX + b$. The PDF of Y is given by*

$$f_Y(y) = \left|\frac{1}{a}\right| f_X\left(\frac{y - b}{a}\right).$$

Proof. First assume that $a > 0$. It turns out to be easier to work out the CDF first:

$$
\begin{aligned}
F_Y(y) &= \mathrm{P}(Y \le y) \\
&= \mathrm{P}(aX + b \le y) \\
&= \mathrm{P}\left(X \le \frac{y-b}{a}\right) \qquad \text{(since } a > 0\text{)} \\
&= F_X\left(\frac{y-b}{a}\right).
\end{aligned}
$$

So,

$$
F_Y(y) = F_X\left(\frac{y-b}{a}\right),
$$

and by differentiating both sides with respect to y we get

$$
f_Y(y) = \frac{1}{a} f_X\left(\frac{y-b}{a}\right).
$$

If $a < 0$, the inequality changes sign, introducing a minus sign in the expression for the PDF. Both cases can therefore be summarised by the result.
□

Proposition 3.13 (PDF of a differentiable 1-1 transformation) *Using a similar argument it is straightforward to deduce the PDF of an arbitrary differentiable invertible transformation $Y = g(X)$ as*

$$
f_Y(y) = f_X\left(g^{-1}(y)\right) \left| \frac{d}{dy} g^{-1}(y) \right|.
$$

Note that the term $\left| \frac{d}{dy} g^{-1}(y) \right|$ is known as the 'Jacobian' of the transformation.

3.7.3 The law of large numbers

In many scenarios involving stochastic simulation, it is desirable to approximate the expectation of a random quantity, X, by the sample mean of a collection of independent observations of the random quantity X. Suppose we have a computer program that can simulate independent realisations of X (we will see in the next chapter how to do this); $X_1, X_2, \ldots, X_n$. The *sample mean* is given by

$$
\bar{X} = \frac{1}{n} \sum_{i=1}^{n} X_i.
$$

If in some appropriate sense the sample mean converges to the expectation, $\mathrm{E}(X)$, then for large n, we can use $\bar{X}$ as an estimate of $\mathrm{E}(X)$. It turns out that for random quantities with finite variance, this is indeed true, but a couple of lemmas are required before this can be made precise.

Lemma 3.6 *If X has finite mean and variance, $\mathrm{E}(X) = \mu$ and $\mathrm{Var}(X) = \sigma^2$, and $X_1, X_2, \ldots, X_n$ is a collection of independent realisations of X used to form the* *sample mean $\bar{X}$, then*

$$E(\bar{X}) = \mu \quad and \quad Var(\bar{X}) = \frac{\sigma^2}{n}.$$

Proof. The result follows immediately from Propositions 3.5 and 3.6.
□

Lemma 3.7 (Markov's inequality) *If $X \geq 0$ is a non-negative random quantity with finite expectation $E(X) = \mu$, we have*

$$P(X \geq a) \leq \frac{\mu}{a}, \quad \forall a > 0.$$

Proof. Note that this result is true for arbitrary non-negative random quantities, but we present here the proof for the continuous case.

$$\frac{\mu}{a} = \frac{E(X)}{a}$$
$$= \frac{1}{a} \int_0^\infty x\, f(x)\, dx$$
$$= \int_0^\infty \frac{x}{a} f(x)dx$$
$$\geq \int_a^\infty \frac{x}{a} f(x)dx$$
$$\geq \int_a^\infty \frac{a}{a} f(x)dx$$
$$= \int_a^\infty f(x)dx$$
$$= P(X \geq a).$$

□

Lemma 3.8 (Chebyshev's inequality) *If X is a random quantity with finite mean $E(X) = \mu$ and variance $Var(X) = \sigma^2$, we have*

$$P(|X - \mu| < k\sigma) \geq 1 - \frac{1}{k^2}, \quad \forall k > 0.$$

Proof. Since $(X - \mu)^2$ is positive with expectation σ^2, Markov's inequality gives

$$P([X - \mu]^2 \geq a) \leq \frac{\sigma^2}{a}.$$

Putting $a = k^2\sigma^2$ then gives

$$P([X - \mu]^2 \geq k^2\sigma^2) \leq \frac{1}{k^2}$$
$$\Rightarrow P(|X - \mu| \geq k\sigma) \leq \frac{1}{k^2}$$
$$\Rightarrow P(|X - \mu| < k\sigma) \geq 1 - \frac{1}{k^2}.$$

□

We are now in a position to state the main result.

Proposition 3.14 (weak law of large numbers, WLLN) *For X with finite mean $E(X) = \mu$ and variance $\mathrm{Var}(X) = \sigma^2$, if $X_1, X_2, \ldots, X_n$ is an independent sample from X used to form the sample mean $\bar{X}$, we have*

$$P\big(|\bar{X} - \mu| < \varepsilon\big) \geq 1 - \frac{\sigma^2}{n\varepsilon^2} \xrightarrow[\infty]{n} 1, \quad \forall \varepsilon > 0.$$

In other words, the WLLN states that no matter how small one chooses the positive constant ε, the probability that $\bar{X}$ is within a distance ε of μ tends to 1 as the sample size n increases. This is a precise sense in which the sample mean 'converges' to the expectation.

Proof. Using Lemmas 3.6 and 3.8 we have

$$P\left(|\bar{X} - \mu| < k\frac{\sigma}{\sqrt{n}}\right) \geq 1 - \frac{1}{k^2}.$$

Substituting $k = \varepsilon\sqrt{n}/\sigma$ gives the result.
□

This result is known as the *weak* law, as there is a corresponding *strong* law, which we state without proof.

Proposition 3.15 (strong law of large numbers, SLLN) *For X with finite mean $E(X) = \mu$ and variance $\mathrm{Var}(X) = \sigma^2$, if $X_1, X_2, \ldots, X_n$ is an independent sample from X used to form the sample mean $\bar{X}$, we have*

$$P\left(\bar{X} \xrightarrow[\infty]{n} \mu\right) = 1.$$

3.8 The uniform distribution

Now that we understand the basic properties of continuous random quantities, we can look at some of the important standard continuous probability models. The simplest of these is the uniform distribution. This distribution turns out to be central to the theory of stochastic simulation that will be developed in the next chapter.

The random quantity X has a *uniform distribution* over the range $[a, b]$, written

$$X \sim U(a, b)$$

if the PDF is given by

$$f_X(x) = \begin{cases} \dfrac{1}{b - a}, & a \leq x \leq b, \\ 0, & \text{otherwise.} \end{cases}$$

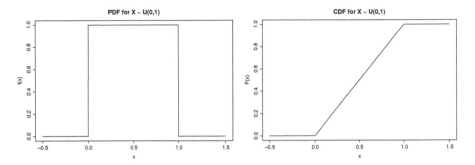

Figure 3.4 *PDF and CDF for a $U(0, 1)$ distribution.*

Thus if $x \in [a, b]$, then

$$F_X(x) = \int_{-\infty}^{x} f_X(y)\, dy$$

$$= \int_{-\infty}^{a} f_X(y)\, dy + \int_{a}^{x} f_X(y)\, dy$$

$$= 0 + \int_{a}^{x} \frac{1}{b - a}\, dy$$

$$= \frac{x - a}{b - a}.$$

Therefore,

$$F_X(x) = \begin{cases} 0, & x < a, \\ \dfrac{x - a}{b - a}, & a \leq x \leq b, \\ 1, & x > b. \end{cases}$$

We can plot the PDF and CDF in order to see the 'shape' of the distribution. Plots for $X \sim U(0, 1)$ are shown in Figure 3.4.

Clearly the lower quartile, median, and upper quartile of the uniform distribution are

$$\frac{3}{4}a + \frac{1}{4}b, \quad \frac{a + b}{2}, \quad \frac{1}{4}a + \frac{3}{4}b,$$

respectively. The expectation of a uniform random quantity is

$$
\begin{aligned}
\mathrm{E}(X) &= \int_{-\infty}^{\infty} x\, f_X(x)\, dx \\
&= \int_{-\infty}^{a} x\, f_X(x)\, dx + \int_{a}^{b} x\, f_X(x)\, dx + \int_{b}^{\infty} x\, f_X(x)\, dx \\
&= 0 + \int_{a}^{b} \frac{x}{b-a}\, dx + 0 \\
&= \left[\frac{x^2}{2(b-a)} \right]_{a}^{b} \\
&= \frac{a+b}{2}.
\end{aligned}
$$

Note that in this case the mean is equal to the median, and that both correspond to the mid-point of the interval. The mean and median often coincide for symmetric distributions — see the end of chapter exercises.

We can also calculate the variance of X. First we calculate $\mathrm{E}(X^2)$ as follows:

$$
\begin{aligned}
\mathrm{E}(X^2) &= \int_{a}^{b} \frac{x^2}{b-a}\, dx \\
&= \frac{b^2 + ab + a^2}{3}.
\end{aligned}
$$

Now,

$$
\begin{aligned}
\mathrm{Var}(X) &= \mathrm{E}(X^2) - \mathrm{E}(X)^2 \\
&= \frac{b^2 + ab + a^2}{3} - \frac{(a+b)^2}{4} \\
&= \frac{(b-a)^2}{12}.
\end{aligned}
$$

The uniform distribution is too simple to realistically model actual experimental data, but is very useful for computer simulation, as random quantities from many different distributions can be obtained from standard uniform random quantities.

3.8.1 The standard uniform distribution

A random variable, usually denoted U, has a standard uniform distribution if $U \sim U(0,1)$. In the next chapter we will see how standard uniform random numbers are used as a starting point for random number generation and Monte Carlo simulation. For now it is sufficient to note that $\mathrm{E}(U) = 1/2$, $\mathrm{Var}(U) = 1/12$, and $F(u) = u$, $\forall 0 \le u \le 1$.

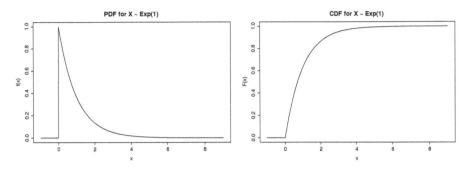

Figure 3.5 *PDF and CDF for an Exp(1) distribution.*

3.9 The exponential distribution

For reasons still to be explored, it turns out that the exponential distribution is the most important continuous distribution in the theory of discrete-event stochastic simulation. It is therefore vital to have a good understanding of this distribution and its many useful properties. We will begin by introducing this distribution in the abstract, but we will then go on to see why it arises so naturally by exploring its relationship with the Poisson process.

The random variable X has an *exponential distribution* with parameter $\lambda > 0$, written

$$X \sim Exp(\lambda)$$

if it has PDF,

$$f_X(x) = \begin{cases} \lambda e^{-\lambda x}, & x \geq 0, \\ 0, & \text{otherwise.} \end{cases}$$

The distribution function, $F_X(x)$ is therefore given by

$$F_X(x) = \begin{cases} 0, & x < 0, \\ 1 - e^{-\lambda x}, & x \geq 0. \end{cases}$$

The PDF and CDF for an $Exp(1)$ are shown in Figure 3.5.

The expectation of the exponential distribution is

$$\begin{aligned} E(X) &= \int_0^\infty x\lambda e^{-\lambda x} dx \\ &= \left[-xe^{-\lambda x}\right]_0^\infty + \int_0^\infty e^{-\lambda x} dx \qquad \text{(by parts)} \\ &= 0 + \left[\frac{e^{-\lambda x}}{-\lambda}\right]_0^\infty \\ &= \frac{1}{\lambda}. \end{aligned}$$

Also,

$$E(X^2) = \int_0^\infty x^2 \lambda e^{-\lambda x} dx = \frac{2}{\lambda^2},$$

and so

$$Var(X) = \frac{2}{\lambda^2} - \frac{1}{\lambda^2} = \frac{1}{\lambda^2}.$$

Note that this means the expectation and standard deviation are both $1/\lambda$.

Notes

1. As λ increases, the probability of small values of X increases and the mean decreases.

2. The median m is given by

$$m = \frac{\log 2}{\lambda} = \log 2 \, E(X) < E(X).$$

3. The exponential distribution is often used to model times between random events (such as biochemical reactions). Some of the reasons are given below.

Proposition 3.16 (memoryless property) *If $X \sim Exp(\lambda)$, then for any $s, t \geq 0$ we have*

$$P(X > (s+t)|X > t) = P(X > s).$$

If we think of the exponential random quantity as representing the time to an event, we can regard that time as a 'lifetime'. Then the proposition states that the probability of 'surviving' a further time s, having survived time t, is the same as the original probability of surviving a time s. This is called the 'memoryless' property of the distribution (as the distribution 'forgets' that it has survived to time t). It is therefore the continuous analogue of the geometric distribution, which is the (unique) discrete distribution with such a property.

Proof.

$$P(X > (s+t)|X > t) = \frac{P([X > (s+t)] \cap [X > t])}{P(X > t)}$$
$$= \frac{P(X > (s+t))}{P(X > t)}$$
$$= \frac{1 - P(X \le (s+t))}{1 - P(X \le t)}$$
$$= \frac{1 - F_X(s+t)}{1 - F_X(t)}$$
$$= \frac{1 - [1 - e^{-\lambda(s+t)}]}{1 - [1 - e^{-\lambda t}]}$$
$$= e^{-\lambda s}$$
$$= 1 - [1 - e^{-\lambda s}]$$
$$= 1 - F_X(s)$$
$$= 1 - P(X \le s)$$
$$= P(X > s).$$

□

Proposition 3.17 *Consider a Poisson process with rate λ. Let T be the time to the first event (after zero). Then $T \sim Exp(\lambda)$.*

Proof. Let N_t be the number of events in the interval $(0, t]$ (for given fixed $t > 0$). We have seen previously that (by definition) $N_t \sim Po(\lambda t)$. Consider the CDF of T,

$$F_T(t) = P(T \le t)$$
$$= 1 - P(T > t)$$
$$= 1 - P(N_t = 0)$$
$$= 1 - \frac{(\lambda t)^0 e^{-\lambda t}}{0!}$$
$$= 1 - e^{-\lambda t}.$$

This is the distribution function of an $Exp(\lambda)$ random quantity, and so $T \sim Exp(\lambda)$.
□

So the time to the *first* event of a Poisson process is an exponential random variable. But then using the independence properties of the Poisson process, it should be reasonably clear that the time between any two such events has the same exponential distribution. Thus the times between events of the Poisson process are exponential.

There is another way of thinking about the Poisson process that this result makes clear. For an infinitesimally small time interval dt we have

$$P(T \le dt) = 1 - e^{-\lambda dt} = 1 - (1 - \lambda dt) = \lambda dt,$$

and due to the independence property of the Poisson process, this is the probability for any time interval of length dt. The Poisson process can therefore be thought of as a process with constant event 'hazard' λ, where the 'hazard' is essentially a measure of 'event density' on the time axis. The exponential distribution with parameter λ can therefore also be reinterpreted as the time to an event of constant hazard λ.

The two properties above are probably the most fundamental. However, there are several other properties that we will require of the exponential distribution when we come to use it to simulate discrete stochastic models of biochemical networks, and so they are mentioned here for future reference. The first describes the distribution of the minimum of a collection of independent exponential random quantities.

Proposition 3.18 *If $X_i \sim Exp(\lambda_i)$, $i = 1, 2, \ldots, n$, are independent random variables, then*

$$X_0 \equiv \min_i\{X_i\} \sim Exp(\lambda_0), \text{ where } \lambda_0 = \sum_{i=1}^{n} \lambda_i.$$

Proof. First note that for $X \sim Exp(\lambda)$ we have $P(X > x) = e^{-\lambda x}$. Then

$$P(X_0 > x) = P\left(\min_i\{X_i\} > x\right)$$
$$= P([X_1 > x] \cap [X_2 > x] \cap \cdots \cap [X_n > x])$$
$$= \prod_{i=1}^{n} P(X_i > x)$$
$$= \prod_{i=1}^{n} e^{-\lambda_i x}$$
$$= e^{-x \sum_{i=1}^{n} \lambda_i}$$
$$= e^{-\lambda_0 x}.$$

So $P(X_0 \leq x) = 1 - e^{-\lambda_0 x}$ and hence $X_0 \sim Exp(\lambda_0)$. □

The next lemma is for the following proposition.

Lemma 3.9 *Suppose that $X \sim Exp(\lambda)$ and $Y \sim Exp(\mu)$ are independent random variables. Then*

$$P(X < Y) = \frac{\lambda}{\lambda + \mu}.$$

Proof.

$$P(X < Y) = \int_0^\infty P(X < Y|Y = y) f(y)\, dy \qquad \text{(Proposition 3.11)}$$
$$= \int_0^\infty P(X < y) f(y)\, dy$$
$$= \int_0^\infty (1 - e^{-\lambda y})\mu e^{-\mu y}\, dy$$
$$= \frac{\lambda}{\lambda + \mu}$$

□

This next result gives the likelihood of a particular exponential random quantity of an independent collection being the smallest.

Proposition 3.19 *If $X_i \sim Exp(\lambda_i)$, $i = 1, 2, \ldots, n$ are independent random variables, let j be the index of the smallest of the X_i. Then j is a discrete random variable with PMF*

$$\pi_i = \frac{\lambda_i}{\lambda_0}, \quad i = 1, 2, \ldots, n, \ where \ \lambda_0 = \sum_{i=1}^{n} \lambda_i.$$

Proof.

$$\pi_j = \mathrm{P}\left(X_j < \min_{i \neq j}\{X_i\}\right)$$

$$= \mathrm{P}(X_j < Y),$$

where $Y = \min_{i \neq j}\{X_i\}$, so that $Y \sim Exp(\lambda_{-j})$, where $\lambda_{-j} = \sum_{i \neq j} \lambda_i$

$$= \frac{\lambda_j}{\lambda_j + \lambda_{-j}} \quad \text{(by the lemma)}$$

$$= \frac{\lambda_j}{\lambda_0}.$$

□

This final result is a trivial consequence of Proposition 3.12.

Proposition 3.20 *Consider $X \sim Exp(\lambda)$. Then for $\alpha > 0$, $Y = \alpha X$ has distribution*

$$Y \sim Exp(\lambda/\alpha).$$

3.10 The normal/Gaussian distribution

3.10.1 Definition and properties

Another fundamental distribution in probability theory is the normal or Gaussian distribution. It turns out that sums of random quantities often approximately follow a normal distribution. Since the change of state of a biochemical network can sometimes be represented as a sum of random quantities, it turns out that normal distributions are useful in this context.

Definition 3.15 *A random quantity X has a normal distribution with parameters μ and σ^2, written*

$$X \sim N(\mu, \sigma^2)$$

if it has probability density function

$$f_X(x) = \frac{1}{\sigma\sqrt{2\pi}} \exp\left\{-\frac{1}{2}\left(\frac{x-\mu}{\sigma}\right)^2\right\}, \quad -\infty < x < \infty,$$

for $\sigma > 0$.

Note that $f_X(x)$ is symmetric about $x = \mu$, and so (provided the density integrates to 1), the median of the distribution will be μ. Checking that the density integrates to 1 requires the computation of a difficult integral. However, it follows directly from the known 'Gaussian' integral

$$\int_{-\infty}^{\infty} e^{-\alpha x^2} \, dx = \sqrt{\frac{\pi}{\alpha}}, \qquad \alpha > 0,$$

since then

$$\int_{-\infty}^{\infty} f_X(x) \, dx = \int_{-\infty}^{\infty} \frac{1}{\sigma\sqrt{2\pi}} \exp\left\{-\frac{1}{2}\left(\frac{x-\mu}{\sigma}\right)^2\right\} dx$$

$$= \frac{1}{\sigma\sqrt{2\pi}} \int_{-\infty}^{\infty} \exp\left\{-\frac{1}{2\sigma^2} z^2\right\} dz \qquad \text{(putting } z = x - \mu)$$

$$= \frac{1}{\sigma\sqrt{2\pi}} \sqrt{\frac{\pi}{1/2\sigma^2}}$$

$$= 1.$$

Now that we know that the given PDF represents a valid density, we can calculate the expectation and variance of the normal distribution as follows:

$$E(X) = \int_{-\infty}^{\infty} x \, f_X(x) \, dx$$

$$= \int_{-\infty}^{\infty} x \, \frac{1}{\sigma\sqrt{2\pi}} \exp\left\{-\frac{1}{2}\left(\frac{x-\mu}{\sigma}\right)^2\right\} dx$$

$$= \mu.$$

The last line follows after a little algebra and calculus, though the result should be clear due to the symmetry and the fact that the integral is clearly convergent. Similarly,

$$\text{Var}(X) = \int_{-\infty}^{\infty} (x-\mu)^2 f_X(x) \, dx$$

$$= \int_{-\infty}^{\infty} (x-\mu)^2 \frac{1}{\sigma\sqrt{2\pi}} \exp\left\{-\frac{1}{2}\left(\frac{x-\mu}{\sigma}\right)^2\right\} dx$$

$$= \sigma^2.$$

The PDF and CDF for a $N(0,1)$ are shown in Figure 3.6.

3.10.2 The standard normal distribution

A standard normal random quantity is a normal random quantity with zero mean and variance equal to 1. It is usually denoted Z, so that

$$Z \sim N(0,1).$$

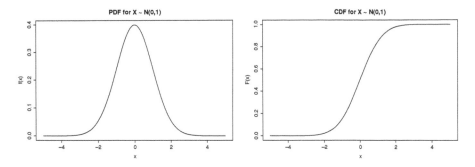

Figure 3.6 *PDF and CDF for a $N(0, 1)$ distribution.*

Therefore, the density of Z, which is usually denoted $\phi(z)$, is given by

$$\phi(z) = \frac{1}{\sqrt{2\pi}} \exp\left\{ -\frac{1}{2}z^2 \right\}, \qquad -\infty < z < \infty.$$

It is important to note that the PDF of the standard normal is symmetric about zero. The distribution function of a standard normal random quantity is denoted $\Phi(z)$, that is

$$\Phi(z) = \int_{-\infty}^{z} \phi(x)dx.$$

There is no neat analytic expression for $\Phi(z)$, so tables of the CDF are used. Of course, we do know that $\Phi(-\infty) = 0$ and $\Phi(\infty) = 1$, as it is a distribution function. Also, because of the symmetry of the PDF about zero, it is clear that we must also have $\Phi(0) = 1/2$ and $\Phi(-z) = 1 - \Phi(z)$, and these can prove useful in calculations. The standard normal distribution is important because it is easy to transform any normal random quantity to a standard normal random quantity by means of a simple linear scaling. Consider $Z \sim N(0, 1)$ and put

$$X = \mu + \sigma Z,$$

for $\sigma > 0$. Then $X \sim N(\mu, \sigma^2)$. To show this, we must show that the PDF of X is the PDF for a $N(\mu, \sigma^2)$ random quantity. Using Proposition 3.12 we have

$$f_X(x) = \frac{1}{\sigma} \phi\left(\frac{x - \mu}{\sigma} \right)$$

$$= \frac{1}{\sigma\sqrt{2\pi}} \exp\left\{ -\frac{1}{2}\left(\frac{x - \mu}{\sigma} \right)^2 \right\},$$

which is the PDF of a $N(\mu, \sigma^2)$ distribution. Conversely, if

$$X \sim N(\mu, \sigma^2)$$

then

$$Z = \frac{X - \mu}{\sigma} \sim N(0, 1).$$

Even more importantly, the distribution function of X is given by

$$F_X(x) = \Phi\left(\frac{x - \mu}{\sigma}\right),$$

and so the cumulative probabilities for any normal random quantity can be calculated using tables for the standard normal distribution.

It is a straightforward generalisation of the above result to see that any linear rescaling of a normal random quantity is another normal random quantity. However, the normal distribution also has an additive property similar to that of the Poisson distribution.

Proposition 3.21 *If $X_1 \sim N(\mu_1, \sigma_1^2)$ and $X_2 \sim N(\mu_2, \sigma_2^2)$ are independent normal random quantities, then their sum $Y = X_1 + X_2$ is also normal, and*

$$Y \sim N(\mu_1 + \mu_2, \sigma_1^2 + \sigma_2^2).$$

The proof is straightforward, but omitted. Putting the above results together, we can see that any linear combination of independent normal random quantities will be normal. We can then use the results for the mean and variance of a linear combination to deduce the mean and variance of the resulting normal distribution.

3.10.3 The central limit theorem

Now that we know about the normal distribution and its properties, we need to understand how and why it arises so frequently. The answer is contained in the central limit theorem, which is stated without proof; see (for example) Miller and Miller (2004) for a proof and further details.

Theorem 3.3 (central limit theorem) *If $X_1, X_2, \ldots, X_n$ are independent realisations of an arbitrary random quantity with mean μ and variance σ^2, let $\bar{X}_n$ be the sample mean and define*

$$Z_n = \frac{\bar{X}_n - \mu}{\sigma/\sqrt{n}}.$$

Then the limiting distribution of Z_n as $n \longrightarrow \infty$ is the standard normal distribution. In other words, $\forall z$,

$$P(Z_n \leq z) \xrightarrow[\infty]{n} \Phi(z).$$

Note that Z_n is just $\bar{X}_n$ linearly scaled so that it has mean zero and variance 1. By rescaling it this way, it is then possible to compare it with the standard normal distribution. Typically, Z_n will not have a normal distribution for any finite value of n, but what the CLT tells us is that the distribution of Z_n becomes more like that of a standard normal as n increases. Then since $\bar{X}_n$ is just a linear scaling of Z_n, $\bar{X}_n$ must also become more normal as n increases. And since $S_n \equiv \sum_{i=1}^{n} X_i$ is just a linear scaling of $\bar{X}_n$, this must also become more normal as n increases. In practice, values of n as small as 10 or 20 are often quite adequate for a normal approximation to work reasonably well.

3.10.4 Normal approximation of binomial and Poisson

A *Bernoulli* random quantity with parameter p, written $Bern(p)$, is a $Bin(1, p)$ random quantity. It therefore can take only the values zero or 1, and has mean p and variance $p(1 - p)$. It is of interest because it can be used to construct the binomial distribution. Let $I_k \sim Bern(p)$, $k = 1, 2, \ldots, n$, where the I_k are mutually independent and identically distributed (IID). Then put

$$X = \sum_{k=1}^{n} I_k.$$

It is clear that X is a count of the number of successes in n trials, and therefore $X \sim Bin(n, p)$. Then, because of the central limit theorem, this will be well approximated by a normal distribution if n is large and p is not too extreme (if p is very small or very large, a Poisson approximation will be more appropriate). A useful guide is that if

$$0.1 \leq p \leq 0.9 \quad \text{and} \quad n > \max \left[\frac{9(1 - p)}{p}, \frac{9p}{1 - p} \right]$$

then the binomial distribution may be adequately approximated by a normal distribution. It is important to understand exactly what is meant by this statement. No matter how large n is, the binomial will always be a discrete random quantity with a PMF, whereas the normal is a continuous random quantity with a PDF. These two distributions will always be qualitatively different. The similarity is measured in terms of the CDF, which has a consistent definition for both discrete and continuous random quantities. It is the CDF of the binomial which can be well approximated by a normal CDF. Fortunately, it is the CDF which matters for typical computations involving cumulative probabilities.

When the n and p of a binomial distribution are appropriate for approximation by a normal distribution, the approximation is achieved by matching expectation and variance. That is,

$$Bin(n, p) \simeq N(np, np[1 - p]).$$

Since the Poisson is derived from the binomial, it is unsurprising that in certain circumstances, the Poisson distribution may also be approximated by the normal. It is generally considered appropriate to apply the approximation if the mean of the Poisson is bigger than 20. Again the approximation is done by matching mean and variance:

$$X \sim Po(\lambda) \simeq N(\lambda, \lambda) \text{ for } \lambda > 20.$$

3.11 The gamma distribution

The *gamma function*, $\Gamma(x)$, is defined by the integral

$$\Gamma(x) = \int_0^\infty y^{x-1} e^{-y} \, dy.$$

By integrating by parts, it is easy to see that $\Gamma(x + 1) = x\Gamma(x)$, and it is similarly clear that $\Gamma(1) = 1$. Together these give $\Gamma(n + 1) = n!$ for integer $n > 0$, and so

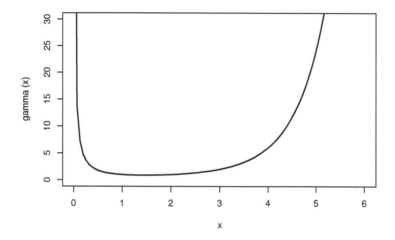

Figure 3.7 *Graph of* $\Gamma(x)$ *for small positive values of* x.

the gamma function can be thought of as a generalisation of the factorial function to non-integer values, x.[§] A graph of the function for small positive values is given in Figure 3.7. It is also worth noting that $\Gamma(1/2) = \sqrt{\pi}$. The gamma function is used in the definition of the *gamma distribution*.

Definition 3.16 *The random variable* X *has a* gamma distribution *with parameters* $\alpha,\ \beta > 0$, *written* $Ga(\alpha, \beta)$ *if it has PDF*

$$f(x) = \begin{cases} \dfrac{\beta^{\alpha}}{\Gamma(\alpha)} x^{\alpha-1} e^{-\beta x} & x > 0 \\ 0 & x \leq 0. \end{cases}$$

Using the substitution $y = \beta x$ and the definition of the gamma function, it is straight-forward to see that this density must integrate to 1. It is also clear that $Ga(1, \lambda) = Exp(\lambda)$, so the gamma distribution is a generalisation of the exponential distribution. It is also relatively straightforward to compute the mean and variance as

$$\mathrm{E}(X) = \frac{\alpha}{\beta}, \qquad \mathrm{Var}(X) = \frac{\alpha}{\beta^2}.$$

[§] Recall that $n!$ (pronounced 'n-factorial') is given by $n! = 1 \times 2 \times 3 \times \cdots \times (n-1) \times n$.

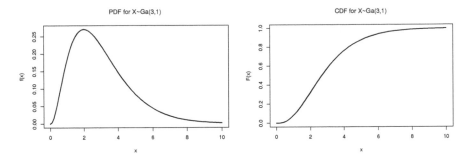

Figure 3.8 *PDF and CDF for a Ga(3, 1) distribution.*

We compute $E(X)$ as follows $(\text{Var}(X)$ is similar):

$$
\begin{aligned}
E(X) &= \int_0^\infty x f(x)\, dx \\
&= \int_0^\infty x \frac{\beta^\alpha}{\Gamma(\alpha)} x^{\alpha-1} e^{-\beta x}\, dx \\
&= \frac{\alpha}{\beta} \int_0^\infty \frac{\beta^{\alpha+1}}{\alpha\Gamma(\alpha)} x^\alpha e^{-\beta x}\, dx \\
&= \frac{\alpha}{\beta} \int_0^\infty \frac{\beta^{\alpha+1}}{\Gamma(\alpha+1)} x^\alpha e^{-\beta x}\, dx \\
&= \frac{\alpha}{\beta}.
\end{aligned}
$$

The last line follows as the integral on the line above is the integral of a $Ga(\alpha+1, \beta)$ density, and hence must be 1. A plot of the PDF and CDF of a gamma distribution is shown in Figure 3.8. The gamma distribution has a very important property.

Proposition 3.22 *If $X_1 \sim Ga(\alpha_1, \beta)$ and $X_2 \sim Ga(\alpha_2, \beta)$ are independent, and $Y = X_1 + X_2$, then*

$$
Y \sim Ga(\alpha_1 + \alpha_2, \beta).
$$

The proof is straightforward but omitted. This clearly generalises to more than two gamma random quantities, and hence implies that the sum of n independent $Exp(\lambda)$ random quantities is $Ga(n, \lambda)$. Since we know that the inter-event times of a Poisson process with rate λ are $Exp(\lambda)$, we now know that the time to the nth event of a Poisson process is $Ga(n, \lambda)$. This close relationship between the gamma distribution, the exponential distribution, and the Poisson process is the reason why the gamma distribution turns out to be important for discrete stochastic modelling.

3.12 Quantifying 'noise'

3.12.1 Variance and standard deviation

Often in stochastic modelling, or when analysing data that is noisy or stochastic in some way, it is desirable to obtain a measure of how 'noisy' some random quantity, X, is. We have already seen that the natural statistical measure of variation is the variance $\sigma^2 = \mathrm{Var}(X)$. However, as previously mentioned, many people prefer to think about the standard deviation $\sigma = \mathrm{SD}(X)$, not least because it is measured in the same units as X. It is therefore the case that the standard deviation scales with the data, so that if $\mathrm{SD}(X) = \sigma$, then for any non-random real number a we have

$$\mathrm{SD}(aX) = |a|\sigma.$$

This is quite natural, but is sometimes inconvenient, especially when dealing with random quantities measured on an arbitrary scale. For example, if X is measured in millimoles, and has a standard deviation of σ, then changing the measurement units to moles will change the standard deviation to $\sigma/1000$. Therefore, interpreting the noise requires knowing and understanding the scale on which the quantity is measured. If X is measured in 'arbitrary fluorescence units', for example, how should the standard deviation be interpreted? Here, a dimensionless measure of noise would be valuable.

3.12.2 The coefficient of variation

The *coefficient of variation* (CV) of a random quantity X is usually defined by

$$\mathrm{CV}(X) = \frac{\mathrm{SD}(X)}{|\mathrm{E}(X)|} = \frac{\sigma}{|\mu|}.$$

Although in principle this definition can be applied to any X such that $\mathrm{E}(X) \neq 0$, in practice it is most often applied to non-negative random quantities (having $\mathrm{E}(X) > 0$). It is clear from this definition that the CV is dimensionless, and has the property

$$\mathrm{CV}(aX) = \mathrm{CV}(X),$$

for $a \neq 0$. This makes the CV independent of the scale on which the variable is measured, and can be desirable in many practical experimental situations. For example, if the coefficient of variation of a random quantity is measured on a machine in one laboratory, it should be comparable with the CV measured by a similar machine in another laboratory, even if the measurements are made in arbitrary units, and the machines are calibrated slightly differently.

The reciprocal of the coefficient of variation, $|\mu|/\sigma$, is known as the *signal-to-noise ratio*, and is often used by people from engineering backgrounds. It clearly has the same dimensionless, scale invariant properties as the CV, and which to use is obviously just a matter of taste. The signal-to-noise ratio is typically used less often than the CV in the study of noise, simply because the coefficient of variation has the intuitive property that more noise leads to a bigger number.

3.12.3 Dispersion index

The *dispersion index*, sometimes also known as the *coefficient of dispersion*, *index of dispersion*, the *variance-to-mean ratio* (VMR), and in certain contexts as the *Fano factor*, is defined by

$$\mathrm{VMR}(X) = \frac{\mathrm{Var}(X)}{|\,\mathrm{E}(X)\,|} = \frac{\sigma^2}{|\mu|}.$$

Like the standard deviation, the VMR has the same units as X, with the same scaling property

$$\mathrm{VMR}(aX) = |a|\,\mathrm{VMR}(X).$$

It should be emphasised that the VMR only has a natural interpretation when X is a discrete random quantity representing some kind of count. In this case we have seen that the Poisson distribution is the natural reference distribution, and this clearly has the property that $\mathrm{VMR}(X) = 1$, irrespective of the expected value (since the mean and variance of a Poisson random quantity are both the same). Therefore the VMR provides a natural way to compare how noisy a discrete random quantity is relative to the a Poisson distribution with the same expectation. Random variables X with the property that $\mathrm{VMR}(X) < 1$ are said to be *under-dispersed* relative to the Poisson, and quantities such that $\mathrm{VMR}(X) > 1$ are said to be *over-dispersed*. Again, it should be stressed that such comparisons only make sense when X is a discrete count, otherwise it is possible to arbitrarily change the value of the VMR simply by changing the measurement units.

3.13 Exercises

1. Prove the addition law (item 4 of Proposition 3.2).
2. Show that if events E and F both have positive probability they cannot be both independent and exclusive.
3. Prove Proposition 3.13.
4. Show that for a continuous distribution that is symmetric about zero, zero must be a median of the distribution. Further show that if the mean (expectation) exists, then it too must be zero. Deduce the corresponding results for a continuous distribution symmetric about a value, μ.
5. The standard Cauchy distribution is a continuous distribution defined on the whole of the real line, with PDF

$$f(x) = \frac{1}{\pi(1 + x^2)}, \qquad x \in \mathbb{R}.$$

   (a) Write down the median and mode of the distribution.
   (b) Does this distribution have a mean?
   (c) Does this distribution have a variance?
   (d) Derive the CDF, and use it to show that the PDF does integrate to one.

6. Suppose that a cell produces mRNA transcripts for a particular gene according to a Poisson process with a rate of 2 per second.

   (a) What is the distribution of the number of transcripts produced in 5 seconds?

   (b) What is the mean and variance of the number produced in 5 seconds?

   (c) What is the probability that exactly the mean number will be produced?

   (d) What is the distribution of the time until the first transcript?

   (e) What is the mean and variance of the time until the first transcript?

   (f) What is the probability that the time is more than 1 second?

   (g) What is the distribution of the time until the third transcript?

   (h) What is the mean and variance of the time until the third transcript?

   (i) What is the probability that the third transcript happens in the first second?

7. Derive from first principles the variance of the gamma distribution.

8. For $X \sim Geom(p)$, $p > 0$, write down, $E(X)$, $Var(X)$, $SD(X)$, $CV(X)$, and $VMR(X)$. For what values of $p > 0$ is the geometric distribution more 'noisy' than the Poisson?

3.14 Further reading

For a more comprehensive and less terse introduction to probability models, classic texts such as Ross (2009) are ideal. However, most introductory statistics textbooks also cover much of the more basic material. More advanced statistics texts, such as Rice (2007) or Miller and Miller (2004), also cover this material in addition to statistical methodology that will be helpful in later chapters.

## CHAPTER 4

Stochastic simulation

4.1 Introduction

As we saw in the previous chapter, probability theory is a powerful framework for understanding random phenomena. For simple random systems, mathematical analysis alone can provide a complete description of all properties of interest. However, for the kinds of random systems that we are interested in (stochastic models of biochemical networks), mathematical analysis is not possible, and the systems are described as *analytically intractable*. However, this does not mean that it is not possible to understand such systems. With the aid of a computer, it is possible to *simulate* the time-evolution of the system dynamics. Stochastic simulation is concerned with the computer simulation of random (or stochastic) phenomena, and it is therefore essential to know something about this topic before proceeding further.

4.2 Monte Carlo integration

The rationale for stochastic simulation can be summarised very easily: to understand a probabilistic model, simulate many realisations from it and study them. To make this more concrete, one way to understand stochastic simulation is to perceive it as a way of numerically solving the difficult integration problems that naturally arise in probability theory.

Suppose we have a (continuous) random variable X, with probability density function (PDF), $f(x)$, and we wish to evaluate $E(g(X))$ for some function $g(\cdot)$. We know that

$$E(g(X)) = \int_X g(x)f(x)\,dx,$$

and so the problem is one of integration. For simple $f(\cdot)$ and $g(\cdot)$ this integral might be straightforward to compute directly. On the other hand, in more complex scenarios, it is likely to be analytically intractable. However, if we can *simulate* realisations $x_1, \dots, x_n$ of X, then we can form realisations of the random variable $g(X)$ as $g(x_1), \dots, g(x_n)$. Then, provided that the variance of $g(X)$ is finite, the laws of large numbers (Propositions 3.14 and 3.15) assure us that for large n we may approximate the integral by

$$E(g(X)) \simeq \frac{1}{n} \sum_{i=1}^{n} g(x_i).$$

In fact, even if we cannot simulate realisations of X, but can simulate realisations $y_1, \dots, y_n$ of Y (a random variable with the same support as X), which has PDF

$h(\cdot)$, then

$$E(g(X)) = \int_X g(x)f(x)\,dx$$

$$= \int_X \frac{g(x)f(x)}{h(x)}h(x)\,dx$$

and so $E(g(X))$ may be approximated by

$$E(g(X)) \simeq \frac{1}{n}\sum_{i=1}^{n} \frac{g(y_i)f(y_i)}{h(y_i)}.$$

This procedure is known as *importance sampling*, and it can be very useful when there is reasonable agreement between $f(\cdot)$ and $h(\cdot)$.

The above examples show that it is possible to compute summaries of interest such as expectations, provided we have a mechanism for simulating realisations of random quantities. It turns out that doing this is a rather non-trivial problem. We begin first by thinking about the generation of uniform random quantities, and then move on to thinking about the generation of random quantities from other standard distributions.

4.3 Uniform random number generation

4.3.1 Discrete uniform random numbers

Most stochastic simulation begins with a uniform random number generator (Gentle, 2003). Computers are essentially deterministic, so algorithms are produced which generate a deterministic sequence of numbers that appears to be random. Consequently such generators are often referred to as *pseudo-random number generators*.

Typically a number theoretic method is used to generate an apparently random integer from 0 to $2^N - 1$ (often, $N = 16$, 32, or 64). The methods employed these days are often very sophisticated, as modern cryptographic algorithms rely heavily on the availability of 'good' random numbers. A detailed discussion would be out of place in the context of this book, but a brief explanation of a very simple algorithm is perhaps instructive.

Linear congruential generators are the simplest class of algorithm used for pseudo-random number generation. The algorithm begins with a *seed* x_0 and then generates new values according to the (deterministic) rule

$$x_{n+1} = (a\,x_n + b \mod 2^N)$$

for carefully chosen a and b. Clearly such a procedure must return to a previous value at some point and then cycle indefinitely. However, if a 'good' choice of a, b, and N are used, this deterministic sequence of numbers will have a very large cycle length (significantly larger than the number of random quantities one is likely to want to generate in a particular simulation study) and give every appearance of being uniformly random. One reasonable choice is

$$N = 59,\ b = 0,\ a = 13^{13},$$

but there are many others. The Wikipedia page for 'Linear congruential generator' has a list of commonly used values and their sources.

4.3.2 Standard uniform random numbers

Most computer programming languages and scientific libraries have built-in functions for returning a pseudo-random integer. The technique used will be similar in principle to the linear congruential generator described above, but should be more sophisticated (and therefore less straightforward for cryptographers to 'crack'). Provided the maximum value is very large, the random number can be divided by the maximum value to give a pseudo-random $U(0,1)$ number (and again, most languages will provide a function which does this). We will not be concerned with the precise mechanisms of how this is done, but rather take it as our starting point to examine how these uniform random numbers can be used to simulate more interesting distributions. Once we have a method for simulating standard uniform random variates, we can use them in order to simulate random quantities from any distribution we want. Large Monte Carlo studies do rely heavily on good quality generators, and so it is always worth choosing a high-quality generator if one is available. The Mersenne Twister (Matsumoto and Nishimura, 1998) is a sophisticated and popular generator with a very long cycle length, and it is often implemented within scientific libraries and software.

4.4 Transformation methods

Suppose that we wish to simulate realisations of a random variable X, with PDF $f(x)$, and that we are able to simulate $U \sim U(0,1)$.

Proposition 4.1 (inverse distribution method) *If $U \sim U(0,1)$ and $F(\cdot)$ is a valid invertible cumulative distribution function (CDF), then*

$$X = F^{-1}(U)$$

has CDF $F(\cdot)$.

Proof.

$$
\begin{aligned}
P(X \le x) &= P\big(F^{-1}(U) \le x\big) \\
&= P(U \le F(x)) \\
&= F_U(F(x)) \\
&= F(x). \qquad \text{(as } F_U(u) = u)
\end{aligned}
$$

□

So for a given $f(\cdot)$, assuming that we also compute the probability distribution function $F(\cdot)$ and its inverse $F^{-1}(\cdot)$, we can simulate a realisation of X using a single $U \sim U(0,1)$ by applying the inverse CDF to the uniform random quantity. Note that the inverse CDF of a random variable is often referred to as the *quantile function*.

4.4.1 Uniform random variates

Recall that the properties of the uniform distribution were discussed in Section 3.8. Given $U \sim U(0,1)$, we can simulate $V \sim U(a,b)$ $(a < b)$ in the 'obvious' way, that is

$$V = a + (b-a)U.$$

We can justify this as V has CDF

$$F(v) = \frac{v-a}{b-a}, \qquad a \leq v \leq b$$

and hence inverse

$$F^{-1}(u) = a + (b-a)u.$$

4.4.2 Exponential random variates

For discrete-event simulation of the time evolution of biochemical networks, it is necessary to have an efficient way to simulate exponential random quantities (Section 3.9). Fortunately, this is very straightforward, as the following result illustrates.

Proposition 4.2 *If $U \sim U(0,1)$ and $\lambda > 0$, then*

$$X = -\frac{1}{\lambda}\log(U)$$

has an $Exp(\lambda)$ distribution.

Proof. Consider $X \sim Exp(\lambda)$. This has density $f(x)$ and distribution $F(x)$, where

$$f(x) = \lambda e^{-\lambda x}, \qquad F(x) = 1 - e^{-\lambda x}, \qquad x \geq 0$$

and so

$$F^{-1}(u) = -\frac{\log(1-u)}{\lambda}, \qquad 0 \leq u \leq 1.$$

The result follows as U and $1-U$ clearly both have a $U(0,1)$ distribution.
□

So to simulate a realisation of X, simulate u from $U(0,1)$, and then put

$$x = -\frac{\log(u)}{\lambda}.$$

4.4.3 Scaling

It is worth noting the scaling issues in the above example, as these become more important for distributions that are more difficult to simulate from. If $U \sim U(0,1)$, then $Y = -\log U \sim Exp(1)$, sometimes referred to as the *standard* or *unit* exponential. The parameter, λ, of an exponential distribution is a *scale parameter* because we can obtain exponential variates with other parameters from a variate with a unit parameter by a simple linear scaling. That is, if

$$Y \sim Exp(1)$$

then

$$X = \frac{1}{\lambda}Y \sim Exp(\lambda).$$

In general, we can spot *location* and *scale* parameters in a distribution as follows. If Y has PDF $f(y)$ and CDF $F(y)$, and $X = aY + b$, then X has CDF

$$\begin{aligned}
F_X(x) &= \mathrm{P}(X \le x) \\
&= \mathrm{P}(aY + b \le x) \\
&= \mathrm{P}\left(Y \le \frac{x-b}{a}\right) \\
&= F\left(\frac{x-b}{a}\right)
\end{aligned}$$

and PDF

$$f_X(x) = \frac{1}{a} f\left(\frac{x-b}{a}\right).$$

Parameters like a are scale parameters, and parameters like b are location parameters.

4.4.4 Gamma random variates

The inverse transformation method is ideal for distributions with tractable CDF and quantile function, but the gamma distribution (Section 3.11) is a good example of a distribution where it is not practical to compute an analytic expression for the inverse of the CDF, and so it is not possible to use the inverse transformation method directly. In this case, it is possible to use numerical techniques to compute the CDF and its inverse, but this is not very efficient. More sophisticated techniques are therefore required in general. One way to simulate $X \sim Ga(n, \lambda)$ random variates for integer n is to use the fact that if

$$Y_i \sim Exp(\lambda),$$

and the Y_i are independent, then

$$X = \sum_{i=1}^{n} Y_i \sim Ga(n, \lambda).$$

So just simulate n exponential random variates and add them up.

The first parameter of the gamma distribution (here n) is known as the *shape* parameter, and the second (here λ) is known as the *scale* parameter. It is important to understand that the second parameter is a scale parameter, because many gamma generation algorithms work by first generating gamma variables with arbitrary shape but unit scale. Once this has been achieved, it can then easily be rescaled to give the precise gamma variate required. This can be done easily because the $Ga(\alpha, \beta)$ PDF is

$$f_X(x) = \frac{\beta^\alpha}{\Gamma(\alpha)} x^{\alpha-1} e^{-\beta x}, \qquad x > 0$$

and so the $Ga(\alpha, 1)$ PDF is

$$f_Y(y) = \frac{1}{\Gamma(\alpha)} y^{\alpha-1} e^{-y}, \quad y > 0.$$

We can see that if $Y \sim Ga(\alpha, 1)$, then $X = Y/\beta \sim Ga(\alpha, \beta)$ because

$$f_X(x) = \beta f_Y(\beta x).$$

Consequently the CDFs must be related by

$$F_X(x) = F_Y(\beta x).$$

Techniques for efficiently generating gamma variates with arbitrary shape parameter are usually based on rejection techniques (to be covered later). Note, however, that for shape parameters which are an integer multiple of 0.5, use can be made of the fact that $\chi_n^2 = Ga(n/2, 1/2)$.* So given a technique for generating χ^2 quantities, gamma variates with a shape parameter that is an integer multiple of 1/2 can be generated by a simple rescaling.

4.4.5 Normal random variates

The ability to efficiently simulate normal (or Gaussian) random quantities is of crucial importance for stochastic simulation. However, the inverse transformation method is not really appropriate due to the analytic intractability of the CDF and inverse CDF of the normal distribution, so another method needs to be found. Note that all we need is a technique for simulating $Z \sim N(0, 1)$ random variables, as then $X = \mu + \sigma Z \sim N(\mu, \sigma^2)$. Also note that standard normal random variables can be used to generate χ^2 random variables. If $Z_i \sim N(0, 1)$ and the Z_i are independent, then

$$C = \sum_{i=1}^{n} Z_i^2$$

has a χ_n^2 distribution.

CLT-based method

One simple way to generate normal random variables is to make use of the Central Limit Theorem (Theorem 3.3). Consider

$$Z = \sum_{i=1}^{12} U_i - 6$$

where $U_i \sim U(0, 1)$, $i = 1, 2, \ldots, 12$ are independent. Clearly $\mathrm{E}(Z) = 0$ and $\mathrm{Var}(Z) = 1$, and by the Central Limit Theorem (CLT), Z is approximately normal. However, this method is not exact. For example, Z only has support on $[-6, 6]$

---

* We did not discuss the χ^2 (or *chi-square*) distribution in Chapter 3, but it is usually defined as the sum of squares of independent standard normal random quantities, which leads directly to the simulation method discussed in the next section. That the χ^2 distribution is a special case of the gamma distribution is demonstrated in Miller and Miller (2004).

and is poorly behaved in the extreme tails. However, for $Z \sim N(0,1)$ we have $P(|Z| > 6) \simeq 2 \times 10^{-9}$, and so the truncation is not much of a problem in practice, and this method is good enough for many purposes. The main problem with this method is that for 'good' random number generators, simulating uniform random numbers is quite slow, and so simulating 12 in order to get one normal random quantity is undesirable.

Box–Muller method

A more efficient (and 'exact') method for generating normal random variates is the following. Simulate

$$\Theta \sim U(0, 2\pi), \quad R^2 \sim Exp(1/2)$$

independently. Then

$$X = R\cos\Theta, \quad Y = R\sin\Theta$$

are two independent standard normal random variables. It is easier to show this the other way around.

Proposition 4.3 *If X and Y are independent standard normal random quantities, and they are regarded as the Cartesian coordinates (X,Y) of a 2-d random variable, then the polar coordinates of the variable (R,Θ) are independent with $R^2 \sim Exp(1/2)$ and $\Theta \sim U(0, 2\pi)$.*

Proof. Suppose that $X, Y \sim N(0,1)$ and that X and Y are independent. Then

$$f_{X,Y}(x,y) = \frac{1}{\sqrt{2\pi}}\exp\{-x^2/2\}\frac{1}{\sqrt{2\pi}}\exp\{-y^2/2\} = \frac{1}{2\pi}\exp\{-(x^2+y^2)/2\}.$$

Put

$$X = R\cos\Theta \quad \text{and} \quad Y = R\sin\Theta.$$

Then,

$$f_{R,\Theta}(r,\theta) = f_{X,Y}(x,y)\left|\frac{\partial(x,y)}{\partial(r,\theta)}\right|$$

$$= \frac{1}{2\pi}e^{-r^2/2}\begin{vmatrix}\cos\theta & -r\sin\theta \\ \sin\theta & r\cos\theta\end{vmatrix}$$

$$= \frac{1}{2\pi} \times re^{-r^2/2}.$$

So, Θ and R are independent, $\Theta \sim U(0, 2\pi)$, and $f_R(r) = re^{-r^2/2}$. It is then easy to show (using Proposition 3.13) that $R^2 \sim Exp(1/2)$.
□

Note that here we have used a couple of results that were not covered directly in Chapter 3: namely, that independence of continuous random quantities corresponds to factorisation of the density, and also that the random variable transformation formula (Proposition 3.13) generalises to higher dimensions (where the 'Jacobian' of the transformation is now the *determinant* of the matrix of partial derivatives). For completeness, the algorithm can be summarised as follows:

1. Generate $u_1, u_2 \sim U(0,1)$, independently

2. Put $\theta = 2\pi u_1$ and $r = \sqrt{-2\log u_2}$

3. Put $x = r\cos\theta$ and $y = r\sin\theta$

4. Return x, y

The returned values of x and y are independent $N(0,1)$ random variates. This algorithm is relatively efficient, as there are not too many mathematical operations involved, and the algorithm produces one standard normal random variate for each standard uniform that is used. It is however a feature of the algorithm that standard uniforms are 'consumed' in pairs, and that standard normal deviates are produced in pairs. This causes slight issues for efficient implementation, as one obviously wants a function which returns a single normal random deviate. One way to achieve this is for the function to actually generate two values, but then 'cache' one ready to return immediately on the next function call, and return the other. This is very efficient, but there can be implications associated with caching a value between calls for 'thread safety' in certain languages and environments, and so some implementations will sacrifice efficiency for thread safety and just generate a single normal variate from two uniform variates at each function call. Algorithms for efficiently simulating normal random quantities are built into most scientific and numerical libraries, and it is often preferable to use these where available.

4.5 Lookup methods

The so-called lookup method is just the discrete version of the inverse transformation method. However, as it looks a little different at first sight, it is worth examining in detail. Suppose one is interested in simulating a discrete random quantity X with outcome space $S = \{0, 1, 2, \ldots\}$, or a subset thereof. Define $p_k = P(X = k)$, $k = 0, 1, 2, \ldots$, the PMF of X. Some of the p_k may be zero. Indeed, if X is finite, then all but finitely many of the p_k will be zero. Next define

$$q_k = P(X \leq k) = \sum_{i=0}^{k} p_i.$$

Realisations of X can be generated by first simulating $U \sim U(0,1)$ and then putting

$$X = \min\{k | q_k \geq U\}.$$

This works because then by definition we must have $q_{k-1} < U \leq q_k$, and so

$$P(X = k) = P(U \in (q_{k-1}, q_k]) = q_k - q_{k-1} = p_k,$$

as required. In practice, for a one-off simulation, first simulate a U and then compute a running sum of the p_i. At each stage, check to see if the running sum is as big as U. If it is, return the index, k of the final p_k. If many realisations are required, more efficient strategies are possible, but in the context of discrete-event simulation of biochemical networks, a one-off simulation is usually all that is required. As we will see, typically one picks one of a finite set of reactions using a lookup method and chooses a time to wait until that reaction occurs by simulating an exponential random quantity.

As this lookup method is so fundamental to many stochastic simulation algorithms, it is worth spending a little time understanding exactly how it works, how it is implemented, and how to make it more efficient. The standard uniform covers a range of width one, going from zero to one. Each possible event has a probability, and the sum of probabilities is one, the width of the standard uniform. The lookup method works by assigning each event to an interval of $[0, 1]$, with width equal to probability. The algorithm described above does that assignment in index order. This may or may not be efficient, depending on exactly how the algorithm is implemented. Many algorithms use the running sum method described above, starting at the beginning of the interval. It should be clear that if such a strategy is to be adopted, then it is going to be much more efficient to have the events arranged so that they are ordered in decreasing probability. This will ensure that the matching event is found with the fewest number of comparisons, on average. However, sorting events by their probability is also an expensive operation, so that will not be worthwhile for a one-off simulation. In the context of the discrete stochastic simulation algorithms we will consider later, the probabilities are different each time, but are typically *not much* different. This raises the possibility of using simple dynamic event re-ordering strategies in order to gradually 'bubble' the high-probability events to the top of the queue, without doing a full sort at each iteration. In fact, even if the probabilities are perfectly ordered (in decreasing order), a linear search through the list for a match is still not optimal. For further information on more efficient strategies for lookup methods, see the discussion of Walker's alias method in Ripley (1987).

4.6 Rejection samplers

We have now examined a range of useful techniques for random number generation, but we still have not seen a technique that can be used for general continuous random quantities where the inverse CDF cannot easily be computed. In this situation, rejection samplers are most often used.

Proposition 4.4 (uniform rejection method) *Suppose that we want to simulate from $f(x)$ with (finite) support on $[a, b]$, and that $f(x) \leq m$, $\forall x \in [a, b]$. Then consider simulating*

$$X \sim U(a, b) \quad and \quad Y \sim U(0, m).$$

Accept X if $Y < f(X)$, otherwise reject *and try again. Then the accepted X values have PDF $f(x)$.*

Intuitively we can see that this will work because it has the effect of scattering points uniformly, first over a rectangle, and then, via the rejection step, just over the region bounded by the PDF and the x-axis. Then there will be more X values where the density is high than where it is low, and in exactly the right proportion.

Proof. Call the acceptance region A, and the accepted value $\tilde{X}$. Then

$$
\begin{aligned}
F_{\tilde{X}}(x) &= \mathrm{P}\left(\tilde{X} \le x\right) \\
&= \mathrm{P}(X \le x | (X,Y) \in A) \\
&= \frac{\mathrm{P}((X \le x) \cap ((X,Y) \in A))}{\mathrm{P}((X,Y) \in A)} \\
&= \frac{\displaystyle\int_a^b \mathrm{P}((X \le x) \cap ((X,Y) \in A)|X = z) \times \frac{1}{b-a}\,dz}{\displaystyle\int_a^b \mathrm{P}((X,Y) \in A|X = z) \times \frac{1}{b-a}\,dz}
\end{aligned}
$$

$$
\begin{aligned}
&= \frac{\dfrac{1}{b-a} \displaystyle\int_a^x \mathrm{P}((X,Y) \in A|X = z)\, dz}{\dfrac{1}{b-a} \displaystyle\int_a^b \mathrm{P}((X,Y) \in A|X = z)\, dz} \\
&= \frac{\displaystyle\int_a^x \frac{f(z)}{m}\,dz}{\displaystyle\int_a^b \frac{f(z)}{m}\,dz} \\
&= \int_a^x f(z)\, dz \\
&= F(x).
\end{aligned}
$$

□

In summary, we simulate a value x uniformly from the support of X and accept this value with probability $f(x)/m$, otherwise we reject and try again. Obviously the efficiency of this method depends on the overall proportion of candidate points that are accepted. The actual acceptance probability for this method is

$$
\begin{aligned}
\mathrm{P}(\text{Accept}) &= \mathrm{P}((X,Y) \in A) \\
&= \int_a^b \mathrm{P}((X,Y) \in A|X = x) \times \frac{1}{b-a}\,dx \\
&= \int_a^b \frac{f(x)}{m} \times \frac{1}{b-a}\,dx \\
&= \frac{1}{m(b-a)} \int_a^b f(x)\,dx \\
&= \frac{1}{m(b-a)}.
\end{aligned}
$$

If this acceptance probability is very low, the procedure will be very inefficient, and a better procedure should be sought — the *envelope method* is one possibility.

4.6.1 Envelope method

Once we have established that scattering points uniformly over the region bounded by the density and the x-axis generates x-values with the required distribution, we can extend it to distributions with infinite support and make it more efficient by choosing our *enveloping* region more carefully.

Suppose that we wish to simulate X with PDF $f(\cdot)$, but that we can already simulate values of Y (with the same support as X), which has PDF $h(\cdot)$. Suppose further that there exists some constant a such that

$$f(x) \le a\, h(x), \quad \forall x.$$

That is, a is an upper bound for $f(x)/h(x)$. Note also that $a \ge 1$, as both $f(x)$ and $h(x)$ integrate to 1.

Consider the following algorithm. Draw $Y = y$ from $h(\cdot)$, and then $U = u \sim U(0, a\, h(y))$. Accept y as a simulated value of X if $u < f(y)$, otherwise reject and try again. This works because it first distributes points uniformly over a region covering $f(x)$ (using the rejection method idea in reverse), and then only keeps points in the required region (under $f(x)$):

$$P\left(\tilde{X} \le x\right) = P(Y \le x | U \le f(Y))$$

$$= \frac{P([Y \le x] \cap [U \le f(Y)])}{P(U \le f(Y))}$$

$$= \frac{\displaystyle\int_{-\infty}^{\infty} P([Y \le x] \cap [U \le f(Y)] | Y = y)\, h(y) dy}{\displaystyle\int_{-\infty}^{\infty} P(U \le f(Y) | Y = y)\, h(y) dy}$$

$$= \frac{\displaystyle\int_{-\infty}^{x} P(U \le f(Y) | Y = y)\, h(y) dy}{\displaystyle\int_{-\infty}^{\infty} P(U \le f(Y) | Y = y)\, h(y) dy}$$

$$= \frac{\displaystyle\int_{-\infty}^{x} \frac{f(y)}{a\, h(y)} h(y) dy}{\displaystyle\int_{-\infty}^{\infty} \frac{f(y)}{a\, h(y)} h(y) dy} = \frac{\displaystyle\int_{-\infty}^{x} \frac{f(y)}{a} dy}{\displaystyle\int_{-\infty}^{\infty} \frac{f(y)}{a} dy}$$

$$= \int_{-\infty}^{x} f(y)\, dy = F(x).$$

To summarise, just simulate a proposed value from $h(\cdot)$ and accept this with probability $f(y)/[a\, h(y)]$, otherwise reject and try again. The accepted values will have PDF $f(\cdot)$.

Obviously, this method will work well if the overall acceptance rate is high, but not otherwise. The overall acceptance probability can be computed as

$$
\begin{aligned}
\mathrm{P}(U < f(Y)) &= \int_{-\infty}^{\infty} \mathrm{P}(U < f(Y)|Y = y)\, h(y)dy \\
&= \int_{-\infty}^{\infty} \frac{f(y)}{a\,h(y)} h(y)dy \\
&= \int_{-\infty}^{\infty} \frac{f(y)}{a}dy \\
&= \frac{1}{a}.
\end{aligned}
$$

Consequently, we want a to be as small as possible (that is, as close as possible to 1). What 'small enough' means is context-dependent, but generally speaking, if $a > 10$, the envelope is not adequate — too many points will be rejected, so a better envelope needs to be found. If this is not practical, then an entirely new approach is required.

4.7 Importance resampling

Importance resampling is an idea closely related to both importance sampling and the envelope rejection method. One of the problems with using the rejection method is finding a good envelope and computing the envelope bounding constant, a. Importance resampling, like importance sampling, can in principle use any proposal distribution $h(\cdot)$ with the same support as the target distribution $f(\cdot)$, and there is no need to calculate any kind of bounding constant. It is therefore widely applicable, but in practice will work *well* only if $h(\cdot)$ is sufficiently similar to $f(\cdot)$. Further, unlike rejection sampling, importance resampling is not exact, so the generated samples are only approximately from $f(\cdot)$, with the approximation improving as the number of generated samples increases.

Importance sampling was introduced in Section 4.2. There it was demonstrated that an expectation of an arbitrary function $g(\cdot)$ with respect to a target distribution $f(\cdot)$ can be approximated using samples y_i from a proposal distribution $h(\cdot)$ by

$$
\mathrm{E}(g(X)) \simeq \frac{1}{n} \sum_{i=1}^{n} \frac{g(y_i)f(y_i)}{h(y_i)}.
$$

We can re-write this as

$$
\mathrm{E}(g(X)) \simeq \frac{1}{n} \sum_{i=1}^{n} w_i g(y_i),
$$

where $w_i = f(y_i)/h(y_i)$. That is, samples from $h(\cdot)$ can be used as if they were samples from $f(\cdot)$, provided that they are re-weighted appropriately by the w_i. This motivates importance resampling: first generate samples from the proposal $h(\cdot)$, then *resample* from the sample, using the weights w_i. Then the new sample is distributed approximately according to $f(\cdot)$. We can describe the algorithm explicitly as follows.

1. Sample $y_1, y_2, \ldots, y_n \sim h(\cdot)$

2. Compute the weights $w_k = f(y_k)/h(y_k)$, $k = 1, 2, \ldots, n$

3. Compute the sum of the weights, $w_0 = \sum_{j=1}^{n} w_j$

4. Compute the *normalised* weights $w_k' = w_k/w_0$, $k = 1, 2, \ldots, n$

5. Sample n times, *with replacement* from the set $\{y_1, y_2, \ldots, y_n\}$ using the probabilities $\{w_1', w_2', \ldots, w_n'\}$ (for example, using the lookup method) to generate a new sample $\{x_1, x_2, \ldots, x_n\}$

6. Return the new sample $\{x_1, x_2, \ldots, x_n\}$ as an approximate sample from $f(\cdot)$

As already discussed, this algorithm is very general and is central to several more advanced Monte Carlo techniques, such as *sequential importance resampling* (SIR) and particle filtering, which will be discussed in the latter chapters.

A rigorous proof that importance resampling works is beyond the scope of this book (see Doucet et al. (2001) for such details), but an informal justification can be given in the case of a univariate continuous distribution as follows. The method is only approximate, and the approximation improves as the number of particles increases, so consider a *very* large number of particles (samples), N. Let us also consider an arbitrary very small interval $[x, x + dx)$ in the support of $f(\cdot)$ and $h(\cdot)$. The probability that a given sample from the proposal $h(\cdot)$ will be contained in this interval is $h(x)dx$, so the expected number of particles is $Nh(x)dx$. The weight of each of these particles is $w(x) = f(x)/h(x)$. On the other hand, the expected weight of an arbitrary random particle is

$$\mathrm{E}(w(X)) = \int_{\mathbb{R}} w(x)h(x)dx = \int_{\mathbb{R}} \frac{f(x)}{h(x)}h(x)dx = \int_{\mathbb{R}} f(x)dx = 1,$$

and so $w_0 = N$. The normalised weights are therefore $w'(x) = f(x)/[Nh(x)]$, and so the combined normalised weight of all particles in the interval $[x, x + dx)$ is

$$\frac{f(x)}{Nh(x)} \times Nh(x)dx = f(x)dx.$$

When we resample from our set of particles N times we therefore expect to get $Nf(x)dx$ particles in our interval, corresponding to a proportion $f(x)dx$ and a density $f(x)$, and so our new set of particles has density $f(x)$, asymptotically, as $N \longrightarrow \infty$. Although there are a number of gaps in this argument, it isn't too difficult to tighten up into a reasonably rigorous proof, at least in the case of one-dimensional continuous distributions. The algorithm also works for very general multi-dimensional distributions, but demonstrating the validity of the algorithm in that case requires more work.

4.8 The Poisson process

Consider a Poisson process with rate λ defined on the interval $[0, T]$. As we know, the inter-event times are $Exp(\lambda)$, and so we have a very simple algorithm for simulating realisations of the process. Initialise the process at time zero. Then simulate $t_1 \sim Exp(\lambda)$, the time to the first event, and define $X_1 = t_1$. Next simulate $t_2 \sim Exp(\lambda)$, the time from the first to the second event, and set $X_2 = X_1 + t_2$. At step k, simulate

$t_k \sim Exp(\lambda)$, the time from the $k-1$ to kth event, and set $X_k = X_{k-1} + t_k$. Repeat until $X_k > T$, and keep the $X_1, X_2, \ldots$ as the realisation of the process.

4.9 Using the statistical programming language R

4.9.1 Introduction

R (sometimes known as GNU S) is a very popular programming language for data analysis and statistics. It is a completely free open-source software application and very widely used by professional statisticians. It is also very popular in certain application areas, including bioinformatics. R is a dynamically typed interpreted language, and it is typically used interactively. It has many built-in functions and libraries and a mature optional–package ecosystem called CRAN. R is extensible, allowing users to define their own functions and procedures using R, C, C++, or Fortran, for example. It also has several object systems. In addition, it is a suitable language for a functional programming style, as it is lexically scoped and has functions as first-class objects. R is a particularly convenient environment to use for stochastic simulation, visualisation, and analysis. It will therefore be used in the forthcoming chapters in order to provide concrete illustrations of the theory being described. It is strongly recommended that readers unfamiliar with R download and install it, then work through this mini-tutorial. Although there are several GUI front-ends for R, it is fundamentally a programming language, and best used by entering commands directly at the R prompt. However, many people find that the RStudio IDE for R increases their productivity.

4.9.2 Vectors

Vectors are a fundamental concept in R, as many functions operate on and return vectors, so it is best to master these as soon as possible. In R, a (numeric) vector is an object consisting of a one-dimensional array of scalars.

```
> rep (1,10)
 [1] 1 1 1 1 1 1 1 1 1 1
>
```

Here **rep** is a function that returns a vector (here, 1 repeated 10 times). You can get documentation for **rep** by typing

```
> ?rep
```

You can assign any object (including vectors) using the assignment operator <- (or =), and combine vectors and scalars with the **c** function.

```
> a <- rep (1,10)
> b <- 1:10
> c (a,b)
 [1]  1  1  1  1  1  1  1  1  1  1  1  2  3  4  5  6
[17]  7  8  9 10
```

```
> a+b
 [1]   2   3   4   5   6   7   8   9  10  11
> a+2*b
 [1]   3   5   7   9  11  13  15  17  19  21
> a/b
 [1] 1.0000000 0.5000000 0.3333333 0.2500000 0.2000000
 [6] 0.1666667 0.1428571 0.1250000 0.1111111 0.1000000
> c(1,2,3)
[1] 1 2 3
>
```

Note that arithmetic operations act element-wise on vectors. To look at any object (function or data), just type its name.

```
> b
 [1]   1   2   3   4   5   6   7   8   9  10
```

To list all of your objects, use **ls** (). Note that because of the existence of a function called **c** (and another called t) it is best to avoid using these as variable names in user-defined functions (this is a common source of bugs).

Vectors can be 'sliced' very simply:

```
> d <- c(3,5,3,6,8,5,4,6,7)
> d
[1] 3 5 3 6 8 5 4 6 7
> d[4]
[1] 6
> d[2:4]
[1] 5 3 6
> d<7
[1]  TRUE   TRUE   TRUE   TRUE FALSE   TRUE   TRUE   TRUE
    FALSE
> d[d<7]
[1] 3 5 3 6 5 4 6
>
```

Vectors can be sorted and randomly sampled. The following command generates some lottery numbers.

```
> sort(sample(1:59,6))
[1] 12 15 17 37 39 48
```

Get help (using ?) on **sort** and **sample** to see how they work. R is also good at stochastic simulation of vectors of quantities from standard probability distributions, so

```
> rpois(20,2)
 [1] 1 4 0 2 2 3 3 3 4 2 2 4 2 1 3 4 1 2 1
```

generates 20 independent Poisson random variates with mean 2, and

```
> rnorm (5,1,2)
[1] -0.01251322 -0.03181018   0.30426031   3.24302197
    -2.04370284
```

generates 5 normal random variates with mean 1 and *standard deviation* (*not* variance) 2. There are many functions that act on vectors.

```
> x<-rnorm (50,2,3)
> x
 [1]    2.04360635    5.01113289  -1.52215979  -0.19789766   1.41945311  -0.08850784
 [7]   -0.91161025    3.47199019   6.13447194   4.62796165   0.07600234  -2.99687943
[13]    1.75153104    8.55000833   3.11921624   3.38411717   3.86860456   0.29103619
[19]    1.25823419    3.88427191  -0.77722215  -0.57774833   2.99937058   4.29042603
[25]    6.10597239    2.83832381   3.73618138   4.12999252   6.23009274   1.07251421
[31]   -0.19645150    1.77581296   2.08783542   1.62948606   2.74911850   0.44028844
[37]    1.80996899    1.86436309   0.29372974   2.37077354   1.54285955   4.40098545
[43]   -3.01913118   -0.23174209   3.58252631   5.18954147   3.61988373   4.08815220
[49]    6.30878505    4.56744882
> mean (x)
[1] 2.361934
> length (x)
[1] 50
> var (x)
[1] 6.142175
> summary (x)
    Min. 1st Qu.   Median    Mean 3rd Qu.    Max.
 -3.0190   0.3304   2.2290   2.3620   4.0370   8.5500
> cumsum (x)
 [1]     2.043606     7.054739     5.532579     5.334682     6.754135     6.665627
 [7]     5.754017     9.226007    15.360479    19.988441    20.064443    17.067563
[13]    18.819095    27.369103    30.488319    33.872436    37.741041    38.032077
[19]    39.290311    43.174583    42.397361    41.819613    44.818983    49.109409
[25]    55.215382    58.053705    61.789887    65.919879    72.149972    73.222486
[31]    73.026035    74.801848    76.889683    78.519169    81.268288    81.708576
[37]    83.518545    85.382908    85.676638    88.047412    89.590271    93.991257
[43]    90.972125    90.740383    94.322910    99.512451   103.132335   107.220487
[49]   113.529272   118.096721
> sum (x)
[1] 118.0967
>
```

Note that the function **var** computes and returns the *sample variance* of a data vector (and the function **sd** returns the square root of this). The sample variance, usually denoted by s^2, is an estimator of the population variance in the same sense that the sample mean, $\bar{X}$, is an estimator of the population mean, μ. It is defined by

$$s^2 = \frac{1}{n-1} \left(\left[\sum_{i=1}^{n} X_i^2 \right] - n\bar{X}^2 \right).$$

It has the property that $\mathrm{E}(s^2) = \mathrm{Var}(X)$, and is therefore referred to as the *unbiased* estimator of the population variance.

4.9.3 Plotting

R has many functions for data visualisation and for producing publication–quality plots — running **demo (graphics)** will give an idea of some of the possibilities. Some more basic commands are given below. Try them in turn and see what they do.

```
> plot (1:50, cumsum (x))
```

```
> plot (1:50, cumsum (x), type="l", col="red")
> plot (x, 0.6*x+rnorm (50, 0.3))
> curve (0.6*x, -5, 10, add=TRUE)
> hist (x)
> hist (x, 20)
> hist (x, freq=FALSE)
> curve (dnorm (x, 2, 3), -5, 10, add=TRUE)
> boxplot (x, 2*x)
> barplot (d)
>
```

Study the Help file for each of these commands to get a feel for the way each can be customised.

4.9.4 User-defined functions

R is a full programming language, and so before long you are likely to want to add your own functions. Consider the following declaration.

```
rchi<-function (n, p=2) {
  x<-matrix (rnorm (n*p), nrow=n, ncol=p)
  y<-x*x
  as.vector (y %*% rep (1, p))
  }
```

The first line declares the object rchi to be a function with two arguments, n and p, the second of which will default to a value of 2 if not specified. Then everything between { and } is the function body, which can use the variables n and p as well as any globally defined objects. The second line declares a local variable x to be a matrix with n rows and p columns, whose elements are standard normal random variables. The next line forms a new matrix y whose elements are the squares of the elements in x. The last line computes the matrix-vector product of y and a vector of p ones, then coerces the resulting n by 1 matrix into a vector. The result of the last line of the function body is the return result of the function. In fact, this function provides a fairly efficient way of simulating Chi-squared random quantities with p degrees of freedom (when p is small), but that is not particularly important. The function is just another R object, and hence can be viewed by entering rchi on a line by itself. It can be edited by doing **fix** (rchi). The function can be called just like any other, so

```
> rchi (10, 3)
 [1] 1.847349 5.590369 3.994036 4.243734 2.104224
 [6] 1.027634 1.119508 6.653095 5.660968 5.384954
> rchi (10)
 [1] 0.09356735 3.63633129 1.34073206 1.79412360
 [5] 1.46038656 2.67362870 0.50413958 6.04307710
 [9] 1.03116671 1.39662895
>
```

generates 10 chi-squared random variates with 3 and 2 degrees of freedom, respectively.

4.9.5 Using R packages

Much of the power of R comes from the huge range of optional packages that are available, most of which have been contributed by the wider R user community. Almost every aspect of statistical modelling and data analysis is covered by at least one R package. The packages are distributed via the Comprehensive R Archive Network (CRAN), and for a computer connected to the internet, the downloading and installation of most packages can be accomplished by executing a single command at the R command prompt.

The use of R packages will be illustrated using the package deSolve, which provides many useful functions related to the numerical solution of (deterministic) differential equation models. This package is pre-installed on many R installations. This can be checked by attempting to load the package with

```
> library(deSolve)
>
```

If the command returns silently, then the package is installed and has successfully loaded. An error probably means that the package is not installed. Installing the package should be as simple as

```
> install.packages("deSolve")
```

This should connect to the default CRAN mirror, download the package, and install it in the default location. Note that various options can be used to control which CRAN server to use, where to install the package, etc. — see ? **install . packages** for further details. All packages should have adequate documentation associated with them. Basic information about the package can be obtained with

```
> help(package="deSolve")
```

This should include an index of all functions associated with the package. For example, it shows that the package includes a function called ode, and further help on this function can be obtained with ?ode in the usual way. Some packages also have additional documentation in the form of 'vignettes', and where they exist they are extremely helpful, as they typically provide an overview of the package and a brief user guide to getting started. The vignettes associated with a package can be listed with

```
> vignette(package="deSolve")
```

which shows that this particular package has two vignettes associated with it, one of which is called deSolve, and this can be viewed with

```
> vignette("deSolve", package="deSolve")
```

This opens the vignette in the form of a PDF document, and provides a short user guide to getting started with the package. We will look again at this package in Chapter 6.

4.9.6 Reading and writing data

Of course, in order to use R for data analysis, it is necessary to be able to read data into R from other sources. It is often also desirable to be able to output data from R in a format that can be read by other applications. Unsurprisingly, R has a range of functions for accomplishing these tasks, but we shall just look here at the two simplest.

The simplest way to get data into R is to read a list of numbers from a text file using the **scan** command. This is most easily illustrated by first writing some data out to a text file and then reading it back into R. A vector of numbers can be output with a command like

```
> write(x,"scandata.txt")
>
```

Then, to load data from the file scandata.txt, use a command like

```
> x<-scan("scandata.txt")
Read 50 items
>
```

In general, you may need to use the **getwd** and **setwd** commands to inspect and change the working directory that R is using.

More often, we will be concerned with loading tabular data output from a spreadsheet or database or even another statistics package. R copes best with whitespace-separated data, but can be persuaded to read other formats with some effort. The key command here is **read.table** (and the corresponding output command **write. table**). So, suppose that mytable.txt is a plain text file containing the following lines:

```
"Name" "Shoe size" "Height"
"Fred" 9 170
"Jim" 10 180
"Bill" 9 185
"Jane" 7 175
"Jill" 6 170
"Janet" 8 180
```

To read this data into R, do

```
> mytab<-read.table("mytable.txt",header=TRUE)
> mytab
    Name Shoe.size Height
1   Fred         9    170
2    Jim        10    180
3   Bill         9    185
4   Jane         7    175
5   Jill         6    170
6  Janet         8    180
>
```

Note that in addition to **read.table**, there are related functions such as **read.csv** for reading comma separated values (CSV) files output from databases and spreadsheets, and also **read.**`delim`.

R also contains some (primitive) functions for editing data frames like this (and other objects), so

```
> mytabnew<-edit(mytab)
```

will pop up a simple editor for `mytab`, and on quitting, the edited version will be stored in `mytabnew`. Data frames like `mytab` are a key object type in R and tend to be used often. Here are some ways to interact with data frames.

```
> mytab$Height
[1] 170 180 185 175 170 180
> mytab[,2]
[1]  9 10  9  7  6  8
> plot(mytab[,2],mytab[,3])
> mytab[4,]
  Name Shoe.size Height
4 Jane         7    175
> mytab[5,3]
[1] 170
> mytab[mytab$Name=="Jane",]
  Name Shoe.size Height
4 Jane         7    175
> mytab$Height[mytab$Shoe.size > 8]
[1] 170 180 185
>
```

Also see the Help on **source** and **dump** for input and output of R objects of other types.

4.9.7 Further reading for R

A great deal of excellent R documentation can be obtained from the Comprehensive R Archive Network — CRAN. The next thing to work through is the official *Introduction to R*, which covers more material in more depth than this very quick introduction. Other useful documents are also available under the 'Manuals' section on CRAN.

4.10 Analysis of simulation output

4.10.1 Analysing Monte Carlo samples

We look now at using R for the analysis of a random quantity using stochastic simulation. Suppose interest lies in $Y = \exp(X)$, where $X \sim N(2, 1)$. In fact, Y is a standard distribution (it is log-normally distributed), and all of the interesting properties of Y can be derived directly, analytically. However, we will suppose that we are not able to do this, and instead study Y using stochastic simulation, using only the ability to simulate normal random quantities. Using R, samples from Y can be generated as follows:

```
> x<-rnorm(10000,2,1)
> y<-exp(x)
```

The variable Y has a skew distribution, which can be visualised with

```
> hist(y,breaks=50,freq=FALSE)
```

and a version of this plot is shown in Figure 4.1. The samples can also be used for computing summary statistics. Basic sample statistics can be obtained with

```
> summary(y)
   Min. 1st Qu.  Median    Mean 3rd Qu.     Max.
  0.134   3.733   7.327  12.220  14.490  728.700
> sd(y)
[1] 17.21968
>
```

and the sample mean, median, and quartiles provide estimates of the true population quantities. Focusing on the sample mean, $\bar{x}$, the value obtained here (12.220) is an estimate of the population mean. Of course, should the experiment be repeated, the estimate will be different. However, we can use the fact that $\bar{X} \sim N(\mu, \sigma^2/n)$ (by the CLT), to obtain $Z \sim N(0,1)$, where $Z \sim \sqrt{n}(\bar{X} - \mu)/\sigma$. Then since $P(|Z| < 2) \simeq 0.95$, we have

$$P\big(|\bar{X} - \mu| < 2\sigma/\sqrt{n}\big) \simeq 0.95,$$

and substituting in n and the estimated value of σ (17.21968), we get

$$P\big(|\bar{X} - \mu| < 0.344\big) \simeq 0.95.$$

We therefore expect that $\bar{X}$ is likely to be within 0.344 of μ (though careful readers will have noted that the conditioning here is really the wrong way around). In fact, in this particular case, the true mean of this distribution can be calculated analytically as $\exp(2.5) = 12.18249$, which is seen to be consistent with the simulation estimate. Thus, in more complex examples where the true population properties are not available, estimated sample quantities can be used as a substitute, provided that enough samples can be generated to keep the 'Monte Carlo error' to a minimum.

4.10.2 Testing an importance resampling algorithm

We finish this chapter with an example implementation of an importance resampling algorithm, and compare the samples generated against the true distribution. Here we will use samples from a standard Cauchy distribution (see the Chapter 3 exercises for details) as a proposal distribution for generating (approximate) standard normal random samples. It is straightforward to develop a transformation method for directly sampling Cauchy random variates, but we will use the built-in R function, **rcauchy**. We will also use the built-in density functions **dnorm** and **dcauchy** to calculate the importance weights.

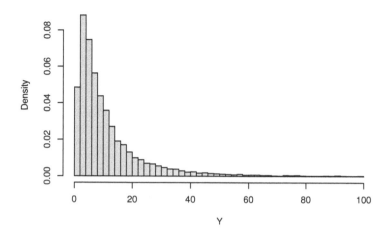

Figure 4.1 *Density of* $Y = \exp(X)$, *where* $X \sim N(2, 1)$.

```
> y=rcauchy(1000)
> w=dnorm(y)/dcauchy(y)
> mean(w)
[1] 1.003260
> x=sample(y, replace=TRUE, prob=w)
> summary(x)
    Min.   1st Qu.   Median     Mean   3rd Qu.     Max.
-3.29600 -0.73190 -0.16310 -0.06087  0.56060  3.18400
> sd(x)
[1] 0.9746765
> qqnorm(x)
> qqline(x)
>
```

Note the check to make sure that the average weight is close to one. Also note that there is no need to explicitly normalise the weights here, as the **sample** function does that by default. The statistics of the sample are close to their theoretical values and the normal Q–Q plot, shown in Figure 4.2, shows the quantiles of the sample matching up very well with the theoretical quantiles of the standard normal distribution.

4.11 Exercises

1. If you have not already done so, follow the links from this book's website and download and install R. Work through the mini-tutorial from this chapter. Also consider installing and using RStudio (linked from this book's website).

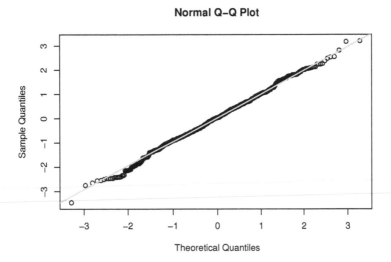

Figure 4.2 *Normal Q–Q plot for the samples resulting from the importance resampling procedure, showing good agreement with the theoretical distribution.*

2. Download the official *Introduction to R* from CRAN (linked from this book's website) and work through the first half at least.

3. The random variable X has PDF

$$f(x) = \begin{cases} \sin(x), & 0 \leq x \leq \pi/2, \\ 0, & \text{otherwise.} \end{cases}$$

(a) Derive a transformation method for simulating values of X based on $U(0,1)$ random variates.

(b) Derive a uniform rejection method for simulating values from X. What is the acceptance probability?

(c) Derive an envelope rejection method for simulating values of X based on a proposal with density

$$h(x) = \begin{cases} kx, & 0 \leq x \leq \pi/2, \\ 0, & \text{otherwise,} \end{cases}$$

for some fixed k. You should use the fact that $\sin(x) \leq x$, $\forall x \geq 0$. What is the acceptance probability?

(d) Implement the above three algorithms in R, and compare them to one another by computing summary statistics, and with the function **qqplot**. If you have a reasonable math background, compare the mean and variance of your samples with the corresponding theoretical values, which you should calculate analytically.

4. Write your own function `myrexp`, which does the same as **rexp**, but does not rely on the built-in version. Test that it really works as it should.

5. Write a function to simulate normal random quantities using the CLT method. Use plots and summary statistics to compare the distribution you obtain with those of the built-in **rnorm** function (which is exact).

6. Write your own function to simulate $Ga(3, 2)$ random quantities (and again, compare with the built-in version). See if you can also write your own function to simulate $Ga(3.5, 5)$ random quantities.

7. Obtain Monte Carlo solutions to the problems posed in Exercise 6. from Chapter 3.

4.12 Further reading

There are several good introductory texts on stochastic simulation, including Gentle (2003), Morgan (1984), and Ripley (1987). Devroye (1986) is the classic reference work on the subject. It is now out of print but is available freely on-line. Gentle et al. (2004) is another useful source available freely on-line. See the website for this book for up-to-date links. The standard reference for R is the R Development Core Team (2005), but other references such as Crawley (2007) are complementary.

CHAPTER 5

Markov processes

5.1 Introduction

We now have a grounding in elementary probability theory and an understanding of stochastic simulation. The only remaining theory required before studying the dynamics of genetic and biochemical networks (and chemical kinetics more generally) is an introduction to the theory of stochastic processes. A stochastic process is a random variable (say, the state of a biochemical network) which evolves through time. The state may be continuous or discrete, and it can evolve through time in a discrete or continuous way. A Markov process is a stochastic process which possesses the property that the future behaviour depends only on the current state of the system. Put another way, given information about the current state of the system, information about the past behaviour of the system is of no help in predicting the time-evolution of the process. It turns out that Markov processes are particularly amenable to both theoretical and computational analyses. Fortunately, the dynamic behaviour of biochemical networks can be effectively modelled by a Markov process, so familiarity with Markov processes is sufficient for studying many problems which arise naturally in systems biology.

5.2 Finite discrete time Markov chains

5.2.1 Introduction

The set $\{\theta^{(t)}|t = 0, 1, 2, \ldots\}$ is a *discrete time stochastic process*. The *state-space S* is such that $\theta^{(t)} \in S$, $\forall t$ and may be discrete or continuous.

A (first-order) *Markov chain* is a stochastic process with the property that the future states are independent of the past states given the present state. Formally, for $A \subseteq S, t = 0, 1, 2, \ldots$, we have

$$P\left(\theta^{(t+1)} \in A|\theta^{(t)} = x, \theta^{(t-1)} = x_{t-1}, \ldots, \theta^{(0)} = x_0\right)$$
$$= P\left(\theta^{(t+1)} \in A|\theta^{(t)} = x\right), \qquad \forall x, x_{t-1}, \ldots, x_0 \in S.$$

The past states provide no information about the future state if the present state is known. The behaviour of the chain is therefore determined by $P(\theta^{(t+1)} \in A|\theta^{(t)} = x)$. In general, this depends on t, A, and x. However, if there is no t dependence, so that

$$P\left(\theta^{(t+1)} \in A|\theta^{(t)} = x\right) = P(x, A), \quad \forall t,$$

then the Markov chain is said to be (time) *homogeneous*, and the *transition kernel*,

$P(x, A)$ determines the behaviour of the chain. Note that $\forall x \in S$, $P(x, \cdot)$ is a probability measure over S.

5.2.2 Notation

When dealing with discrete state-spaces, it is easier to write

$$P(x, \{y\}) = P(x, y) = P\left(\theta^{(t+1)} = y | \theta^{(t)} = x\right).$$

In the case of a finite discrete state-space, $S = \{x_1, \ldots, x_r\}$, we can write $P(\cdot, \cdot)$ as a matrix

$$P = \begin{pmatrix} P(x_1, x_1) & \cdots & P(x_1, x_r) \\ \vdots & \ddots & \vdots \\ P(x_r, x_1) & \cdots & P(x_r, x_r) \end{pmatrix}.$$

The matrix P is a *stochastic matrix*.

Definition 5.1 *A real $r \times r$ matrix P is said to be a* stochastic matrix *if its elements are all non-negative and its rows sum to 1.*

Note that if we define $\mathbb{1}$ to be an r-dimensional row vector of ones, we can write the row-sum convention as $P\mathbb{1}^\mathsf{T} = \mathbb{1}^\mathsf{T}$, or equivalently, as $\mathbb{1}P^\mathsf{T} = \mathbb{1}.^*$

Proposition 5.1 *The product of two stochastic matrices is another stochastic matrix. Every eigenvalue λ of a stochastic matrix satisfies $|\lambda| \le 1.^\dagger$ Also, every stochastic matrix has at least one eigenvalue equal to 1.*

The proof of this proposition is straightforward and left to the end-of-chapter exercises.

Suppose that at time t, we have

$$P\left(\theta^{(t)} = x_1\right) = \pi^{(t)}(x_1)$$

$$P\left(\theta^{(t)} = x_2\right) = \pi^{(t)}(x_2)$$

$$\vdots \qquad \vdots$$

$$P\left(\theta^{(t)} = x_r\right) = \pi^{(t)}(x_r).$$

We can write this as an r-dimensional row vector

$$\pi^{(t)} = (\pi^{(t)}(x_1), \pi^{(t)}(x_2), \ldots, \pi^{(t)}(x_r)).$$

---

* There is a widely adopted convention in the theory of Markov chains that the stochastic matrix is defined to have rows (not columns) summing to one, and that all vectors are row-vectors, operated on the right by (stochastic) matrices. This can be quite confusing at first, but the convention is so widespread that we will persist with it.

† When considering a matrix A, the vector v is called a (column) *eigenvector* of A if $Av = \lambda v$ for some scalar (real or complex number) λ, which is known as an *eigenvalue* of A, corresponding to the eigenvector v. The row eigenvectors of A are the column eigenvectors of A^T. Although row and column eigenvectors are different, the corresponding eigenvalues are the same. That is, A and A^T have the same (column) eigenvalues.

The probability distribution at time $t + 1$ can be computed using Theorem 3.1, as

$$P\left(\theta^{(t+1)} = x_1\right) = P(x_1, x_1)\,\pi^{(t)}(x_1) + P(x_2, x_1)\,\pi^{(t)}(x_2) +$$
$$\cdots + P(x_r, x_1)\,\pi^{(t)}(x_r),$$

and similarly for $P\left(\theta^{(t+1)} = x_2\right)$, $P\left(\theta^{(t+1)} = x_3\right)$, *etc.* We can write this in matrix form as

$$(\pi^{(t+1)}(x_1), \pi^{(t+1)}(x_2), \ldots, \pi^{(t+1)}(x_r)) = (\pi^{(t)}(x_1), \pi^{(t)}(x_2), \ldots, \pi^{(t)}(x_r))$$
$$\times \begin{pmatrix} P(x_1, x_1) & \cdots & P(x_1, x_r) \\ \vdots & \ddots & \vdots \\ P(x_r, x_1) & \cdots & P(x_r, x_r) \end{pmatrix}$$

or equivalently

$$\pi^{(t+1)} = \pi^{(t)} P.$$

So,

$$\pi^{(1)} = \pi^{(0)} P$$
$$\pi^{(2)} = \pi^{(1)} P = \pi^{(0)} P P = \pi^{(0)} P^2$$
$$\pi^{(3)} = \pi^{(2)} P = \pi^{(0)} P^2 P = \pi^{(0)} P^3$$
$$\vdots \quad = \quad \vdots$$
$$\pi^{(t)} = \pi^{(0)} P^t.$$

That is, the initial distribution $\pi^{(0)}$, together with the transition matrix P, determine the probability distribution for the state at all future times. Further, if the one-step transition matrix is P, then the n-step transition matrix is P^n. Similarly, if the m–step transition matrix is P^m and the n–step transition matrix is P^n, then the $(m+n)$–step transition matrix is $P^m P^n = P^{m+n}$. The set of linear equations corresponding to this last statement are known as the *Chapman–Kolmogorov equations*.

These equations are so fundamental to much of the theory of Markov chains that we now consider a direct derivation. If we define $P(n)$ to be the n–step transition matrix, with elements $p_{ij}(n)$, then by the theorem of total probability we have

$$p_{ij}(m + n) = P\left(\theta^{(m+n)} = j \middle| \theta^{(0)} = i\right)$$
$$= \sum_{k=1}^{r} P\left(\theta^{(m+n)} = j \middle| \theta^{(m)} = k\right) P\left(\theta^{(m)} = k \middle| \theta^{(0)} = i\right)$$
$$= \sum_{k=1}^{r} p_{kj}(n) p_{ik}(m).$$

That is,

$$p_{ij}(m+n) = \sum_{k=1}^{r} p_{ik}(m)p_{kj}(n), \tag{5.1}$$

and these are the Chapman–Kolmogorov equations. If we prefer, we can write this in matrix form as $P(m+n) = P(m)P(n)$. The advantage of deriving the Chapman–Kolmogorov equations directly in this way is that the construction is very general. In particular, it is clear that they remain valid in the case of a countably infinite state-space ($r = \infty$). This construction also generalises when we move from discrete to continuous time, as we will see later.

5.2.3 Stationary distributions

A distribution π is said to be a *stationary distribution* of the homogeneous Markov chain governed by the transition matrix P if

$$\pi = \pi P. \tag{5.2}$$

Note that π is a row eigenvector of the transition matrix, with corresponding eigenvalue equal to 1. It is also a fixed point of the linear map induced by P. The stationary distribution is so-called because if at some time n, we have $\pi^{(n)} = \pi$, then $\pi^{(n+1)} = \pi^{(n)}P = \pi P = \pi$, and similarly $\pi^{(n+k)} = \pi$, $\forall k \geq 0$. That is, if a chain has a stationary distribution, it retains that distribution for all future time. Note that

$$\pi = \pi P \iff \pi - \pi P = 0$$
$$\iff \pi(I - P) = 0,$$

where I is the $r \times r$ identity matrix. Hence the stationary distribution of the chain may be found by solving

$$\pi(I - P) = 0. \tag{5.3}$$

Note that the trivial solution $\pi = 0$ is not of interest here, as it does not correspond to a probability distribution (its elements do not sum to 1). However, there are always infinitely many solutions to (5.3), so proper solutions can be found by finding a positive solution and then imposing the unit-sum constraint. In the case of a unique stationary distribution (just one eigenvalue of P equal to 1), then there will be a one–dimensional set of solutions to (5.3), and the unique stationary distribution will be the single solution with positive elements summing to 1.

5.2.4 Convergence

Convergence of Markov chains is a rather technical topic, which we will not examine in detail here. This short section presents a very informal explanation of why Markov chains often do converge to their stationary distribution and how the rate of convergence can be understood.

By convergence to stationary distribution, we mean that irrespective of the starting distribution, $\pi^{(0)}$, the distribution at time t, $\pi^{(t)}$, will converge to the stationary distribution, π, as t tends to infinity. If the limit of $\pi^{(t)}$ exists, it is referred to as the *equilibrium* distribution of the chain (sometimes referred to as the *limiting* distribution). Clearly the equilibrium distribution will be a stationary distribution, but a stationary distribution is not guaranteed to be an equilibrium distribution.[‡] The relationship between stationary and equilibrium distributions is therefore rather subtle.

Let π be a (row) eigenvector of P with corresponding eigenvalue λ. Then

$$\pi P = \lambda \pi.$$

Also $\pi P^t = \lambda^t \pi$. It is easy to show that for stochastic P we must have $|\lambda| \leq 1$ (see Exercises). We also know that at least one eigenvector is equal to 1 (the corresponding eigenvector is a stationary distribution). Let

$$(\pi_1, \lambda_1), \ (\pi_2, \lambda_2), \ldots, (\pi_r, \lambda_r)$$

be the full eigen-decomposition of P, with $|\lambda_i|$ in decreasing order, so that $\lambda_1 = 1$, and π_1 is a (rescaled) stationary distribution. To keep things simple, let us now make the assumption that the initial distribution $\pi^{(0)}$ can be written in the form

$$\pi^{(0)} = a_1 \pi_1 + a_2 \pi_2 + \cdots + a_r \pi_r$$

for appropriate choice of a_i (this might not always be possible, but making the assumption keeps the mathematics simple). Then

$$\begin{aligned}
\pi^{(t)} &= \pi^{(0)} P^t \\
&= (a_1 \pi_1 + a_2 \pi_2 + \cdots + a_r \pi_r) P^t \\
&= a_1 \pi_1 P^t + a_2 \pi_2 P^t + \cdots + a_r \pi_r P^t \\
&= a_1 \lambda_1^t \pi_1 + a_2 \lambda_2^t \pi_2 + \cdots + a_r \lambda_r^t \pi_r \\
&\to a_1 \pi_1, \qquad \text{as } t \to \infty,
\end{aligned}$$

provided that $|\lambda_2| < 1$. The rate of convergence is governed by the second eigenvalue, λ_2. Therefore, provided that $|\lambda_2| < 1$, the chain eventually converges to an equilibrium distribution, which corresponds to the unique stationary distribution, irrespective of the initial distribution. If there is more than one unit eigenvalue, then there is an infinite family of stationary distributions, and convergence to any particular distribution is not guaranteed. For more details on the theory of Markov chains and their convergence, it is probably a good idea to start with texts such as Ross (1996) and Norris (1998), and then consult the references therein as required. For the rest of this section, we will assume that an equilibrium distribution exists and that it corresponds to a unique stationary distribution.

5.2.5 Reversible chains

If $\theta^{(0)}, \theta^{(1)}, \ldots, \theta^{(N)}$ is a Markov chain, then the reversed sequence of states, $\theta^{(N)}$, $\theta^{(N-1)}, \ldots, \theta^{(0)}$ is also a Markov chain. To see this, consider the conditional distribution of the current state given the future:

[‡] This is clear if there is more than one stationary distribution, but in fact, even in the case of a unique stationary distribution, there might not exist an equilibrium distribution at all.

$$P\left(\theta^{(t)} = y | \theta^{(t+1)} = x_{t+1}, \ldots, \theta^{(N)} = x_N\right)$$

$$= \frac{P\left(\theta^{(t+1)} = x_{t+1}, \ldots, \theta^{(N)} = x_N | \theta^{(t)} = y\right) P\left(\theta^{(t)} = y\right)}{P(\theta^{(t+1)} = x_{t+1}, \ldots, \theta^{(N)} = x_N)}$$

$$= \frac{P\left(\theta^{(t+1)} = x_{t+1} | \theta^{(t)} = y\right) \cdots \cdots P\left(\theta^{(N)} = x_N | \theta^{(N-1)} = x_{N-1}\right) P\left(\theta^{(t)} = y\right)}{P(\theta^{(t+1)} = x_{t+1}) P(\theta^{(t+2)} = x_{t+2} | \theta^{(t+1)} = x_{t+1}) \cdots P(\theta^{(N)} = x_N | \theta^{(N-1)} = x_{N-1})}$$

$$= \frac{P\left(\theta^{(t+1)} = x_{t+1} | \theta^{(t)} = y\right) P\left(\theta^{(t)} = y\right)}{P(\theta^{(t+1)} = x_{t+1})}$$

$$= P\left(\theta^{(t)} = y | \theta^{(t+1)} = x_{t+1}\right).$$

This is exactly the condition required for the reversed sequence of states to be Markovian.

Now let $P_t^*(x, y)$ be the transition kernel for the reversed chain. Then

$$P_t^*(x, y) = P\left(\theta^{(t)} = y | \theta^{(t+1)} = x\right)$$

$$= \frac{P\left(\theta^{(t+1)} = x | \theta^{(t)} = y\right) P\left(\theta^{(t)} = y\right)}{P\left(\theta^{(t+1)} = x\right)} \qquad \text{(by Theorem 3.2)}$$

$$= \frac{P(y, x) \, \pi^{(t)}(y)}{\pi^{(t+1)}(x)}.$$

Therefore, in general, the reversed chain is not homogeneous. However, if the chain has reached its equilibrium distribution, then

$$P^*(x, y) = \frac{P(y, x) \, \pi(y)}{\pi(x)},$$

and so the reversed chain is homogeneous, and has a transition matrix which may be determined from the transition matrix for the forward chain (and its stationary distribution). Note that we can write this expression for the reverse transition kernel in matrix form as

$$P^* = \text{diag}\{\pi\}^{-1} P^{\mathsf{T}} \text{diag}\{\pi\},$$

where for any r-dimensional vector v, $\text{diag}\{v\}$ is the $r \times r$ matrix with the elements of v along the diagonal and zeroes elsewhere,[§] and for any matrix A, the inverse of A, written A^{-1}, is the matrix such that $A^{-1}A = AA^{-1} = I$.[¶]
If

$$P^*(x, y) = P(x, y), \qquad \forall x, y$$

[§] The usual convention with the diag{} operator is that when it is applied to a vector it returns a square matrix with the elements of the vector along the diagonal, and when it is applied to a (square) matrix, it returns a vector consisting of the diagonal elements. So in particular, $\text{diag}\{\mathbb{1}\} = I$ and $\text{diag}\{I\} = \mathbb{1}$.

[¶] The inverse of a diagonal matrix is obtained by replacing each element along the diagonal with its reciprocal.

then the chain is said to be (time) *reversible* (as then the sequence of states appear the same if time is reversed), and we have the *detailed balance equations*:

$$\pi(x)\,\mathrm{P}(x,y) = \pi(y)\,\mathrm{P}(y,x), \quad \forall x, y. \tag{5.4}$$

The detailed balance equations capture symmetry in the flow of probability between pairs of states. The left-hand side is the probability that a particular transition will be a move from x to y, and the right-hand side is the probability that a particular transition will be a move from y to x. We can write the detailed balance equations in matrix form as

$$\mathrm{diag}\{\pi\}\,P = P^{\mathsf{T}}\,\mathrm{diag}\{\pi\}$$

or equivalently as

$$\mathrm{diag}\{\pi\}\,P = [\mathrm{diag}\{\pi\}\,P]^{\mathsf{T}}.$$

In other words, *the chain satisfies detailed balance if the matrix* $\mathrm{diag}\{\pi\}\,P$ *is symmetric.*

If we have a chain with transition kernel $\mathrm{P}(x,y)$ and a distribution $\pi(\cdot)$ satisfying (5.4), then it follows that the chain is reversible with stationary distribution $\pi(\cdot)$. The chain also has other nice properties (such as positive recurrence) which make it comparatively well behaved.

Proposition 5.2 *Consider a Markov chain with transition matrix P, satisfying (5.4) for some probability vector π. Then the chain has π as a stationary distribution and is time reversible.*

Proof. Summing both sides of (5.4) with respect to x gives

$$\sum_x \pi(x)\,\mathrm{P}(x,y) = \sum_x \pi(y)\,\mathrm{P}(y,x)$$
$$= \pi(y)\sum_x \mathrm{P}(y,x)$$
$$= \pi(y), \quad \forall y.$$

In other words, $\pi P = \pi$. Hence π is a stationary distribution. We then immediately conclude that the chain is reversible by dividing (5.4) through by $\pi(x)$.
□

For completeness, we can deduce the stationarity of a chain satisfying detailed balance directly from the matrix form as follows:

$$\pi P = \mathbb{1}\,\mathrm{diag}\{\pi\}\,P = \mathbb{1}\,[\mathrm{diag}\{\pi\}\,P]^{\mathsf{T}}$$
$$= \mathbb{1}\,P^{\mathsf{T}}\,\mathrm{diag}\{\pi\}$$
$$= \mathbb{1}\,\mathrm{diag}\{\pi\}$$
$$= \pi.$$

The first line is just the detailed balance equation pre-multiplied by $\mathbb{1}$ (to form the column sums), and the third line follows as P is a stochastic matrix.

```
rfmc <- function(n, P, pi0)
{
        v = vector("numeric",n)
        r = length(pi0)
        v[1] = sample(r,1,prob=pi0)
        for (i in 2:n) {
                v[i] = sample(r,1,prob=P[v[i-1],])
        }
        ts(v)
}
```

Figure 5.1 *An R function to simulate a sample path of length n from a Markov chain with transition matrix P and initial distribution pi0.*

5.2.6 Stochastic simulation and analysis

We next turn our attention to the problem of stochastic simulation of Markov chains on a computer and analysis of the simulation results. The key requirement is the ability to simulate a new state randomly, with probabilities given by an arbitrary probability vector p. The solution is given in Section 4.5. For simulation of a Markov chain to take place, a transition matrix P and an initial distribution $\pi^{(0)}$ are required. Simulation begins with sampling an initial state $\theta^{(0)}$ from $\pi^{(0)}$ using a lookup method. Once the initial state has been obtained, a value for $\theta^{(1)}$ can be sampled using the set of probabilities from the $\theta^{(0)}$th row of P. Indeed, at time t, the state $\theta^{(t)}$ can be sampled using the probabilities from the $\theta^{(t-1)}$th row of P. A simple R function for simulating the path of a Markov chain is given in Figure 5.1, and an example R session illustrating its use is given in Figure 5.2. The example session shows how to plot and summarise the simulated sample path and how to calculate the proportion of time spent in each state. These latter proportions approximate the equilibrium distribution of the chain.

5.3 Markov chains with continuous state-space

Mostly we will regard biochemical networks as having a discrete state, but sometimes it is helpful to regard the state of certain quantities as continuous. In this case, we have to understand how the concept of a discrete state Markov chain extends to the continuous state case. In fact, this extension is exactly analogous to the generalisation of discrete random quantities to that of continuous random quantities.

Here we are still working with discrete time, but we are allowing the state-space S of the Markov chain to be continuous (e.g., $S \subseteq \mathbb{R}$).

Example — First-order auto-regressive model

Consider the AR(1) model, which arises in elementary time-series analysis. Here, AR stands for auto-regressive, and the (1) means that the order of the auto-regression

```
> P=matrix(c(0.9,0.1,0.2,0.8),ncol=2,byrow=TRUE)
> P
      [,1] [,2]
[1,]  0.9  0.1
[2,]  0.2  0.8
> pi0=c(0.5,0.5)
> pi0
[1] 0.5 0.5
> samplepath=rfmc(200,P,pi0)
> samplepath
Time Series:
Start = 1
End = 200
Frequency = 1
  [1] 1 1 1 1 1 1 2 2 2 2 2 2 1 1 1 1 1 1 1 1 1 2 2 2 1 1 1 1 1 1 1 1 1 2 1 1 1
 [38] 1 1 2 2 2 2 2 2 2 2 2 2 2 2 2 2 2 2 2 1 2 1 1 1 1 1 1 1 1 1 1 1 1 1
 [75] 1 1 2 1 1 1 1 1 1 1 1 1 2 2 1 1 1 1 1 1 1 1 1 1 1 1 1 1 1 1 1 2 2 2 2 2
[112] 2 2 2 2 2 2 2 2 1 1 1 1 1 1 1 1 1 1 2 1 1 1 1 1 1 1 1 1 1 1 1 1 2 2 1
[149] 1 1 1 1 1 1 1 1 2 2 2 2 2 2 2 2 2 2 2 2 2 2 2 2 2 2 2 2 2 2 2 2 1 2 2 2
[186] 2 2 2 2 2 2 2 2 2 1 1 1 1 1
> plot(samplepath)
> hist(samplepath)
> summary(samplepath)
   Min. 1st Qu.  Median   Mean 3rd Qu.    Max.
   1.00    1.00    1.00   1.44    2.00    2.00
> table(samplepath)
samplepath
  1   2
112  88
> table(samplepath)/length(samplepath)
samplepath
   1    2
0.56 0.44
> # now compute the exact stationary distribution ...
> e=eigen(t(P))$vectors[,1]
> e/sum(e)
[1] 0.6666667 0.3333333
>
```

Figure 5.2 *A sample R session to simulate and analyse the sample path of a finite Markov chain. The last two commands show how to use R to directly compute the stationary distribution of a finite Markov chain.*

is 1. The essential structure of the model for an AR(1) process $\{Z_t | t = 1, 2, \ldots\}$ can be summarised as follows:

$$Z_t = \alpha Z_{t-1} + \varepsilon_t, \qquad \varepsilon_t \sim N(0, \sigma^2).$$

This model captures the idea that the value of the stochastic process at time t, Z_t, depends on the value of the stochastic process at time $t - 1$, and that the relationship is non-deterministic, with the non-determinism captured in the 'noise' term, ε_t. It is assumed that the noise process, $\{\varepsilon_t | t = 1, 2, \ldots\}$ is independent (that is, the pair of random quantities ε_i and ε_j will be independent of one another $\forall i \neq j$). However, it is clear that the stochastic process of interest $\{Z_t | t = 1, 2, \ldots\}$ is not independent due to the dependence introduced by auto-regressing each value on the value at the previous time.

It is clear that the conditional distribution of Z_t given $Z_{t-1} = z_{t-1}$ is just

$$Z_t|(Z_{t-1} = z_{t-1}) \sim N(\alpha z_{t-1}, \sigma^2),$$

and that it does not depend (directly) on any other previous time points. Thus, the AR(1) is a Markov chain and its state-space is the real numbers, so it is a continuous state-space Markov chain.$^{\|}$

5.3.1 Transition kernels

Again, for a homogeneous chain, we can define

$$P(x, A) = P\left(\theta^{(t+1)} \in A | \theta^{(t)} = x\right).$$

For continuous state-spaces we always have $P(x, \{y\}) = 0$, so in this case we define $P(x, y)$ by

$$P(x, y) = P\left(\theta^{(t+1)} \leq y | \theta^{(t)} = x\right)$$
$$= P\left(\theta^{(1)} \leq y | \theta^{(0)} = x\right), \ \forall x, y \in S,$$

the conditional cumulative distribution function (CDF). This is the distributional form of the transition kernel for continuous state space Markov chains, but we can also define the corresponding conditional density

$$p(x, y) = \frac{\partial}{\partial y} P(x, y), \quad x, y, \in S.$$

We can use this to define the density form of the *transition kernel* of the chain. Note that $p(x, y)$ is just the conditional density for the next state (with variable y) given that the current state is x, so it could also be written $p(y|x)$. The density form of the kernel can be used more conveniently than the CDF form for vector Markov chains, where the state-space is multidimensional (say $S \subseteq \mathbb{R}^n$).

Example

If we write the AR(1) process in the form

$$\theta^{(t+1)} = \alpha \theta^{(t)} + \epsilon_t, \qquad \epsilon_t \sim N(0, \sigma^2)$$

then

$$(\theta^{(t+1)}|\theta^{(t)} = x) \sim N(\alpha x, \sigma^2),$$

and so the density form of the transition kernel is just the normal density

$$p(x, y) = \frac{1}{\sigma\sqrt{2\pi}} \exp\left\{-\frac{1}{2}\left(\frac{y - \alpha x}{\sigma}\right)^2\right\},$$

and this is to be interpreted as a density for y for a given fixed value of x.

$^{\|}$ Note, however, that other classical time series models such as MA(1) and ARMA(1,1) are *not* Markov chains. The AR(2) is a *second order* Markov chain, but we will not be studying these.

5.3.2 Stationarity and reversibility

Let the state at time t, $\theta^{(t)}$ be represented by a probability density function, $\pi^{(t)}(x)$, $x \in S$. By the continuous version of the theorem of total probability (Proposition 3.11), we have

$$\pi^{(t+1)}(y) = \int_S p(x,y)\pi^{(t)}(x)\,dx. \tag{5.5}$$

We see from (5.5) that a stationary distribution must satisfy

$$\pi(y) = \int_S p(x,y)\pi(x)\,dx, \tag{5.6}$$

which is clearly just the continuous version of the discrete matrix equation $\pi = \pi P$.

Again, we can use Bayes' theorem to get the transition density for the reversed chain

$$p_t^*(x,y) = \frac{p(y,x)\pi^{(t)}(y)}{\pi^{(t+1)}(x)},$$

which homogenises in the stationary limit to give

$$p^*(x,y) = \frac{p(y,x)\pi(y)}{\pi(x)}.$$

So, if the chain is (time) reversible, we have the continuous form of the detailed balance equations

$$\pi(x)p(x,y) = \pi(y)p(y,x), \quad \forall x,y \in S. \tag{5.7}$$

Proposition 5.3 *Any homogeneous Markov chain satisfying (5.7) is reversible with stationary distribution $\pi(\cdot)$.*

Proof. We can see that detailed balance implies stationarity of $\pi(\cdot)$ by integrating both sides of (5.7) with respect to x and comparing with (5.6). Once we know that $\pi(\cdot)$ is the stationary distribution, reversibility follows immediately.
□

This result is of fundamental importance in the study of Markov chains with continuous state-space, as it allows us to verify the stationary distribution $\pi(\cdot)$ of a (reversible) Markov chain with transition kernel $p(\cdot,\cdot)$ without having to directly verify the integral equation (5.6). Note, however, that a given density $\pi(\cdot)$ will fail detailed balance (5.7) regardless of whether or not it is a stationary distribution of the chain if the chain itself is not time reversible.

Example

We know that linear combinations of normal random variables are normal (Section 3.10), so we expect the stationary distribution of our example AR(1) to be normal. The normal distribution is characterised by its mean and variance, so if we can deduce the mean and variance of the stationary distribution, we can check to see if it really is normal. At convergence, successive distributions are the same. In particular, the first and second moments at successive time points remain constant. First,

$E\big(\theta^{(t+1)}\big) = E\big(\theta^{(t)}\big)$, and so

$$E\left(\theta^{(t)}\right) = E\left(\theta^{(t+1)}\right)$$
$$= E\left(\alpha\theta^{(t)} + \epsilon_t\right)$$
$$= \alpha\, E\left(\theta^{(t)}\right)$$
$$\Rightarrow E\left(\theta^{(t)}\right) = 0.$$

Similarly,

$$\text{Var}\left(\theta^{(t)}\right) = \text{Var}\left(\theta^{(t+1)}\right)$$
$$= \text{Var}\left(\alpha\theta^{(t)} + \epsilon_t\right)$$
$$= \alpha^2\, \text{Var}\left(\theta^{(t)}\right) + \sigma^2$$
$$\Rightarrow \text{Var}\left(\theta^{(t)}\right) = \frac{\sigma^2}{1 - \alpha^2}.$$

So we think the stationary distribution is normal with mean zero and variance $\sigma^2/(1-\alpha^2)$. That is, we think the stationary density is

$$\pi(x) = \frac{1}{\sqrt{\frac{2\pi\sigma^2}{1-\alpha^2}}} \exp\left\{-\frac{1}{2}\frac{x^2}{\frac{\sigma^2}{1-\alpha^2}}\right\} = \sqrt{\frac{1-\alpha^2}{2\pi\sigma^2}} \exp\left\{-\frac{x^2(1-\alpha^2)}{2\sigma^2}\right\}.$$

Since we know the transition density for this chain, we can see if this density satisfies detailed balance:

$$\pi(x)p(x,y) = \sqrt{\frac{1-\alpha^2}{2\pi\sigma^2}} \exp\left\{-\frac{x^2(1-\alpha^2)}{2\sigma^2}\right\} \times \frac{1}{\sigma\sqrt{2\pi}} \exp\left\{-\frac{1}{2}\left(\frac{y-\alpha x}{\sigma}\right)^2\right\}$$
$$= \frac{\sqrt{1-\alpha^2}}{2\pi\sigma^2} \exp\left\{-\frac{1}{2\sigma^2}[x^2 - 2\alpha xy + y^2]\right\}$$

after a little algebra. But this expression is exactly symmetric in x and y, and so

$$\pi(x)p(x,y) = \pi(y)p(y,x).$$

So we see that $\pi(\cdot)$ does satisfy detailed balance, and so the AR(1) is a *reversible* Markov chain and *does* have stationary distribution $\pi(\cdot)$.

5.3.3 Stochastic simulation and analysis

Simulation of Markov chains with a continuous state in discrete time is easy provided that methods are available for simulating from the initial distribution, $\pi^{(0)}(x)$, and from the conditional distribution represented by the transition kernel, $p(x,y)$.

1. First $\theta^{(0)}$ is sampled from $\pi^{(0)}(\cdot)$, using one of the techniques discussed in Chapter 4.

2. We can then simulate $\theta^{(1)}$ from $p(\theta^{(0)}, \cdot)$, as this is just a density.

3. In general, once we have simulated a realisation of $\theta^{(t)}$, we can simulate $\theta^{(t+1)}$ from $p(\theta^{(t)}, \cdot)$, using one of the standard techniques from Chapter 4.

Example

Let us start our AR(1) off at $\theta^{(0)} = 0$, so we do not need to simulate anything for the initial value. Next we want to simulate $\theta^{(1)}$ from $p(\theta^{(0)}, \cdot) = p(0, \cdot)$, that is, we simulate $\theta^{(1)}$ from $N(0, \sigma^2)$. Next we simulate $\theta^{(2)}$ from $p(\theta^{(1)}, \cdot)$, that is, we simulate $\theta^{(2)}$ from $N(\alpha\theta^{(1)}, \sigma^2)$. In general, having simulated $\theta^{(t)}$, we simulate $\theta^{(t+1)}$ from $N(\alpha\theta^{(t)}, \sigma^2)$.

As n gets large, the distribution of $\theta^{(n)}$ tends to the distribution with density $\pi(\cdot)$, the equilibrium distribution of the chain. All values sampled after convergence has been reached are draws from $\pi(\cdot)$. There is a 'burn-in' period before convergence is reached, so if interest is in $\pi(\cdot)$, these values should be discarded before analysis takes place.

If we are interested in an integral

$$\int_S g(x)\pi(x)dx = \mathrm{E}_\pi\left(g(\Theta)\right),$$

then if $\theta^{(1)}, \theta^{(2)}, \ldots, \theta^{(n)}$ are draws from $\pi(\cdot)$, this integral may be approximated by

$$\mathrm{E}_\pi\left(g(\Theta)\right) \simeq \frac{1}{n}\sum_{i=1}^{n} g(\theta^{(i)}).$$

However, *draws from a Markov chain are not independent*, so the variance of the sample mean cannot be computed in the usual way. In fact, when we introduced Monte Carlo simulation ideas in Chapter 4, we relied quite heavily on the independence of draws to establish convergence properties of Monte Carlo estimates, using the law of large numbers and central limit theorem from Chapter 3, both of which required independence of samples in the form in which they were presented. Fortunately, the conditions required for a LLN or CLT are much weaker than those assumed in Chapter 3. A thorough exploration of this topic is beyond the scope of this text, but we will look briefly at how we can establish the convergence properties of Monte Carlo estimates for dependent samples in the context of this AR(1) example.

Suppose $\theta^{(i)} \sim \pi(\cdot)$, $i = 1, 2, \ldots$. Then

$$\mathrm{E}_\pi\left(\Theta\right) \simeq \frac{1}{n}\sum_{i=1}^{n} \theta^{(i)} = \bar{\theta}_n.$$

Let $\mathrm{Var}(\Theta) = \mathrm{Var}\left(\theta^{(i)}\right) = \nu^2$. Then if the θ_i were independent, we would have

$$\mathrm{Var}\left(\bar{\theta}_n\right) = \frac{\nu^2}{n}.$$

However, if the $\theta^{(i)}$ are *not* independent (e.g., sampled from a non-trivial Markov chain), then

$$\text{Var}(\bar{\theta}_n) \neq \frac{\nu^2}{n}.$$

Example

This example relies on some results from multivariate probability theory not covered in the text. It can be skipped without significant loss.

$$\text{Var}\left(\theta^{(i)}\right) = \frac{\sigma^2}{1 - \alpha^2} = \nu^2$$

and

$$\gamma(k) = \text{Cov}\left(\theta^{(i)}, \theta^{(i+k)}\right) = \nu^2 \alpha^k,$$

so

$$\text{Var}(\bar{\theta}_n) = \frac{1}{n^2} \text{Var}\left(\sum_{i=1}^{n} \theta^{(i)}\right)$$

$$= \frac{\nu^2}{n^2}\left(n + \sum_{i=1}^{n-1} 2(n-i)\alpha^i\right),$$

which sums to give

$$\text{Var}(\bar{\theta}_n) = \frac{1}{n^2}\frac{\sigma^2}{1-\alpha^2}\frac{n + 2\alpha^{n+1} - 2\alpha - n\alpha^2}{(1-\alpha)^2}$$

$$= \frac{1}{n}\frac{\sigma^2}{1-\alpha^2}\left[\frac{1+\alpha}{1-\alpha} - \frac{2\alpha(1-\alpha^n)}{n(1-\alpha)^2}\right].$$

To a first approximation, we have

$$\text{Var}(\bar{\theta}_n) \simeq \frac{1}{n}\frac{\sigma^2}{1-\alpha^2}\frac{1+\alpha}{1-\alpha},$$

and so the 'correction factor' for the naive calculation based on an assumption of independence is $(1+\alpha)/(1-\alpha)$. For α close to 1, this can be very large; for example, for $\alpha = 0.95$, $(1+\alpha)/(1-\alpha) = 39$, and so the variance of the sample mean is actually around 40 times bigger than calculations based on assumptions of independence would suggest. Similarly, confidence intervals should be around six times wider than calculations based on independence would suggest. However, the crucial property we need is that the variance of the mean tends to zero as the number of samples increases, and in fact here, we are fortunate that the variance decreases as order $1/n$, just as in the case of independent samples. In this case, it is fairly straightforward to directly establish a LLN for Monte Carlo estimates based on dependent samples from the equilibrium distribution of this process.

We can actually use this analysis in order to analyse other Markov chains. If the Markov chain is reasonably well approximated by an AR(1) (and many are), then we can estimate the variance of our sample estimates by the AR(1) variance estimate. For an AR(1), α is just the lag 1 auto-correlation of the chain $(\mathrm{Corr}(\theta^{(i)}, \theta^{(i+1)}) = \alpha)$, and so we can estimate the α of any simulated chain by the sample auto-correlation at lag 1. We can then use this to compute or correct sample variance estimates based on sample means of chain values.

5.4 Markov chains in continuous time

We have now looked at Markov chains in discrete time with both discrete and continuous state-spaces. However, biochemical processes evolve continuously in time, and so we now turn our attention to the continuous time case. We begin by studying chains with a finite number of states, but relax this assumption in due course.

Before we begin we should try to be explicit about what exactly a Markov process is in the continuous time case. Intuitively it is a straightforward extension of the discrete time definition. In continuous time, we can write this as

$$
\begin{aligned}
\mathrm{P}(X(t+dt) = x | \{X(t) = x(t) | t \in [0, t]\}) \\
= \mathrm{P}(X(t+dt) = x | X(t) = x(t)), \quad \forall t \in [0, \infty), \; x \in S.
\end{aligned}
$$

Again, this expresses the idea that the future behaviour of the process depends on the past behaviour of the process only via the current state.

5.4.1 Finite state-space

Consider first a process which can take on one of r states, which we label $S = \{1, 2, \ldots, r\}$. If at time t the process is in state $x \in S$, its future behaviour can be characterised by the transition kernel

$$
p(x, t, x', t') \equiv \mathrm{P}(X(t + t') = x' | X(t) = x).
$$

If this function does not depend explicitly on t, the process is said to be *homogeneous*, and the kernel can be written $p(x, x', t')$. For each value of t', this kernel can be expressed as an $r \times r$ transition matrix, $P(t')$. It is clear that $P(0) = I$, the $r \times r$ identity matrix, as no transitions will take place in a time interval of length zero. Also note that since $P(\cdot)$ is a transition matrix for each value of t, we can multiply these matrices together to give combined transition matrices in the usual way. In particular, we have $P(t + t') = P(t)P(t') = P(t')P(t)$, just as in the discrete time case. Note in particular that our derivation of the Chapman–Kolmogorov equations (5.1) did not rely in any way on the discreteness of time, and hence are equally valid in the continuous time case, where we can write them out in full component form as

$$
p(i, j, t + t') = \sum_{k=1}^{r} p(i, k, t) p(k, j, t'), \quad i, j = 1, \ldots, r.
$$

Now define the *transition rate matrix*, Q, to be the derivative of $P(t')$ at $t' = 0$. Then

$$Q = \frac{d}{dt'}P(t')\bigg|_{t'=0}$$
$$= \lim_{\delta t \to 0} \frac{P(\delta t) - P(0)}{\delta t}$$
$$= \lim_{\delta t \to 0} \frac{P(\delta t) - I}{\delta t}.$$

The elements of the Q matrix give the 'hazards' of moving to different states. Rearranging gives the infinitesimal transition matrix

$$P(dt) = I + Q\,dt.$$

Note that for $P(dt)$ to be a stochastic matrix (with non-negative elements and rows summing to 1), the above implies several constraints which must be satisfied by the rate matrix Q. Since the off-diagonal elements of I are zero, the off-diagonal elements of $P(dt)$ and $Q\,dt$ must be the same, and so the off-diagonal elements of Q must be non-negative. Also, since the diagonal elements of $P(dt)$ are bounded above by one, the diagonal elements of Q must be non-positive. Finally, since the rows of $P(dt)$ and I both sum to 1, the rows of Q must each sum to zero. These properties must be satisfied by *any* rate matrix Q.

The above rearrangement gives us a way of computing the stationary distribution of the Markov chain, as a probability row vector π will be stationary only if

$$\pi P(dt) = \pi$$
$$\Rightarrow \pi(I + Qdt) = \pi$$
$$\Rightarrow \pi Q = 0.$$

Solving this last equation (subject to the constraint that the elements of π sum to 1) will give a stationary distribution for the system.

If $P(t)$ is required for finite t, it may be computed by solving a matrix differential equation. This can be derived by considering the derivative of $P(t)$ for arbitrary times t.

$$\frac{d}{dt}P(t) = \frac{P(t + dt) - P(t)}{dt}$$
$$= \frac{P(dt)P(t) - P(t)}{dt}$$
$$= \frac{P(dt) - I}{dt}P(t)$$
$$= Q\,P(t).$$

Therefore, $P(t)$ is a solution to the matrix differential equation

$$\frac{d}{dt}P(t) = QP(t),$$

subject to the initial condition $P(0) = I$. This differential equation has solution

$$P(t) = \exp\{Q\,t\},$$

where $\exp\{\cdot\}$ denotes the matrix exponential function (Golub and Van Loan, 1996). Working with the matrix exponential function is straightforward, but beyond the scope of this book.* Note that if we prefer, we do not have to work with the differential equation in matrix form. If we write it out in component form, we have

$$\frac{d}{dt}p(i,j,t) = \sum_{k=1}^{r} q_{ik}p(k,j,t), \quad i,j = 1, 2, \ldots, r. \tag{5.8}$$

These are known as *Kolmogorov's backward equations* (Allen, 2011). It is also worth noting that had we expanded $P(t + dt)$ as $P(t)P(dt)$ rather than $P(dt)P(t)$, we would have ended up with the matrix differential equation

$$\frac{d}{dt}P(t) = P(t)Q,$$

which we can write out in component form as

$$\frac{d}{dt}p(i,j,t) = \sum_{k=1}^{r} q_{kj}p(i,k,t), \quad i,j = 1, 2, \ldots, r. \tag{5.9}$$

These are known as *Kolmogorov's forward equations*.

Example

Consider a very simple model for the activation of a single prokaryotic gene. In this model, the gene will be activated unless a repressor protein is bound to its regulatory region. We will consider just two states in our system: state 0 (inactive), and state 1 (active). In the inactive state (0), we will assume a constant hazard of $\alpha > 0$ for activation. In the active state, we will assume a constant hazard of $\beta > 0$ for inactivation. Given that the rows of Q must sum to zero, it is now completely specified as

$$Q = \begin{pmatrix} -\alpha & \alpha \\ \beta & -\beta \end{pmatrix}.$$

Solving $\pi Q = 0$ gives the stationary distribution

$$\pi = \left(\frac{\beta}{\alpha + \beta} \,, \quad \frac{\alpha}{\alpha + \beta} \right).$$

We can also compute the infinitesimal transition matrix

$$P(dt) = I + Q\,dt = \begin{pmatrix} 1 - \alpha\,dt & \alpha\,dt \\ \beta\,dt & 1 - \beta\,dt \end{pmatrix}.$$

It is straightforward to encode this model in SBML. The SBML-shorthand for it is given in Figure 5.3 (and the full SBML can be downloaded from this book's website). A simulated realisation of this process is shown in Figure 5.4. Note that this process

---

* Note that the exponential of a matrix is not the matrix obtained by applying the exponential function to each element of the matrix. In fact, it is usually defined by its series expansion, $\exp\{A\} = \sum_{k=0}^{\infty} A^k/k! = I + A + A^2/2 + \cdots$, but this expansion does not lead to an efficient way of computing the function. Note also that many scientific libraries for numerical linear algebra provide a function for computing the matrix exponential.

```
@model:3.1.1=Activation
 s=item, t=second, v=litre, e=item
@compartments
 Cell
@species
 Cell:A=0 s
 Cell:I=1 s
@reactions
@r=Activation
 I -> A
 alpha : alpha=0.5
@r=Inactivation
 A -> I
 beta : beta=1
```

Figure 5.3 *SBML-shorthand for the simple gene activation process with $\alpha = 0.5$ and $\beta = 1$.*

is sometimes known as the *telegraph process* and can also form the basis for a very simple model of ion channels.

5.4.2 Stochastic simulation

There are three straightforward approaches one can take to simulating this process on a computer. The first is based on a fine time-discretisation of the process, similar in spirit to the first-order Euler method for integrating ordinary differential equations. Given the definition of the infinitesimal transition matrix

$$P(dt) = I + Q\,dt,$$

for small time steps Δt we will have

$$P(\Delta t) \simeq I + Q\,\Delta t.$$

$P(\Delta t)$ can then be regarded as the transition matrix of a discrete time Markov chain, and a simulated sequence of states at times $0,\ \Delta t,\ 2\,\Delta t,\ 3\,\Delta t, \ldots$ may be generated in the usual way.

The above method can be easily improved by replacing the above approximation for $P(\Delta t)$ by its *exact* value

$$P(\Delta t) = \exp\{Q\,\Delta t\},$$

provided that a method for computing the matrix exponential is available. Then it does not matter how small Δt is chosen to be, provided it is small enough to clearly show the behaviour of the process and not so large that interesting transitions are 'missed'.

A third approach to simulation may be taken by simulating each transition event and its corresponding time sequentially, rather than simply looking at the processes only at times on a given lattice. Like the previous method, this gives an exact realisation of the process and offers the additional advantage that recording every event ensures none will be 'missed'. Such an approach is known as *discrete event simulation*, and in the statistics literature the technique dates back to at least Kendall (1950). If the process is currently in state x, then the xth row of Q gives the hazards for the transition to other states. As the row sums to zero, $-q_{xx}$ gives the combined hazard for moving away from the current state — a discrete transition event (note that q_{xx} is non-positive). So, the time to a transition event is exponential with rate $-q_{xx}$. When that transition event occurs, the new state will be random with probabilities proportional to the xth row of Q (with q_{xx} omitted). The above intuitive explanation can be formalised as follows.

To understand how to simulate the process we must consider being in state i at time t, and think about the probability that the next event will be in the time interval $(t + t', t + t' + dt]$, *and* will consist of a move to state j. Let this probability divided by dt be denoted by $f(t', j|t, i)$, so that the probability is $f(t', j|t, i)dt$. It is clear that as the Markov process is homogeneous, there will be no explicit dependence on t in this probability, but we will include it to be clear about exactly what we mean. Then

$$f(t', j|t, i)dt = \text{P(Next event in } (t + t', t + t' + dt]|t, i)$$
$$\times \text{P}(j|\text{Next event in } (t + t', t + t' + dt], t, i).$$

Thinking about the first term, we know that the hazards for the individual transitions are given by the off-diagonal elements of the ith row of Q. The combined hazard is the sum of these off-diagonal elements, which is $-q_{ii}$ (as the row sums to zero). Combined hazards can always be computed as the sum of hazards in this way because the probability that two events occur in the interval $(t, t + dt]$ is of order dt^2 and can therefore be neglected. Now we know from our consideration of the exponential distribution as the time to an event of constant hazard that the time to the next event is $Exp(-q_{ii})$, and so the first term must be $-q_{ii}e^{q_{ii}t'} dt$. The second term is

$$\text{P}(X(t + t' + dt) = j|[X(t + t') = i] \cap [X(t + t' + dt) \neq i])$$
$$= \frac{\text{P}(X(t + t' + dt) = j|X(t + t') = i)}{\text{P}(X(t + t' + dt) \neq i|X(t + t') = i)} = \frac{q_{ij}dt}{\sum_{k \neq i} q_{ik}dt} = \frac{q_{ij}}{-q_{ii}}.$$

Taking the two terms together we have

$$f(t', j|t, i) = -q_{ii}e^{q_{ii}t'} \times \frac{q_{ij}}{-q_{ii}}.$$

The fact that this function factorises into the form of a probability density for the time to the next event and a probability mass function for the type of that event means that we can simulate the next event with the generation of two random variables. Note also that there is no j dependence in the PDF for t' and no t' dependence in the PMF for j, so the two random variables are independent of one another and hence can be simulated independently.

It is the consideration of $f(t', j|t, i)$ that leads to the standard discrete event simulation algorithm, which could be stated as follows:

1. Initialise the process at $t = 0$ with initial state i;

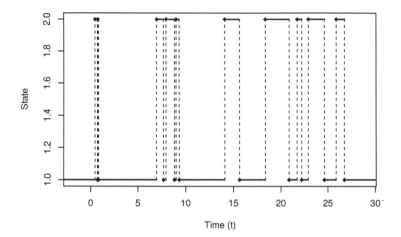

Figure 5.4 *A simulated realisation of the simple gene activation process with* $\alpha = 0.5$ *and* $\beta = 1$.

2. Call the current state i. Simulate the time to the next event, t', as an $Exp(-q_{ii})$ random quantity;

3. Put $t := t + t'$;

4. Simulate new state j as a discrete random quantity with PMF $-q_{ik}/q_{ii}$, $k \neq i$;

5. Output the time t and state j;

6. If $t < T_{max}$, return to step 2.

This particular discrete event simulation technique is known as the *direct method*. A simple R function to implement this algorithm is given in Figure 5.5. The function returns a step-function object, which is easy to plot. Using this function, a plot similar to Figure 5.4 can be obtained with the following command:

```
plot(rcfmc(20,matrix(c(-0.5,0.5,1,-1),ncol=2,byrow=TRUE)
    ,c(1,0)))
```

All of these simulation methods give a single realisation of the Markov process. Now obviously, just as one would not study a normal distribution by looking at a single simulated value, the same is true with Markov processes. Many realisations must be simulated in order to get a feel for the *distribution* of values at different times. In the case of a finite number of states, this distribution is relatively straightforward to compute directly without any simulation at all, but for the stochastic kinetic models we will consider later, simulation is likely to be the only tool we have available to us for gaining insight into the behaviour of the process.

```
rcfmc <- function (n,Q,pi0)
{
        xvec = vector ("numeric",n+1)
        tvec = vector ("numeric",n)
        r = length (pi0)
        x = sample (r,1,prob=pi0)
        t = 0
        xvec[1] = x
        for (i in 1:n) {
                t = t+rexp (1,-Q[x,x])
                weights = Q[x,]
                weights [x] = 0
                x = sample (r,1,prob=weights)
                xvec[i+1] = x
                tvec[i] = t
        }
        stepfun (tvec,xvec)
}
```

Figure 5.5 *An* R *function to simulate a sample path with* n *events from a continuous time Markov chain with transition rate matrix* Q *and initial distribution* pi0.

5.4.3 Countable state-space

Before moving on to thinking about continuous state-spaces, it is worth spending a little time looking at the case of a countably infinite state-space. Rather than attempting to present the theory in generality, we will concentrate on a simple example, which illustrates many of the interesting features. The model is known as the *immigration–death process*. In this model, individuals arrive into the population with constant hazard λ, and each individual dies independently with constant hazard μ. Consequently, the population of individuals increases by one when an immigration event occurs and decreases by one when a death event occurs. There is no reproduction in this model. Figure 5.6 gives the SBML-shorthand corresponding to this model. The key transition equations are:

$$P(X(t + dt) = x + 1 | X(t) = x) = \lambda \, dt$$
$$P(X(t + dt) = x - 1 | X(t) = x) = x\mu \, dt$$
$$P(X(t + dt) = x | X(t) = x) = 1 - (\lambda + x\mu)dt$$
$$P(X(t + dt) = y | X(t) = x) = 0, \, \forall y \notin \{x - 1, x, x + 1\}.$$

These equations clearly define a homogeneous Markov process, but with infinite state-space $S = 0, 1, 2, \ldots$. We therefore cannot easily write down a set of matrix equations for the process, as the matrices are infinite dimensional, but this does not prevent us from working with the process or from simulating it on a computer.

First let's think about understanding this process theoretically. Although the Q

```
@model:3.1.1=ImmigrationDeath
  s=item, t=second, v=litre, e=item
@compartments
  Cell
@species
  Cell:X=0 s
@reactions
@r=Immigration
  -> X
  lambda : lambda=1
@r=Death
  X ->
  mu*X : mu=0.1
```

Figure 5.6 *SBML-shorthand for the immigration–death process with $\lambda = 1$ and $\mu = 0.1$.*

matrix is infinite in extent, we can write its general form as follows:

$$
Q = \begin{pmatrix}
-\lambda & \lambda & 0 & 0 & 0 & \cdots \\
\mu & -\lambda - \mu & \lambda & 0 & 0 & \cdots \\
0 & 2\mu & -\lambda - 2\mu & \lambda & 0 & \\
0 & 0 & 3\mu & -\lambda - 3\mu & \lambda & \ddots \\
\vdots & & \ddots & & \ddots & \ddots
\end{pmatrix}.
$$

Then for an infinite dimensional $\pi = (\pi_0, \pi_1, \pi_2, \ldots)$ we can solve $\pi Q = 0$ to get the stationary distribution one equation at a time, expressing each π_k in terms of π_0 to find the general form

$$
\pi_k = \frac{\lambda^k}{k! \mu^k} \pi_0, \quad k = 1, 2, \ldots.
$$

But these are terms in the expansion of $\pi_0 e^{\lambda/\mu}$, and so imposing the unit-sum constraint we get $\pi_0 = e^{-\lambda/\mu}$, giving the general solution

$$
\pi_k = \frac{(\lambda/\mu)^k e^{-\lambda/\mu}}{k!}, \quad k = 0, 1, 2, \ldots.
$$

This is easily recognised as the PMF of a Poisson random quantity with mean λ/μ (Section 3.6). Hence, the stationary distribution of this process is Poisson with mean λ/μ.

We can also simulate realisations of this process on a computer. Here it is easiest to use the technique of discrete event simulation. If the current state of the process is x, the combined hazard for moving away from the current state is $\lambda + x\mu$, and so the time to the next event is an exponentially distributed random quantity with rate $\lambda + x\mu$. When that event occurs, the process will move up or down with probabilities proportional to their respective hazards, λ and $x\mu$. That is, the state will increase by 1 with probability $\lambda/(\lambda + x\mu)$ and decrease by 1 with probability $x\mu/(\lambda + x\mu)$. This

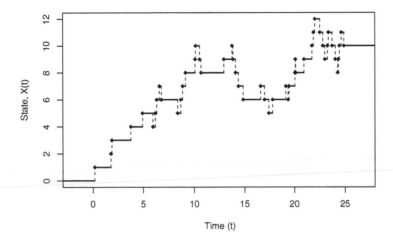

Figure 5.7 *A single realisation of the immigration–death process with parameters $\lambda = 1$ and $\mu = 0.1$, initialised at $X(0) = 0$. Note that the stationary distribution of this process is Poisson with mean 10.*

sequence can be easily simulated on a computer to give a set of states and event times which can be plotted, summarised, etc. A simulated realisation of this immigration–death process is shown in Figure 5.7. An R function to simulate the process is given in Figure 5.8.

5.4.4 Inhomogeneous Poisson process

Our treatment of the Poisson process has so far been fairly low level and intuitive. The relationship between the (homogeneous) Poisson process, the Poisson distribution, and the exponential distribution was made explicit in Proposition 3.17. The Poisson process is described as *homogeneous* because the event hazard λ is constant throughout time. In this section we will see that the Poisson process can be regarded as a Markov process and understand how it may be generalised to the *inhomogeneous* case. Understanding the inhomogeneous Poisson process will be necessary for some of the fast, accurate hybrid stochastic simulation algorithms considered in Chapter 8.

Recall that for a (homogeneous) Poisson process with rate λ, we previously defined N_t to be the number of events of the Poisson process in the interval $(0, t]$, noting that we therefore have $N_t \sim Po(\lambda t)$. The process N_t is the *counting process* associated with the *point process* that is the Poisson process itself (represented by a collection of random event times). It turns out that the counting process N_t is a

```
imdeath <- function (n=20, x0=0, lambda=1, mu=0.1)
{
        xvec = vector ("numeric",n+1)
        tvec = vector ("numeric",n)
        t = 0
        x = x0
        xvec[1] <- x
        for (i in 1:n) {
                t = t+rexp (1,lambda+x*mu)
                if ( runif (1,0,1) < lambda/ (lambda+x*mu)
                     )
                        x <- x+1
                else
                        x <- x-1
                xvec[i+1] <- x
                tvec[i] <- t
        }
        stepfun (tvec, xvec)
}
```

Figure 5.8 *R function for discrete-event simulation of the immigration–death process.*

Markov process, governed by the homogeneous Markovian transition equations

$$P(N_{t+dt} = x + 1 | N_t = x) = \lambda \, dt$$
$$P(N_{t+dt} = x | N_t = x) = 1 - \lambda \, dt$$
$$P(N_{t+dt} = y | N_t = x) = 0, \quad \forall y \notin \{x, x + 1\}.$$

The Poisson process is special because its transition rates do not depend on the current state. It is this lack of dependence on the current state that leads to the analytical tractability of the Poisson process. Associating the Poisson (point) process with the Markov (counting) process N_t gives a new and powerful way of thinking about the relation between the two. Indeed, many standard texts choose to *define* the Poisson process in terms of the associated counting process, as the latter is much more convenient mathematically (it is a state-independent time-homogeneous Markov process). However, in my opinion, the distinction between a point process and its associated counting process is conceptually quite important.

All of the Markov processes that we have considered so far have been homogeneous in the sense that the transition equations do not depend explicitly on the current time, t. Consider now a generalisation of the Poisson process where the event hazard is not a constant λ, but a function, $\lambda(t)$. That is, the probability of an event in the interval $(t, t + dt]$ is $\lambda(t)dt$, so the inhomogeneous Markovian transition equations for the associated counting process N_t are

$$P(N_{t+dt} = x + 1 | N_t = x) = \lambda(t)dt$$
$$P(N_{t+dt} = x | N_t = x) = 1 - \lambda(t)dt$$
$$P(N_{t+dt} = y | N_t = x) = 0, \quad \forall y \notin \{x, x + 1\}.$$

A formal analysis of this process is fairly straightforward, and the reader is referred to a standard text such as Ross (1996) for the technical details. Intuitively, as the hazard is approximately constant in a sufficiently small interval, the number of events in that interval will be approximately Poisson. In the limit, the number of events in the interval $(t, t + dt]$ will be $Po(\lambda(t)dt)$, independent of all other intervals. The number of events in the interval $(0, t]$, N_t will then be the sum (integral) over all such intervals. As the sum of independent Poissons is Poisson, we get

$$N_t \sim Po\left(\int_0^t \lambda(s)ds\right).$$

It is helpful to define the *cumulative hazard*

$$\Lambda(t) = \int_0^t \lambda(s)ds,$$

which then gives $N_t \sim Po(\Lambda(t))$. Similarly, the number of events in the interval $(s, t]$, $0 < s < t$ is $Po(\Lambda(t) - \Lambda(s))$.

Time-change of a Poisson process

An important concept in the study of the Poisson process (and in the study of stochastic processes in continuous time, more generally) is the relationship between the rate of the process and the rate at which time changes. For example, suppose we start with a unit Poisson process (rate 1) having counting process $M(t)$, and define a new process $N(t)$ which speeds up time by a factor of λ. What we mean here is that $N(t) = M(\lambda t)$. For example, if the time unit for the process M is minutes and the time unit for process N is hours, then $\lambda = 60$. It is immediately apparent that $N(t)$ is the counting process of a homogeneous Poisson process of rate λ. So the time-change $t' = \lambda t$ gives a Poisson process with rate λ. This generalises to non-linear time-changes in the obvious way. Clearly time-changes need to be strictly monotonic, and we can ensure this by representing our time-change as the integral of a positive function. That is, we will assume a time-change

$$t' = \int_0^t \lambda(s)\, ds,$$

where $\lambda(t) > 0$, $\forall t > 0$. If we again start with a unit process $M(t)$ and define a new process $N(t) = M(t')$, then we clearly have

$$N(t) = M(t') = M\left(\int_0^t \lambda(s)\, ds\right) \sim Po\left(\int_0^t \lambda(s)\, ds\right),$$

and hence $N(t)$ is an inhomogeneous Poisson process with rate $\lambda(t)$. So a unit Poisson process can be transformed to an arbitrary Poisson process by distorting time appropriately.

Proposition 5.4 *For the inhomogeneous Poisson process with rate function $\lambda(t)$, the time, T, to the first event has distribution function*

$$F(t) = 1 - \exp\{-\Lambda(t)\},$$

and hence has density function

$$f(t) = \lambda(t)\exp\{-\Lambda(t)\},\ t > 0,$$

where $\Lambda(t)$ is the cumulative hazard defined above.

Proof.

$$
\begin{aligned}
F(t) &= \mathrm{P}(T \le t)\\
&= 1 - \mathrm{P}(T > t)\\
&= 1 - \mathrm{P}(N_t = 0)\\
&= 1 - \frac{\Lambda(t)^0 \exp\{-\Lambda(t)\}}{0!}\\
&= 1 - \exp\{-\Lambda(t)\}.
\end{aligned}
$$

□

In stochastic simulation, one will often want to simulate the time to the first (or next) event of such a process. Using the inverse distribution method (Proposition 4.1) we can simulate $u \sim U(0,1)$ and then solve $F(t) = u$ for t. Rearranging gives $\Lambda(t) = -\log(1-u)$. However, as observed previously, $1-u$ has the same distribution as u, so we just want to solve

$$\Lambda(t) = -\log u \tag{5.10}$$

for t. For simple hazards it will often be possible to solve this analytically, but in general a numerical procedure will be required. Clearly if $\Lambda(t)$ is analytically invertible, the solution is given by $t = \Lambda^{-1}(-\log u)$.

Note that by construction the function $\Lambda(t)$ is monotonically increasing. So for a given $u \in (0,1)$, (5.10) will have at most one solution. It is also clear that unless the function $\Lambda(t)$ has the property that it tends to infinity as t tends to infinity, there may not be a solution to (5.10) at all. That is, the first event may not happen at all, ever. Consequently, when dealing with the inhomogeneous Poisson process, attention is usually restricted to the case where this is true. In practice this means that the event hazard $\lambda(t)$ is not allowed to decay faster than $1/t$. Assuming this to be the case, there will always be exactly one solution to (5.10). This turns out to be useful if rather than knowing the exact event time, one simply needs to know whether or not the event has occurred before a given time t. For then it is clear that if $\Lambda(t) + \log u$ is negative, the event has not yet occurred, and hence the event time is greater than t, and if it is positive, the event time is less than t. The event time itself is clearly the unique root of the expression $\Lambda(t) + \log u$ (regarded as a function of t). Then if the event time really is required, it can either be found analytically, or failing this, an interval bisection method can be used to find it extremely quickly. Although this discussion might currently seem a little theoretical, it turns out to be a very important practical part of several hybrid stochastic simulation algorithms for biochemical network simulation, so is important to understand.

Example

Consider the inhomogeneous Poisson process with rate function $\lambda(t) = \lambda t$ for some constant $\lambda > 0$. This process has an event hazard that linearly increases with time. The cumulative hazard is clearly given by $\Lambda(t) = \lambda t^2/2$. From this we can immediately deduce that $N_t \sim Po(\lambda t^2/2)$ and that the number of events in the interval $(s, t]$ is $Po(\lambda(t^2 - s^2)/2)$. The time to the first event has PDF

$$f(t) = \lambda t \exp\left\{-\frac{\lambda t^2}{2}\right\}, \quad t > 0$$

and this time can be simulated by sampling $u \sim U(0, 1)$ and solving $\lambda t^2/2 = -\log u$ for t to get

$$t = \sqrt{-\frac{2 \log u}{\lambda}}.$$

5.4.5 Exact sampling of jump times

In the context of biochemical network simulation, the cumulative hazard, $\Lambda(t)$, is often known analytically, which makes the procedure just discussed fairly efficient. However, if only the hazard function, $\lambda(t)$ is known, and the cumulative hazard cannot be evaluated without numerical integration, then the procedure is not at all efficient. It turns out that as long as one can establish an upper bound for $\lambda(t)$ over the time interval of interest (usually trivial), it is possible to use *exact sampling* techniques to definitively decide if an event has occurred and if so at what time, based only on the ability to evaluate the hazard $\lambda(t)$ at a small number of time points. Such techniques are used frequently in applied probability and computational statistics, but have not yet seen widespread adoption in the stochastic kinetics literature.

It turns out that we can simulate an inhomogeneous Poisson process by first simulating from a unit (homogeneous) Poisson process defined on a particular region of $\mathbb{R}^2$. To understand how this is done, some basic properties of the two-dimensional (2d) Poisson process are required. Note that the results presented here are brief and informal — for further information, see an appropriate text on stochastic processes, such as Kingman (1993).

The 2d Poisson process is defined on a set $\Omega \subseteq \mathbb{R}^2$, and has the property that for any (measurable) subset $A \subseteq \Omega$ with *area* $\mu(A)$, the number of events in A will have a $Po(\mu(A))$ distribution, independently of all other sets that do not intersect with A. It is then clear that the number of events in an infinitesimal square $(x, x + dx] \times (y, y + dy]$ will have a $Po(dx\,dy)$ distribution, and so the probability of an event in this region will be $dx\,dy$, independently of all other regions. It is therefore the natural 2d generalisation of the (1d) unit Poisson process.

There is a fundamental uniformity property of the (2d) homogeneous Poisson process that is exceptionally useful for simulation purposes. That is, conditional on the number of points in a region, Ω, those points are uniformly distributed over that region. This means that it is possible to simulate a 2d Poisson process on a finite region Ω by first sampling the number of points from a $Po(\mu(\Omega))$ distribution and then scattering these points uniformly over Ω. Then provided that one can evaluate the area $\mu(\Omega)$ and then scatter points uniformly over Ω, sampling of the Poisson process is straightforward.

It should also be reasonably clear that if it is possible to simulate a unit Poisson process on a region $\bar{\Omega} \supseteq \Omega$, then the points of the originally sampled Poisson process that lie within Ω form a Poisson process on Ω. Again, this fact has considerable utility, as for any given Ω, scattering points uniformly over it may not be practical. However, it will usually be straightforward to find a bounding region $\bar{\Omega}$ (say, a rectangle containing Ω) on which it is easy to uniformly scatter points. This then provides us with a rejection sampling algorithm for generating a Poisson process on Ω.

We are now in a position to see how the 2d (unit) Poisson process gives us an alternative way of understanding the (1d) inhomogeneous Poisson process and, more importantly, an alternative way of sampling event times. Consider an inhomogeneous Poisson process with (bounded, non-negative) hazard function $\lambda(t)$ satisfying $0 \leq \lambda(t) \leq U_\lambda$ for some (known) upper-bound U_λ. Suppose that we wish to simulate from this process on $(0, T]$ for some $T > 0$. Let Ω be the region of $\mathbb{R}^2$ bounded by $y = 0$, $x = 0$, $x = T$ and $y = \lambda(x)$. Let $\bar{\Omega}$ be the rectangle $[0, T] \times [0, U_\lambda]$. Clearly by construction we have $\Omega \subseteq \bar{\Omega} \subseteq \mathbb{R}^2$.

It is clear that we can sample a Poisson process on $\bar{\Omega}$ by first simulating the number of points $m \sim Po(U_\lambda T)$. Then for $i = 1, \ldots, m$ independently sample an x-coordinate from $U(0, T)$ and a y-coordinate from $U(0, U_\lambda)$. If we keep only the points in Ω (that is, the points (x,y) satisfying $y \leq \lambda(x)$), then we have a Poisson process on Ω. However, it is now the case that marginally the x-coordinates of a Poisson process on Ω give the event times of the required 1d inhomogeneous Poisson process.

This forms one alternate algorithm for sampling an inhomogeneous Poisson process. Although it relies on a significant amount of stochastic process theory, implementing it is entirely straightforward. We can summarise the algorithm as follows:

1. Sample $m \sim Po(U_\lambda T)$

2. For $i := 1, \ldots, m$

   (a) Sample $x \sim U(0, T)$

   (b) Sample $y \sim U(0, U_\lambda)$

   (c) If $y \leq \lambda(x)$, print x

The printed values represent a single realisation of the inhomogeneous Poisson process on the time interval $(0, T]$. The order in which the values are printed is insignificant.

The output of the above algorithm is a collection of times. If (as is typically the case in the context of this text) only the first time is required, the algorithm above must be run to completion with the smallest sampled time representing the value of interest. This is somewhat inefficient, but easily rectified, by ensuring that the times are sampled in ascending order. We will consider two ways of doing this below. Before so doing, it is also worth noting that there is a finite probability that no event will occur on the the given interval $(0, T]$, but again, this is no issue, as due to the Markov property, another interval can be constructed and the procedure repeated.

Returning to the problem of sampling the event times in order, one approach is to use properties of uniform order statistics. For $X_i \sim U(0, T)$, $i = 1, \dots, m$, denote the sorted values $X_{(i)}$, so that $X_{(1)} \leq X_{(2)} \leq \dots \leq X_{(m)}$. Then clearly

$$X_{(1)} = \min_i \{X_i\},$$

which clearly has cumulative distribution function $F_{(1)}(x) = (x/T)^m$. This can be simulated by sampling $u \sim U(0, 1)$, and setting $x_{(1)} = Tu^{1/m}$. Then conditional on $x_{(i-1)}$, $X_{(i)}$ is the minimum of $m - i + 1$ uniforms on $[x_{(i-1)}, T]$. This can be simulated by sampling $u \sim U(0, 1)$ and setting $x_{(i)} = (T - x_{(i-1)})u^{1/(m-i+1)}$. The revised algorithm can therefore be summarised as follows.

1. Set $x_{(0)} := 0$
2. Sample $m \sim Po(U_\lambda T)$
3. For $i := 1, \dots, m$

   (a) Sample $u \sim U(0, 1)$
   (b) Set $x_{(i)} := (T - x_{(i-1)})u^{1/(m-i+1)}$
   (c) Sample $y \sim U(0, U_\lambda)$
   (d) If $y \leq \lambda(x_{(i)})$, print $x_{(i)}$

In fact there is a much better way of sampling the times in order, based again on the relationship between the 1d and 2d process. It is clear that given a unit Poisson process on the rectangle $\bar{\Omega}$, the x-coordinates are from a 1d Poisson process with hazard U_λ. Therefore another approach to sampling from $\bar{\Omega}$ is to sample the inter-event times from an $Exp(U_\lambda)$ distribution in order to get the x-coordinates, and then sample the corresponding y-coordinates from a $U(0, U_\lambda)$ distribution. We then reject the points outside Ω as previously to get a revised algorithm, which would be referred to in the applied probability literature as a 'thinning' approach (Ross, 1996). This approach is used as a basis for a simulation algorithm in standard texts on stochastic simulation (Ripley, 1987) and spatial statistics (Cressie, 1993). The algorithm was first explicitly presented in Lewis and Shedler (1979).

Thinning approach to jump time simulation

Consider an inhomogeneous Poisson process with (bounded, non-negative) hazard function $\lambda(t)$ satisfying $0 \leq \lambda(t) \leq U_\lambda$ for some (known) upper-bound U_λ. Suppose that we wish to simulate from this process on $(0, T]$ for some $T > 0$. We have now shown that if we start with a homogeneous Poisson process with rate U_λ and keep each event with probability $\lambda(t)/U_\lambda$, then the resulting process will be inhomogeneous with rate $\lambda(t)$. We can describe this process algorithmically as follows.

1. Set $t_0 := 0$
2. Set $i := 1$
3. Repeat

   (a) Sample $t \sim Exp(U_\lambda)$
   (b) Set $t_i := t_{i-1} + t$
   (c) If $t_i > T$, stop.
   (d) Sample $u \sim U(0,1)$
   (e) If $u \leq \lambda(t_i)/U_\lambda$, print t_i
   (f) Set $i := i+1$

The above algorithm can be used as the foundation for further development. Note that often only the first event time will be of interest, and so the algorithm can be trivially modified to return as soon as the first event time is accepted.

The key thing to note about this algorithm is that it only requires knowledge of the function $\lambda(t)$ (and only at a finite number of times), and not on its integral or inverse. Thus this algorithm has the potential to be used in the context of a hazard function that is not analytically convenient, or even deterministic. For example, even if the hazard function $\lambda(t)$ is defined by a stochastic process, the above algorithm can be implemented, provided that an upper bound is available, and that values of $\lambda(t)$ can be sampled *retrospectively* at a small, finite, but arbitrary set of time points. This is often the case if $\lambda(t)$ is in fact a *diffusion process* (Beskos and Roberts, 2005).

Also note that here we are using rejection to thin a homogeneous Poisson process in order to give a target inhomogeneous process. This is very analogous to the uniform rejection method introduced in Chapter 4. There it was noted that much better algorithms could be developed by using more sophisticated rejection 'envelopes'. The same is true here, and the above algorithm can be trivially modified to use any inhomogeneous Poisson process with a rate function $\nu(t)$ such that $\nu(t) \geq \lambda(t) \forall t \in [0,T]$, provided that it is possible to directly simulate times from the inhomogeneous process with rate $\nu(t)$. Linear and quadratic forms for $\nu(t)$ are often used in practice.

5.5 Diffusion processes

5.5.1 Introduction

Markov processes with continuous states that evolve continuously in time are often termed *diffusion processes*. A rigorous formal discussion of the theory of such processes is beyond the scope of a book such as this. Nevertheless, it is useful to provide a non-technical introduction at this point, as these processes provide an excellent approximation to biochemical network models in many situations.

A d-dimensional Itô diffusion process Y is governed by a stochastic differential equation (SDE) of the form

$$dY_t = \mu(Y_t)dt + \Psi(Y_t)dW_t, \tag{5.11}$$

where $\mu : \mathbb{R}^d \to \mathbb{R}^d$ is a d-dimensional drift vector and $\Psi : \mathbb{R}^d \to \mathbb{R}^d \times \mathbb{R}^d$ is a

$(d \times d)$-dimensional diffusion matrix. The SDE can be thought of as a recipe for constructing a realisation of Y from a realisation of a d-dimensional Brownian motion (or Wiener process), W. A d-dimensional Brownian motion has d independent components, each of which is a univariate Brownian motion, B. A univariate Brownian motion B is a process defined for $t \geq 0$ in the following way.

1. $B_0 = 0$,

2. $B_t - B_s \sim N(0, t - s)$, $\forall t > s$,

3. The increment $B_t - B_s$ is independent of the increment $B_{t'} - B_{s'}$, $\forall t > s \geq t' > s'$.

It is clear from property 2 that $B_t \sim N(0, t)$ (and so $\mathrm{E}(B_t) = 0$ and $\mathrm{Var}(B_t) = t$). It is also clear that if for some small time increment Δt we define the process increment $\Delta B_t = B_{t+\Delta t} - B_t$, we then have $\Delta B_t \sim N(0, \Delta t)$, $\forall t$, and since we know that the increments are independent of one another, this provides a mechanism for simulating the process on a regular grid of time points.

If we define the increment in the diffusion process Y (and the multivariate Brownian motion W) similarly, then we can interpret the SDE (5.11) as the limit of the difference equation

$$\Delta Y_t = \mu(Y_t)\Delta t + \Psi(Y_t)\Delta W_t, \qquad (5.12)$$

as Δt gets infinitesimally small. For finite Δt, (5.12) is known as the *Euler approximation* (or, more correctly, as the *Euler–Maruyama approximation*) of the SDE, and it provides a simple mechanism for approximate simulation of the process Y on a regular grid of time points.[†]

In the case $d = 1$ we have a univariate diffusion process, and it is clear that then the increments of the process are approximately distributed as

$$\Delta Y_t \sim N(\mu(Y_t)\Delta t, \Psi(Y_t)^2 \Delta t).$$

An R function to simulate a univariate diffusion using an Euler approximation is given in Figure 5.9. Note that more efficient simulation strategies are possible; see Kloeden and Platen (1992) for further details, and Leimkuhler and Matthews (2013) for a simple recent proposal.

We can approximate a discrete Markov process using a diffusion by choosing the functions $\mu(\cdot)$ and $\Psi(\cdot)$ so that the mean and variance of the increments match. This is best illustrated by example.

Example — diffusion approximation of the immigration–death process

Suppose we have an immigration–death process with immigration rate λ and death rate μ, and that at time t the current state of the system is x. Then at time $t + dt$, the state of the system is a discrete random quantity with PMF,

$$P(X_{t+dt} = x - 1) = x\mu\, dt,$$
$$P(X_{t+dt} = x) = 1 - (\lambda + x\mu)dt,$$
$$P(X_{t+dt} = x + 1) = \lambda\, dt.$$

[†] 'Euler' is pronounced 'oil-er'.

```
rdiff <- function(afun, bfun, x0 = 0, t = 50, dt = 0.01,
    ...)
{
        n <- t/dt
        xvec <- vector("numeric", n)
        x <- x0
        sdt <- sqrt(dt)
        for (i in 1:n) {
                t <- i*dt
                x <- x + afun(x,...)*dt +
                                bfun(x,...)*rnorm(1,0,
                                    sdt)
                xvec[i] <- x
        }
        ts(xvec, deltat = dt)
}
```

Figure 5.9 *R function for simulation of a diffusion process using the Euler–Maruyama method.*

So the increment of the process has PMF

$$P(dX_t = -1) = x\mu \, dt, \; P(dX_t = 0) = 1 - (\lambda + x\mu)dt, \; P(dX_t = 1) = \lambda \, dt.$$

From this PMF we can calculate the expectation and variance as

$$E(dX_t) = (\lambda - \mu x)dt, \quad Var(dX_t) = (\lambda + \mu x)dt.$$

We therefore set $\mu(x) = \lambda - \mu x$ and $\Psi(x)^2 = \lambda + \mu x$ to get the diffusion approximation

$$dX_t = (\lambda - \mu x)dt + \sqrt{\lambda + \mu x} \, dW_t.$$

We can use our code for simulating diffusion processes to get sample paths like that shown in Figure 5.10 using the R code shown in Figure 5.11.

5.5.2 *Diffusion processes in one dimension*

Having given a very brief introduction to diffusion processes and stochastic differential equations, it is worth studying the one-dimensional case in more detail.[‡] Generalisations to multiple dimensions are usually straightforward, but the notation is much simpler for the one dimensional case, and most of the important issues can be illustrated in that context. In one dimension, we can write the Itô SDE for a diffusion process X_t as

$$dX_t = \mu(X_t)dt + \sigma(X_t)dB_t.$$

[‡] This section is necessarily more technical than most, and may be skipped by readers without a strong mathematical background without significant loss.

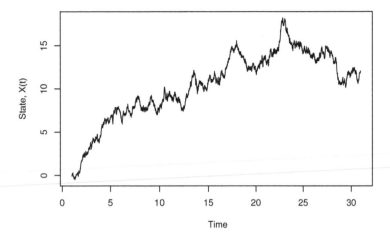

Figure 5.10 *A single realisation of the diffusion approximation to the immigration–death process with parameters $\lambda = 1$ and $\mu = 0.1$, initialised at $X(0) = 0$. Note that this realisation appears to dip below zero near the time origin.*

```
afun <- function (x,lambda,mu)
{
        lambda-mu*x
}
bfun <- function (x,lambda,mu)
{
        sqrt (lambda+mu*x)
}
plot (rdiff (afun,bfun,lambda=1,mu=0.1,t=30))
```

Figure 5.11 *R code for simulating the diffusion approximation to the immigration–death process.*

The process is determined by the deterministic scalar functions $\mu(\cdot)$ and $\sigma(\cdot)$, and an initial condition $X_0 = x_0$. The interpretation of the SDE is that the process is Markovian, that the *infinitesimal mean* of the process is $\mu(X_t)$, and that the *infinitesimal variance* of the process is $\sigma^2(X_t)$. The infinitesimal mean is defined to be

$$\mu(x_t) = \lim_{\delta t \to 0} \frac{1}{\delta t} \, \mathrm{E}(X_{t+\delta t} - x_t),$$

and the infinitesimal variance is defined similarly as

$$\sigma^2(x_t) = \lim_{\delta t \to 0} \frac{1}{\delta t} \, \mathrm{Var}(X_{t+\delta t} - x_t).$$

Note that we also have

$$\sigma^2(x_t) = \lim_{\delta t \to 0} \frac{1}{\delta t} \mathrm{E}\big([X_{t+\delta t} - x_t]^2\big)$$

since $\mathrm{E}(dX_t)^2$ is of order dt^2, and hence will drop out of the limit. There is one further property required of diffusion processes, and that is

$$\lim_{\delta t \to 0} \frac{1}{\delta t} \mathrm{E}(|X_{t+\delta t} - x_t|^q) = 0, \quad \forall q > 2.$$

This technical assumption is required in order to ensure that we have continuous sample paths of the diffusion process. Note that in all of the above we are implicitly conditioning throughout on $X_t = x_t$.

Standard Brownian motion is the very important special case of $\mu(x) = 0$ and $\sigma(x) = 1$, $\forall x \in \mathbb{R}$. To make further progress with the mathematical analysis of diffusion processes, it is helpful to introduce a transition kernel for the process. This can be defined similarly to the case of continuous states in discrete time as

$$P(x, t, x', t') = \mathrm{P}(X_{t+t'} \le x' | X_t = x),$$

with corresponding transition density

$$p(x, t, x', t') = \frac{\partial}{\partial x'} P(x, t, x', t')$$

representing the probability density of $X_{t+t'} | (X_t = x)$. As usual, we will restrict attention to the time homogeneous case where there is no explicit dependence on t in this density, and hence use the notation $p(x, x', t')$ in place of $p(x, t, x', t')$ where this does not cause confusion.

We can now re-write our properties of a diffusion process as integrals with respect to the transition kernel as

$$\mu(x) = \lim_{\delta t \to 0} \frac{1}{\delta t} \int_{\mathbb{R}} (x' - x) p(x, x', \delta t) dx' \qquad (5.13)$$

$$\sigma^2(x) = \lim_{\delta t \to 0} \frac{1}{\delta t} \int_{\mathbb{R}} (x' - x)^2 p(x, x', \delta t) dx' \qquad (5.14)$$

$$0 = \lim_{\delta t \to 0} \frac{1}{\delta t} \int_{\mathbb{R}} |x' - x|^q p(x, x', \delta t) dx', \quad \forall q > 2. \qquad (5.15)$$

These properties turn out to be useful for mathematical analysis. Since the process is Markovian, we can again use the theorem of total probability to write down the Chapman–Kolmogorov equation for a diffusion process as

$$p(x, x', t + t') = \int_{\mathbb{R}} p(z, x', t') p(x, z, t) dz.$$

Just as for the case of discrete state Markov chains in continuous time, we can use the Chapman–Kolmogorov equation in order to derive differential equations representing the Kolmogorov backward and forward equations for a diffusion process. It turns out that the backward equation is slightly easier to derive, so we will start with that.

For the derivation of the backward equation, we will write the transition kernel as $p(x, x_e, t)$, which represents the density of an endpoint state, x_e, as a function of an initial state x and the time prior to the endpoint, t. We begin by thinking about the time derivative of the transition density

$$\frac{\partial}{\partial t} p(x, x_e, t) = \lim_{\delta t \to 0} \frac{p(x, x_e, t + \delta t) - p(x, x_e, t)}{\delta t}. \tag{5.16}$$

Clearly the term $p(x, x_e, t + \delta t)$ can be expanded using the Chapman–Kolmogorov equation in two obvious ways. Expanding as

$$p(x, x_e, t + \delta t) = \int_{\mathbb{R}} p(z, x_e, \delta t) p(x, z, t) dz$$

will give rise to the forward equation, so here instead we use

$$p(x, x_e, t + \delta t) = \int_{\mathbb{R}} p(z, x_e, t) p(x, z, \delta t) dz.$$

If we Taylor expand $p(z, x_e, t)$ in z about x as

$$p(z, x_e, t) = p(x, x_e, t) + (z - x) \frac{\partial}{\partial x} p(x, x_e, t) + \frac{1}{2}(z - x)^2 \frac{\partial^2}{\partial x^2} p(x, x_e, t)$$
$$+ \frac{1}{6}(z - x)^3 \frac{\partial^3}{\partial x^3} p(x, x_e, t) + \cdots$$

and substitute back in to the integral, we get

$$p(x, x_e, t + \delta t) = p(x, x_e, t) \int_{\mathbb{R}} p(x, z, \delta t) dz$$
$$+ \left[\frac{\partial}{\partial x} p(x, x_e, t) \right] \int_{\mathbb{R}} (z - x) p(x, z, \delta t) dz$$
$$+ \frac{1}{2} \left[\frac{\partial^2}{\partial x^2} p(x, x_e, t) \right] \int_{\mathbb{R}} (z - x)^2 p(x, z, \delta t) dz$$
$$+ \frac{1}{6} \left[\frac{\partial^3}{\partial x^3} p(x, x_e, t) \right] \int_{\mathbb{R}} (z - x)^3 p(x, z, \delta t) dz + \cdots$$

At this point we can use our diffusion assumptions, (5.13) – (5.15) to see that for small δt this will approach

$$p(x, x_e, t) + \mu(x) \frac{\partial}{\partial x} p(x, x_e, t) \, \delta t + \frac{1}{2} \sigma^2(x) \frac{\partial^2}{\partial x^2} p(x, x_e, t) \, \delta t,$$

with all higher order terms going to zero. Substituting this back into (5.16) gives the Kolmogorov backward equation as

$$\frac{\partial}{\partial t} p(x, x_e, t) = \mu(x) \frac{\partial}{\partial x} p(x, x_e, t) + \frac{1}{2} \sigma^2(x) \frac{\partial^2}{\partial x^2} p(x, x_e, t).$$

The backward equation can be useful in applications, but is slightly less useful than the forward equation, and also less intuitive. Consequently, much mathematical analysis of diffusion processes relies on the forward equation, which we are now in a position to derive. Again, it is a differential equation for the transition density of the process, but here we will use the notation $p(x_0, x, t)$ to emphasise that it is a density for x at a time t after an initial state x_0.

Theorem 5.1 *The Kolmogorov forward equation for the transition density $p(x_0, x, t)$ of a univariate diffusion process governed by an Itô SDE of the form*

$$dX_t = \mu(X_t)dt + \sigma(X_t)dB_t$$

is

$$\frac{\partial}{\partial t}p(x_0, x, t) = -\frac{\partial}{\partial x}[\mu(x)p(x_0, x, t)] + \frac{1}{2}\frac{\partial^2}{\partial x^2}[\sigma^2(x)p(x_0, x, t)].$$

This equation is commonly referred to as the Fokker–Planck equation. *It is common to drop the dependence on an initial condition and write the equation as*

$$\frac{\partial}{\partial t}p(x, t) = -\frac{\partial}{\partial x}[\mu(x)p(x, t)] + \frac{1}{2}\frac{\partial^2}{\partial x^2}[\sigma^2(x)p(x, t)]$$

for the transition density $p(x, t)$ as a function of x at time t.

Proof. We begin by expressing the time derivative of the transition density as

$$\frac{\partial}{\partial t}p(x_0, x, t) = \lim_{\delta t \to 0} \frac{\int_{\mathbb{R}} p(z, x, \delta t)p(x_0, z, t)dz - p(x_0, x, t)}{\delta t}.$$

At this point it is tempting to follow the derivation of the backward equation and Taylor expand $p(x_0, z, t)$ in z about x. However, this turns out to lead to non-trivial integrals.

The standard way to proceed is to use a technique from analysis, and integrate against a test function by Taylor-expanding the test function and integrating by parts. The idea is reasonably intuitive: if you want to show that two functions $f(x)$ and $g(x)$ are identical, you can just show that $\int_{\mathbb{R}} \varphi(x)f(x)dx = \int_{\mathbb{R}} \varphi(x)g(x)dx$ for *all* functions $\varphi(x)$ in a sufficiently large class. The class of functions typically used is the set of *test functions*, which are smooth, compactly supported functions, and we write $\varphi \in D(\mathbb{R})$. Due to their properties, they are sometimes known as *bump functions*. Note that due to their compact support, integration by parts is just $\int_{\mathbb{R}} \varphi'(x)f(x)dx = -\int_{\mathbb{R}} \varphi(x)f'(x)dx$, and this is very useful in calculations.

We want to evaluate

$$\int_Z p(z, x, \delta t)p(x_0, z, t)dz,$$

and so we integrate against an arbitrary test function

$$\int_X \varphi(x)dx \int_Z p(z, x, \delta t)p(x_0, z, t)dz$$

and change the order of integration to get

$$\int_Z dz\, p(x_0, z, t) \int_X dx\, \varphi(x)p(z, x, \delta t).$$

We now Taylor expand $\varphi(x)$ about z and substitute in to get

$$
\int_Z dz\, p(x_0, z, t) \int_X dx\, \left[\varphi(z) + (x - z)\frac{d}{dz}\varphi(z) + \frac{1}{2}(x - z)^2 \frac{d^2}{dz^2}\varphi(z) \right.
$$

$$
\left. + \frac{1}{6}(x - z)^3 \frac{d^3}{dz^3}\varphi(z) + \cdots \right] p(z, x, \delta t).
$$

At this point we can use our diffusion assumptions, (5.13) – (5.15) to see that for small δt this will approach

$$
\int_Z dz\, p(x_0, z, t) \left[\varphi(z) + \mu(z)\frac{d}{dz}\varphi(z)\delta t + \frac{1}{2}\sigma^2(z)\frac{d^2}{dz^2}\varphi(z)\delta t \right],
$$

with higher-order terms dropping out. As we have now completed the integration with respect to x, we can switch the dummy variable of integration from z to x and integrate terms by parts to get

$$
\int_X dx\, \varphi(x) \left[p(x_0, x, t) - \frac{\partial}{\partial x}\{\mu(x)p(x_0, x, t)\}\,\delta t + \frac{1}{2}\frac{\partial^2}{\partial x^2}\{\sigma^2(x)p(x_0, x, t)\}\,\delta t \right].
$$

By comparing this expression against our starting point, we have demonstrated that

$$
\int_Z p(z, x, \delta t)p(x_0, z, t)dz = p(x_0, x, t) - \frac{\partial}{\partial x}\{\mu(x)p(x_0, x, t)\}\,\delta t
$$

$$
+ \frac{1}{2}\frac{\partial^2}{\partial x^2}\{\sigma^2(x)p(x_0, x, t)\}\,\delta t,
$$

asymptotically, and the result follows.
□

The Fokker–Planck equation tells us how to compute the forward transition kernel of the diffusion process, and as such, has many important applications in the mathematical analysis of diffusion processes. Unfortunately it is analytically intractable except in a few simple special cases. Also note that again we can set the left-hand side of the Kolmogorov forward equation to zero in order to identify stationary distributions of the diffusion. Since the stationary distribution by definition is not time dependent, we can write it as $p(x)$, and identify it as a solution to the second-order ordinary differential equation (ODE)

$$
\frac{d}{dx}[\mu(x)p(x)] = \frac{1}{2}\frac{d^2}{dx^2}[\sigma^2(x)p(x)]. \tag{5.17}
$$

The equation is easiest to remember in the above form, but in applications it can often be useful to expand and simplify to write in the more conventional form of an ODE as

$$
\sigma^2 \frac{d^2 p}{dx^2} + 2\left[2\sigma\frac{d\sigma}{dx} - \mu \right]\frac{dp}{dx} + 2\left[\sigma\frac{d^2\sigma}{dx^2} + \left(\frac{d\sigma}{dx}\right)^2 - \frac{d\mu}{dx} \right]p = 0. \tag{5.18}
$$

It is clear that this expression will be slightly simpler if we parametrise the diffusion directly in terms of infinitesimal variance rather than infinitesimal standard deviation.

For example, defining $\nu(x) = \sigma^2(x)$ gives

$$\nu\frac{d^2p}{dx^2} + 2\left[\frac{d\nu}{dx} - \mu\right]\frac{dp}{dx} + \left[\frac{d^2\nu}{dx^2} - 2\frac{d\mu}{dx}\right]p = 0. \qquad (5.19)$$

Standard processes

We can generalise the briefly introduced concept of Brownian motion to an arbitrary linear scaling of standard Brownian motion. Thus, a *generalised Brownian motion*, X_t, is defined by

$$X_t = \mu t + \sigma B_t,$$

where B_t is (standard) Brownian motion. It is immediately clear from the Gaussian properties of Brownian motion and results on the linear transformation of normal random variables that we have the following key properties of generalised Brownian motion:

- $X_t \sim N(\mu t, \sigma^2 t)$
- $\Delta X_t \equiv X_{t+\Delta t} - X_t \sim N(\mu\Delta t, \sigma^2\Delta t)$
- $\Delta X_t = \mu\Delta t + \sigma\Delta B_t.$

In particular, it is clear that the infinitesimal mean and variance of this process are μ and σ^2, respectively, and so we can write this in the form of the SDE for a diffusion process as

$$dX_t = \mu\,dt + \sigma\,dB_t.$$

The special case $Y_t = t + B_t$, or $dY_t = dt + dB_t$, giving $Y_t \sim N(t,t)$, is the natural *diffusion approximation* for a unit Poisson process.

Time-change of a Brownian motion

When we studied the inhomogeneous Poisson process we considered the relationship between the rate of the process and the rate at which time is measured. We can similarly consider the *time-change of a standard Brownian motion*. Suppose we define a process $X_t = B_{t'}$, using the time-change

$$t' = \int_0^t \lambda(s)\,ds, \ \lambda(t) > 0, \ \forall t \geq 0.$$

Then

$$dX_t = X_{t+dt} - X_t$$
$$= B\left(\int_0^{t+dt}\lambda(s)\,ds\right) - B\left(\int_0^t\lambda(s)\,ds\right)$$
$$\sim N\left(0, \int_t^{t+dt}\lambda(s)\,ds\right) = N(0, \lambda(t)\,dt)$$
$$= \sqrt{\lambda(t)}N(0, dt),$$

and so X_t satisfies the SDE

$$dX_t = \sqrt{\lambda(t)}dB_t.$$

Clearly, for a simple linear change of time units, we get a generalised Brownian motion, but unsurprisingly, for a time-inhomogeneous time transformation the result is a time-inhomogeneous Itô process.

Itô transformations

Theorem 5.2 (Itô's formula for change of variable) *For the diffusion*

$$dX_t = \mu(X_t)\,dt + \sigma(X_t)\,dB_t,$$

define $Y_t = f(X_t)$ for a (twice-differentiable) deterministic real-valued function $f(\cdot)$. Then Y_t satisfies the SDE

$$dY_t = \left(\frac{df}{dx}\mu + \frac{1}{2}\frac{d^2f}{dx^2}\sigma^2\right)dt + \frac{df}{dx}\sigma\,dB_t.$$

Itô's formula can be used to apply (non-linear) transformations to SDEs.
Proof. Here we give a short, informal, sketch proof, which indicates why the result is true, but doesn't really rigorously establish it. We begin with Taylor's theorem

$$\begin{aligned}
dy &= \frac{df}{dx}dx + \frac{1}{2}\frac{d^2f}{dx^2}dx^2 + \cdots \\
&= \frac{df}{dx}[\mu\,dt + \sigma\,dB] + \frac{1}{2}\frac{d^2f}{dx^2}[\mu\,dt + \sigma\,dB]^2 + \cdots \\
&= \frac{df}{dx}\mu\,dt + \frac{df}{dx}\sigma\,dB + \frac{1}{2}\frac{d^2f}{dx^2}\sigma^2 dt,
\end{aligned}$$

as higher-order terms drop out, and the result follows.
□

Here we have used the rules $dB^2 = dt$, $dt^2 = dt\,dB = 0$ and all terms of order 3 and higher are zero. These rules are easily justified by considering the infinitesimal mean and variance of each term.

Example: Square of a generalised Brownian motion

Starting with generalised Brownian motion

$$dX_t = \mu\,dt + \sigma\,dB_t,$$

define $Y_t = X_t^2$. We will use Itô's formula to deduce the SDE satisfied by this process. First note that the transformation is $f(x) = x^2$, and then that

$$\frac{\partial f}{\partial x} = 2x = 2\sqrt{Y_t} \quad \text{and} \quad \frac{\partial^2 f}{\partial x^2} = 2,$$

as $X_t = \sqrt{Y_t}$. Substituting these terms into Itô's formula gives us

$$dY_t = (2\mu\sqrt{Y_t} + \sigma^2)\,dt + 2\sigma\sqrt{Y_t}dB_t.$$

By definition this process is non-negative. Note how the noise term (but not the drift term) tends to zero as Y_t approaches zero.

We can understand another standard process, known as *geometric Brownian motion*, using similar techniques. Starting with generalised Brownian motion

$$dX_t = \mu \, dt + \sigma \, dB_t,$$

define $Y_t = \exp(X_t)$ and apply Itô's formula to get

$$dY_t = \left(\mu + \frac{1}{2}\sigma^2\right) Y_t \, dt + \sigma Y_t dB_t.$$

By definition this process is strictly positive. Note how the noise term (and the drift term) tends to zero as Y_t approaches zero. It is clear that we can calculate any required properties of this process from the corresponding properties of the generalised Brownian motion that it can be transformed to.

Another very important process in physical applications is the *Ornstein–Uhlenbeck (OU) process*,

$$dX_t = -\alpha X_t dt + \sigma \, dB_t. \tag{5.20}$$

The parameters α and σ are assumed to be positive, and the process has support on the whole real line. This is the only Markov process with Gaussian increments which admits a non-trivial stationary distribution (which is also Gaussian). It is the continuous time generalisation of the AR(1) process we considered earlier, in that the values the OU process takes at integer times form an AR(1) process. The expected value of the stationary distribution is zero, but we can use the transformation $Y_t = X_t + \theta$ together with Itô's formula to get the generalised OU process

$$dY_t = \alpha(\theta - Y_t)dt + \sigma \, dB_t,$$

and this has expectation θ. Since the two processes are trivially related, we will restrict attention to the simpler case, X_t.

Analysis of the infinitesimal behaviour of the process and comparison against the AR(1) process suggest a transition distribution of the form

$$X_t|(X_0 = x_0) \sim N\left(x_0 e^{-\alpha t}, \frac{\sigma^2(1 - e^{-2\alpha t})}{2\alpha}\right),$$

leading to transition density

$$p(x_0, x, t) = \sqrt{\frac{\alpha}{\pi\sigma^2(1 - e^{-2\alpha t})}} \exp\left\{\frac{-\alpha(x - x_0 e^{-\alpha t})^2}{\sigma^2(1 - e^{-2\alpha t})}\right\}.$$

This transition density can be confirmed by substitution into the Fokker–Planck equation for the process,

$$\frac{\partial}{\partial t}p(x_0, x, t) = \alpha\frac{\partial}{\partial x}[xp(x_0, x, t)] + \frac{\sigma^2}{2}\frac{\partial^2}{\partial x^2}p(x_0, x, t).$$

Once the transition distribution is established, most interesting properties follow. In particular, by letting t become large, we confirm that the stationary mean of the process is zero, and see that the stationary variance is $\sigma^2/(2\alpha)$.

The *Cox–Ingersoll–Ross* (CIR) process is another standard (and tractable) process. It is defined by

$$dX_t = \alpha(\theta - X_t)\, dt + \sigma\sqrt{X_t}\, dB_t.$$

It is a non-negative process (for positive α, θ, σ). However, it is not a simple transformation of generalised Brownian motion, and it is therefore more difficult to analyse. However, it turns out that increments of the process are non-central χ^2 random variables, and its stationary distribution is gamma. We will look now at how to identify the stationary distribution of the process using the Fokker–Planck equation. We start with (5.19) and substitute in $\mu = \alpha(\theta - x)$ and $\nu = \sigma^2 x$ to get

$$\sigma^2 x\frac{d^2p}{dx^2} + 2[\sigma^2 - \alpha(\theta - x)]\frac{dp}{dx} + 2\alpha p = 0.$$

There are many standard approaches to solving second-order ODEs of this form, but we suspect that the solution is a gamma distribution, and so we can look directly for a solution of the form $p = kx^{a-1}\exp\{-bx\}$. If we manage to find a solution of this form, we will know that the stationary distribution is $Ga(a, b)$. So substituting this in and simplifying gives

$$\sigma^2[b^2x^2 - 2(a-1)bx + (a-1)(a-2)] + 2[\sigma^2 - \alpha(\theta - x)](a - 1 - bx) + 2\alpha x = 0.$$

This is a quadratic in x, and so by equating coefficients of x we get 3 equations in our 2 unknowns, a and b. The coefficients of x^2 give

$$\sigma^2 b^2 - 2\alpha b = 0 \Rightarrow b = \frac{2\alpha}{\sigma^2},$$

similarly, the coefficients of x^0 give

$$a = \frac{2\alpha\theta}{\sigma^2},$$

and as a check, one of these values can be substituted into the equation derived from the coefficients of x^1 to ensure that the equations are consistent (they are). So we conclude that the equilibrium distribution of the CIR process is $Ga(2\alpha\theta/\sigma^2, 2\alpha/\sigma^2)$. This distribution has expectation θ and variance $\sigma^2\theta/(2\alpha)$.

Example: diffusion approximation

Let us now return to the diffusion approximation to the *immigration–death process*. We approximated the process by the diffusion process

$$dX_t = (\lambda - \mu X_t)dt + \sqrt{\lambda + \mu X_t}\, dB_t,$$

which exactly matches the infinitesimal mean and variance of the true process. It is also reasonably easy to show directly that the mean and variance of the stationary distribution exactly match that of the true process (see the end of chapter exercises). However, the plotted trajectory shown in Figure 5.10 suggests that the process is not non-negative, unlike the true process. If we compare the SDE describing this process with the standard processes that we have examined, the source of the problem is clear. All of the non-negative processes we have seen have the property that the infinitesimal variance tends to zero as the state tends to zero. Indeed it is clear that that has to be the case, otherwise fluctuations around zero due to a non-negligible noise term near zero will send the process below zero. Furthermore, we see that the reason the diffusion term does not go to zero is λ, and so, somewhat ironically, it is the noise associated with the immigration process that is causing the process to go negative. It is interesting to try to understand this process, and in particular, its equilibrium distribution, assuming that one exists.

Inspection of the diffusion term shows that the noise actually hits zero at $X_t = -\lambda/\mu$, and so this suggests looking at the transformed process

$$Y_t = X_t + \frac{\lambda}{\mu}.$$

Using Itô's formula, we find that Y_t satisfies

$$dY_t = \mu \left(\frac{2\lambda}{\mu} - Y_t \right) dt + \sqrt{\mu Y_t} \, dB_t.$$

We can see immediately that this shifted process is a special case of a CIR process, and we know that the CIR process has a gamma equilibrium distribution. Consequently we may conclude that X_t has support on $(-\lambda/\mu, \infty)$, with a shifted-Gamma equilibrium.

This example serves to illustrate a more general point; that diffusion approximations to Markov jump processes do not always preserve all desirable features of the process they are approximating. We will revisit this issue in Chapter 8.

5.5.3 Multivariate diffusion processes

All of the concepts discussed in the context of univariate diffusions generalise to multivariate diffusion processes. We have already seen that the concept of a diffusion process has a natural multivariate SDE representation as

$$dX_t = \mu(X_t)dt + \Psi(X_t)dW_t,$$

where $\mu(X_t)$ is a d-vector representing the infinitesimal mean of the diffusion, and $\Sigma(X_t) = \Psi(X_t)\Psi(X_t)^\mathsf{T}$ is a $d \times d$ matrix representing the infinitesimal variance.[§] That is, increments of the process are distributed as

$$dX_t \sim N(\mu(X_t)dt, \Sigma(X_t)dt).$$

It is generally straightforward, in principle, to work with multivariate diffusions, but in practice most are cumbersome to work with algebraically, and very few are mathematically tractable. Once again, the Fokker–Planck equation and Itô transformations

---

[§] Note that both dB_t and dW_t are routinely used to denote increments of both univariate and multivariate standard Brownian motions (Wiener processes). I am tending to use dB_t for univariate and dW_t for multivariate, but that is not a widely adopted convention. In general, dW_t is used more in the physical sciences, and dB_t is used more in the mathematical sciences.

are the two most powerful and commonly used tools. We state without proof the multivariate version of the first of these results.

Theorem 5.3 (multivariate Fokker–Planck equation) *The Kolmogorov forward equation for the d-dimensional diffusion process*

$$dX_t = \mu(X_t)dt + \Psi(X_t)dW_t,$$

is given by

$$\frac{\partial}{\partial t}p(x,t) = -\sum_{i=1}^{d} \frac{\partial}{\partial x_i}\{\mu_i(x)p(x,t)\} + \frac{1}{2}\sum_{i=1}^{d}\sum_{j=1}^{d} \frac{\partial^2}{\partial x_i \partial x_j}\{\Sigma_{ij}(x)p(x,t)\},$$

where $\Sigma(x) = \Psi(x)\Psi(x)^T$.

Using vector calculus notation[¶], we could write this as

$$\frac{\partial p}{\partial t} = -\nabla \cdot [\mu(x)p] + \frac{1}{2}\nabla \cdot (\nabla \cdot [\Sigma(x)p]). \tag{5.21}$$

Again, stationary distributions can be obtained by solving

$$\sum_{i=1}^{d} \frac{\partial}{\partial x_i}\{\mu_i(x)p(x)\} = \frac{1}{2}\sum_{i=1}^{d}\sum_{j=1}^{d} \frac{\partial^2}{\partial x_i \partial x_j}\{\Sigma_{ij}(x)p(x)\},$$

which we could re-write in vector notation as

$$\nabla \cdot [\mu(x)p] = \frac{1}{2}\nabla \cdot (\nabla \cdot [\Sigma(x)p]).$$

The multivariate version of Itô's formula is typically stated for a scalar (one dimensional) function of a multivariate diffusion, but generalisations to multivariate functions are straightforward.

Theorem 5.4 (multivariate Itô formula) *For the d-dimensional diffusion process*

$$dX_t = \mu(X_t)dt + \Psi(X_t)dW_t,$$

define $\Sigma(X_t) = \Psi(X_t)\Psi(X_t)^T$ *and* $Y_t = f(X_t)$ *for twice-differentiable scalar function* $f(\cdot)$. *Then the process* Y_t *satisfies the univariate SDE*

$$dY_t = \left(\nabla f \cdot \mu + \frac{1}{2}\operatorname{tr}([\nabla^2 f]\Sigma)\right)dt + \sqrt{(\nabla f)^T\Sigma\nabla f}\,dB_t.$$

Note that here $\nabla^2 f$ denotes the full Hessian matrix $\nabla\nabla^T f$, and not its trace[‖], $\nabla \cdot \nabla f = \nabla^T\nabla f = \operatorname{tr}(\nabla\nabla^T f)$, known as the Laplacian, and also often denoted $\nabla^2 f$. Before going on to give a sketch proof of this result, it is important to be clear about the implications of this result, which may not be immediately apparent. Since $\nabla f(\cdot)$ and $\Sigma(\cdot)$ will both (typically) depend on aspects of X_t that cannot be captured by

---

[¶] We use $\nabla = \left(\frac{\partial}{\partial x_1}, \frac{\partial}{\partial x_2}, \dots, \frac{\partial}{\partial x_d}\right)^T$ and for vectors v_1 and v_2, $v_1 \cdot v_2 = v_1^T v_2$. We also use the convention $v \cdot M = (v^T M)^T$ for v a vector and M a (conformable) matrix.

[‖] The trace of a (square) matrix A, denoted $\operatorname{tr}(A)$, is the sum of the diagonal elements, $\sum_i a_{ii}$.

the functional $f(\cdot)$, it will (typically) not be possible to write the infinitesimal mean and variance of Y_t so that they are only a function of Y_t. In this case, the marginal behaviour of Y_t will not be Markovian, and certainly will not be marginally a one–dimensional diffusion process. The formula is useful because it gives an insight into the stochastic behaviour of functionals of X_t. To form a transformation that is a diffusion process, one typically requires a multivariate transformation based on an invertible function $f(\cdot)$. It is possible to develop an Itô formula in this case (see Chapter 4 of Øksendal (2003)), but it is beyond the scope of an introductory text such as this.

Proof. Here, as for the univariate case, we give a brief, informal sketch proof. From the multivariate Taylor expansion we have

$$dy = \nabla f^\mathsf{T} \, dx + \frac{1}{2} dx^\mathsf{T} \nabla^2 f \, dx + \cdots$$

$$= \nabla f^\mathsf{T} \left[\mu \, dt + \Psi \, dW \right] + \frac{1}{2} [\mu \, dt + \Psi \, dW]^\mathsf{T} \nabla^2 f \left[\mu \, dt + \Psi \, dW \right] + \cdots$$

$$= \nabla f^\mathsf{T} \mu \, dt + \nabla f^\mathsf{T} \Psi \, dW + \frac{1}{2} [\Psi \, dW]^\mathsf{T} \nabla^2 f \, \Psi \, dW,$$

as higher-order terms will drop out.

$$= \nabla f^\mathsf{T} \mu \, dt + \nabla f^\mathsf{T} \Psi \, dW + \frac{1}{2} \operatorname{tr} \left(\nabla^2 f \, \Psi \, dW [\Psi \, dW]^\mathsf{T} \right)$$

$$= \nabla f^\mathsf{T} \mu \, dt + \nabla f^\mathsf{T} \Psi \, dW + \frac{1}{2} \operatorname{tr} \left(\nabla^2 f \, \Psi \Psi^\mathsf{T} \right) dt$$

$$= \left(\nabla f^\mathsf{T} \mu + \frac{1}{2} \operatorname{tr}(\nabla^2 f \, \Sigma) \right) dt + \nabla f^\mathsf{T} \Psi \, dW.$$

But the increments of this process clearly have variance

$$\operatorname{Var}\left(\nabla f^\mathsf{T} \Psi \, dW \right) = \nabla f^\mathsf{T} \Psi \operatorname{Var}(dW) \, \Psi^\mathsf{T} \nabla f = \nabla f^\mathsf{T} \Sigma \nabla f \, dt$$

and the result follows.

☐

5.6 Exercises

1. (a) Show that the product of two stochastic matrices is stochastic.

   (b) Show that for stochastic P, and *any* row vector π, we have $\|\pi P\|_1 \le \|\pi\|_1$, where $\|v\|_1 = \sum_i |v_i|$. Deduce that all eigenvalues, λ, of P must satisfy $|\lambda| \le 1$. As an aside, note that as a consequence of (a), it is clear that for all *probability* row vectors, π, we have $\|\pi P\|_1 = \|\pi\|_1$.

   (c) Show that a stochastic matrix always has a row eigenvector with eigenvalue 1. Show that this row eigenvector can be chosen to correspond to a stationary distribution of the induced Markov chain. *Hint: It is an easy consequence of the Brouwer fixed point theorem, which says, inter alia, that any continuous map from a compact convex set to itself must have a fixed point.*

2. For the AR(1) process

$$Z_t = \alpha Z_{t-1} + \varepsilon_t, \quad \varepsilon_t \sim Exp(\lambda), \quad \alpha \in (0,1), \ \lambda > 0,$$

   (a) what is the transition kernel, $p(x, y)$, of the chain?

   (b) Hence, deduce an integral equation satisfied by the stationary distribution, $\pi(x)$.

   (c) If Z has density $\pi(x)$, calculate the expected value of Z and show that the variance of Z is given by

$$\text{Var}(Z) = \frac{1}{\lambda^2 (1 - \alpha^2)}.$$

   (d) Is this chain reversible? *Hint: You do not need to work out the stationary density!*

   (e) Simulate the time evolution of the chain by writing appropriate functions for R. Discard the initial burn-in phase from a long run and then use the rest to get empirical estimates of the mean and variance of the stationary distribution. Use this to verify your calculations from (c) for a couple of different combinations of λ and α.

3. Starting with the R function for the simulation of the immigration–death process, modify it to include 'births' in addition to immigrations and deaths. Use your modified function to simulate a long realisation from the birth-immigration–death process where the birth, death, and immigration rates are all one, starting from an initial condition of zero.

4. Use stochastic simulation to investigate the stationary distribution of the diffusion approximation to the immigration–death process. How well does it approximate the Poisson distribution of the original discrete process? By solving $E(dX_t) = 0$, show that the stationary mean of the diffusion approximation is λ/μ. Then (this one is slightly tricky) by solving $\text{Var}(X_t) = \text{Var}(X_t + dX_t)$, show that the stationary variance is also λ/μ.

5. Form a diffusion approximation of the birth–death process, which has transition equations:

$$P(X_{t+dt} = x + 1 | X_t = x) = \lambda x \, dt,$$
$$P(X_{t+dt} = x - 1 | X_t = x) = \mu x \, dt,$$
$$P(X_{t+dt} = x | X_t = x) = 1 - (\lambda + \mu)x \, dt.$$

Assume an non-zero initial condition $X_0 = x_0 > 0$. Does the diffusion approximation define a non-negative stochastic process?

6. Write down the Fokker–Planck equation for standard Brownian motion. Show that the $N(0, t)$ density is a solution for $p(0, x, t)$.

7. If you haven't already done so, install the R package associated with this book, smfsb. See Appendix B.1 for details. The R package includes functions such as rcfmc and rdiff. Reproduce figures showing sample paths of Markov processes, analogous to those given in this chapter.

5.7 Further reading

The literature on the theory of Markov processes is vast. For further information, it is sensible to start with classic texts such as Cox and Miller (1977) and Ross (1996), before moving on to applied texts such as Allen (2011), Gillespie (1992a), Van Kampen (1992), and references therein. The theory of diffusion processes and stochastic differential equations is particularly technical. Iacus (2008) is a good introduction with minimal technical details. An excellent overview of this area, which is more accessible than most, is Øksendal (2003), but it is not appropriate for novices. For numerical methods related to SDEs, the standard text is Kloeden and Platen (1992).

## PART III

Stochastic chemical kinetics

CHAPTER 6

Chemical and biochemical kinetics

6.1 Classical continuous deterministic chemical kinetics

6.1.1 Introduction

Chemical kinetic modelling is concerned with understanding the time-evolution of a reaction system specified by a given set of coupled chemical reactions. In particular, it is concerned with system behaviour away from equilibrium. In order to introduce the concepts it is helpful to use a very simple model system. Consider the 'Lotka–Volterra' (LV) system introduced in Section 1.6,

$$Y_1 \longrightarrow 2Y_1$$
$$Y_1 + Y_2 \longrightarrow 2Y_2$$
$$Y_2 \longrightarrow \emptyset.$$

Although the reaction equations capture the key interactions between the competing species, on their own they are not enough to determine the full dynamic behaviour of the system. For that, we need to know the *rates* at which each of the reactions occurs (together with some suitable initial conditions).

6.1.2 Mass-action kinetics

The above model encourages us to think about the number of prey (Y_1) and predators (Y_2) as integers, which can change only by discrete (integer) amounts when a reaction *event* occurs. This picture is entirely correct, and we will study the implications of such an interpretation later in this chapter. However, we will introduce the study of kinetics by thinking about a more classical chemical reaction setting of macroscopic amounts of chemicals reacting in a 'beaker of water'. There, the amount of each chemical is generally regarded as a *concentration*, measured in (say) moles per litre, M, which can vary continuously as the reaction progresses. Conventionally, the concentration of a chemical species X is denoted $[X]$.

It is generally the case that the instantaneous rate of a reaction is directly proportional to the concentration (in turn directly proportional to mass) of each reactant raised to the power of its stoichiometry. We will see the reason behind this when we study stochastic kinetics later, but for now we will accept it as an empirical law. This kinetic 'law' is known as *mass-action kinetics*. So, for the LV system, the second reaction will proceed at a rate proportional to $[Y_1][Y_2]$. Consequently, due to the effect of this reaction, $[Y_1]$ will decrease at instantaneous rate $k_2[Y_1][Y_2]$ (where k_2 is the constant of proportionality for this reaction), and $[Y_2]$ will increase at the same rate (since the overall effect of the reaction is to decrease $[Y_1]$ at the same rate $[Y_2]$ increases). $k_2[Y_1][Y_2]$ is known as the *rate law* of the reaction, and k_2 is the *rate constant*. Considering all three reactions, we can write down a set of ordinary differential equations (ODEs) for the system:

$$\frac{d[Y_1]}{dt} = k_1[Y_1] - k_2[Y_1][Y_2]$$
$$\frac{d[Y_2]}{dt} = k_2[Y_1][Y_2] - k_3[Y_2].$$

The three rate constants, k_1, k_2, and k_3 (measured in appropriate units) must be specified, as well as the initial concentrations of each species. Once this has been done, the entire dynamics of the system are completely determined and can be revealed by 'solving' the set of ODEs, either analytically (in the rare cases where this is possible) or numerically using a computer.

It is instructive to rewrite the above ODE system in matrix form as

$$\frac{d}{dt}\begin{pmatrix}[Y_1]\\[Y_2]\end{pmatrix} = \begin{pmatrix}1 & -1 & 0\\0 & 1 & -1\end{pmatrix}\begin{pmatrix}k_1[Y_1]\\k_2[Y_1][Y_2]\\k_3[Y_2]\end{pmatrix},$$

where the 2×3 matrix is just the stoichiometry matrix, S, of the reaction system (Definition 2.4). This leads to a general strategy for constructing ODE models from the Petri net reaction network representation discussed in Chapter 2. If we write $r([Y]) = (r_1([Y]), r_2([Y]), \ldots, r_v([Y]))^\mathsf{T}$ for the vector of rate laws of the v different chemical reactions, then the ODE model may be derived from the state updating equation as

$$\frac{d}{dt}[Y] = S\,r([Y]).$$

We will see later that this ODE model may be regarded as a continuous deterministic approximation of the natural discrete stochastic Markov process model description given by the theory of stochastic chemical kinetics.

6.1.3 Equilibrium

Even when the set of ODEs is analytically intractable, it may be possible to discover an 'equilibrium' solution of the system by analytic (or simple numerical) means. An equilibrium solution is a set of concentrations which will not change over time, and hence can be found by solving the set of simultaneous equations formed by setting the RHS of the ODEs to zero. In general this is achieved by solving the equations given by

$$S\,r([Y]) = 0.$$

In the context of the LV example, this is:

$$k_1[Y_1] - k_2[Y_1][Y_2] = 0$$
$$k_2[Y_1][Y_2] - k_3[Y_2] = 0.$$

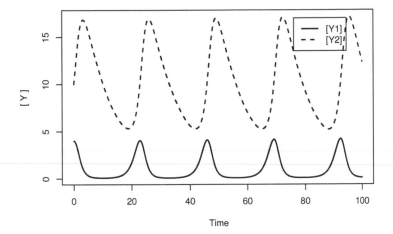

Figure 6.1 *Lotka–Volterra dynamics for* $[Y_1](0) = 4$, $[Y_2](0) = 10$, $k_1 = 1$, $k_2 = 0.1$, $k_3 = 0.1$. *Note that the equilibrium solution for this combination of rate parameters is* $[Y_1] = 1$, $[Y_2] = 10$.

Solving these for $[Y_1]$ and $[Y_2]$ in terms of k_1, k_2, and k_3 gives two solutions. The first is the rather uninteresting

$$[Y_1] = 0, \quad [Y_2] = 0,$$

and the second is

$$[Y_1] = \frac{k_3}{k_2}, \quad [Y_2] = \frac{k_1}{k_2}.$$

Further analysis (beyond the scope of this text and rather tangential to it), reveals that this second solution is not unstable, and hence corresponds to a realistic stable state of the system. However, it is not an 'attractive' stable state, and so there is no reason to suppose that the system will tend to this state irrespective of the starting conditions.

Despite knowing the existence of an equilibrium solution to this system, we have no reason to suppose that any particular set of initial conditions will lead to this equilibrium, and even if we did, it would say nothing (or little) about how the system reaches it. To answer this question we need to specify the initial conditions of the system and integrate the ODEs to uncover the full dynamics. The dynamics for a particular combination of rate parameters and initial conditions are shown in Figure 6.1. An alternative way of displaying these dynamics is as an 'orbit' in 'phase-space' (where the value of one variable is plotted against the others, and time is not shown directly). Figure 6.2 shows the dynamics in this way.

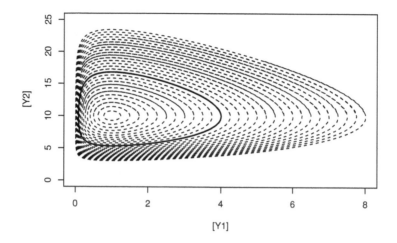

Figure 6.2 *Lotka–Volterra dynamics in phase-space for rate parameters $k_1 = 1$, $k_2 = 0.1$, $k_3 = 0.1$. The dynamics for the initial condition $[Y_1](0) = 4$, $[Y_2](0) = 10$ are shown as the bold orbit. Note that the system moves around this orbit in an anti-clockwise direction. Orbits for other initial conditions are shown as dotted curves. Note that the equilibrium solution for this combination of rate parameters is $[Y_1] = 1$, $[Y_2] = 10$.*

6.1.4 Reversibility

Before going on to examine numerical integration of ODEs (which will explain how plots such as Figure 6.1 are produced), it is worth considering an important special class of reactions, namely reversible reactions. These are reactions that can proceed in both directions. For example, consider a dimerisation reaction,

$$2P \rightleftharpoons P_2.$$

If we make the very strong assumption that *neither of these species are involved in any other reactions*, then we get the ODE system

$$\frac{d}{dt}\begin{pmatrix} [P] \\ [P_2] \end{pmatrix} = \begin{pmatrix} -2 & 2 \\ 1 & -1 \end{pmatrix} \begin{pmatrix} k_1[P]^2 \\ k_2[P_2] \end{pmatrix},$$

where k_1 and k_2 are the forward and backward rate constants, respectively. Expanding gives the pair of ODEs

$$\frac{d[P]}{dt} = 2k_2[P_2] - 2k_1[P]^2$$

$$\frac{d[P_2]}{dt} = k_1[P]^2 - k_2[P_2]. \tag{6.1}$$

We clearly have equilibrium whenever

$$k_2[P_2] = k_1[P]^2.$$

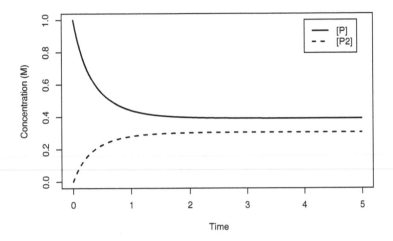

Figure 6.3 *Dimerisation kinetics for* $[P](0) = 1$, $[P_2](0) = 0$, $k_1 = 1$, $k_2 = 0.5$. *This combination of parameters gives* $K_{eq} = 2$, $c = 1$, *and hence equilibrium concentrations of* $[P] = 0.39$, $[P_2] = 0.30$.

Another way of writing this is

$$\frac{[P_2]}{[P]^2} = \frac{k_1}{k_2} \equiv K_{eq}, \qquad (6.2)$$

where K_{eq} is the *equilibrium constant* of the system. It turns out that this equilibrium is stable and attractive, as can be seen from the sample simulated dynamics in Figure 6.3.

For this system, it is possible to make further progress by noting that $[P]$ and $[P_2]$ are deterministically related in this system. One way to see this is to add twice the second ODE to the first to get

$$\frac{d[P]}{dt} + 2\frac{d[P_2]}{dt} = 0$$

$$\Rightarrow \frac{d}{dt}([P] + 2[P_2]) = 0$$

$$\Rightarrow [P] + 2[P_2] = c, \qquad (6.3)$$

where c is the concentration of $[P]$ we would have if the dimers were fully disassociated. Equation (6.3) is a conservation equation, as the value of the LHS is conserved by the reaction system. A more direct approach to finding conservation equations is to use the Petri net theory from Chapter 2. Here, the reaction matrix (Definition 2.4) for the system is given by

$$A = \begin{pmatrix} -2 & 1 \\ 2 & -1 \end{pmatrix}.$$

The conservation equation corresponds to the P-invariant (Definition 2.5) $y = (1,2)^\mathsf{T}$, which can be found directly by looking for non-trivial solutions of the linear system $Ay = 0$.

Conservation equations are useful for reducing the dimension of the system under consideration. Here, for example, we can rearrange (6.3) for $[P_2]$ and substitute back into the equilibrium relation (6.2) in order to find the equilibrium concentration of $[P]$ to be a solution of the quadratic equation

$$2K_{eq}[P]^2 + [P] - c = 0.$$

This clearly has a single positive real root given by

$$[P] = \frac{\sqrt{8cK_{eq} + 1} - 1}{4K_{eq}}.$$

We can therefore find exactly the equilibrium concentrations of $[P]$ and $[P_2]$ in the absence of other reactions.

Alternatively we can use the conservation equation (6.3) to reduce the pair of ODEs to a single first-order ODE

$$\frac{d[P]}{dt} = k_2(c - [P]) - 2k_1[P]^2, \tag{6.4}$$

which can be solved for given initial conditions to give the full dynamical behaviour of the system. It turns out that (6.4) can be solved analytically to give an explicit expression for $[P]$ as a function of t, but the solution is not elegant, and solving ODEs is not the main focus of this book. Note, however, that in order to be able to do this (analytically or otherwise), we must know both the forward and backward rate constants k_1 and k_2, and not just their ratio, K_{eq}.

6.1.5 Numerical integration of ODEs

We will finish this section on deterministic kinetics by examining numerically how we integrate a system of ODEs on a computer. We will just look at this simplest possible technique, known as the first-order Euler method.* It is worth bearing in mind, however, that there are more sophisticated techniques that can be used which give much more accurate dynamics for an equivalent amount of computation time. A good example of this is the fourth-order Runge–Kutta method, which is implemented by many biochemical simulators.

We can write any of the ODE systems that we have considered so far very simply as a vector ODE

$$\frac{dX}{dt} = f(X)$$

where X is a p-dimensional vector and $f(\cdot) : \mathbb{R}^p \to \mathbb{R}^p$ is an arbitrary (non-linear) p-dimensional function of X (for the models we have been considering, we have

---

* In fact, this technique is a special case of the Euler method for numerically solving SDEs that was examined in Section 5.5.

```
simpleEuler <- function(t=50, dt=0.001, fun, ic, ...)
{
        p = length(ic)
        n = t/dt
        xmat = matrix(0,ncol=p,nrow=n)
        x = ic
        t = 0
        xmat[1,] = x
        for (i in 2:n) {
                t = t + dt
                x = x + fun(x,t,...) * dt
                xmat[i,] = x
        }
        ts(xmat, start=0, deltat=dt)
}
```

Figure 6.4 *An R function to numerically integrate a system of coupled ODEs using a simple first-order Euler method.*

$f(X) = Sr(X)$). Recall that the derivative is defined by

$$\frac{dX}{dt}(t) = \lim_{\Delta t \to 0} \frac{X(t + \Delta t) - X(t)}{\Delta t}.$$

So for small Δt we have

$$\frac{X(t + \Delta t) - X(t)}{\Delta t} \simeq f(X(t)),$$

and re-arranging gives

$$X(t + \Delta t) \simeq X(t) + \Delta t f(X(t)). \tag{6.5}$$

Equation (6.5) gives us a simple method for computing $X(t + \Delta t)$ from $X(t)$. If we start off at known $X(0)$, we can compute $X(\Delta t)$, $X(2\Delta t)$, $X(3\Delta t), \ldots$ to get the full dynamics of the system. This is the so-called Euler method.

A very simple R function to implement this algorithm is given in Figure 6.4. The function returns an R time series object, which can easily be plotted. For example, to simulate the dynamics of the LV system, first define the function

```
lv <- function(x,t,k=c(k1=1,k2=0.1,k3=0.1))
{
        with(as.list(c(x,k)),{
                c( k1*x1 - k2*x1*x2 ,
                   k2*x1*x2 - k3*x2 )
        })
}
```

Then typing **plot**(simpleEuler(t=100,fun=lv,ic=**c**(x1=4,x2=10))) gives a plot similar to the one shown in Figure 6.1. For more sophisticated numerical integration strategies, consult a standard numerical analysis text such as Burden and Faires (2010). Note, however, that the R package deSolve (available from CRAN) provides an interface to a fairly sophisticated ODE solving library. We can use that library to integrate the above model as follows:

```
require (deSolve)
times = seq (0,50,by=0.01)
k = c (k1=1,k2=0.1,k3=0.1)
lvlist = function (t,x,k)
         list (lv(x,t,k))
plot (ode(y=c(x1=4,x2=10),times=times,func=lvlist,parms=k
    ))
```

The main difference between using ode and simpleEuler is that here we explicitly provide a sequence of times at which the solution is required. Additionally, the function representing the RHS of the ODE must return its value as a list (and has arguments t and x swapped). Note that the default solver used by the ode function is very much better than a simple Euler method, and so this strategy is worth pursuing if numerical integration of ODE models is of interest. Typing vignette ("deSolve") provides a tutorial introduction to the package.

6.2 Molecular approach to kinetics

The deterministic approach to kinetics fails to capture the discrete and stochastic nature of chemical kinetics at low concentrations. As many intra-cellular processes involve reactions at extremely low concentrations, such discrete stochastic effects are often relevant for systems biology models. We are now in a position to see how chemical kinetics can be modelled in this way.

Consider a bi-molecular reaction of the form

$$X + Y \longrightarrow ?$$

(the RHS is not important). What this reaction really means is that a molecule of X is able to react with a molecule of Y if the pair happen to collide with one another (with sufficient energy), while moving around randomly, driven by Brownian motion. Considering a single pair of such molecules in a container of volume V, it is possible to use statistical mechanical arguments to understand the hazard of molecules colliding. Under fairly weak assumptions regarding the container and its contents (essentially that it is small or well stirred, and in thermal equilibrium), it can be rigorously demonstrated that the collision hazard is *constant*, provided the volume is fixed and the temperature is constant. A comprehensive treatment of this issue is given in Gillespie (1992b), to which the reader is referred for further details. However, the essence of the argument is that as the molecules are uniformly distributed throughout the volume and this distribution does not depend on time, then the probability that the molecules are within reaction distance is also independent of time. In the case of time-varying V (which can be quite relevant in the biological context), the hazard is inversely proportional to V. Again, for the careful statistical mechanical argument, see Gillespie (1992b), but an intuitive explanation can be given as follows. Let the molecules' position in space be denoted by P_1 and P_2, respectively. Then P_1 and P_2 are uniformly and independently distributed over the volume V. This means that for a region of space d with volume v' we have

$$P(P_i \in d) = \frac{v'}{V}, \quad i = 1, 2.$$

Now if we are interested in the probability that X and Y are within a reacting distance (r) of one another at any given instant of time (assuming that r is very small relative to the dimensions of the container, so that boundary effects can be ignored), this probability can be computed as

$$P(|P_1 - P_2| < r) = E(P(|P_1 - P_2| < r|P_2)) \qquad \text{(by Proposition 3.11)}$$

but the conditional probability will be the same for any P_2 away from the boundary, rendering the expectation redundant, and reducing the expression to

$$
\begin{aligned}
&= P(|P_1 - p| < r) && \text{(for any } p \text{ away from the boundary)}\\
&= P(P_1 \in d) && \text{(where } d \text{ is a sphere of radius } r)\\
&= \frac{4\pi r^3}{3V}.
\end{aligned}
$$

This probability is inversely proportional to V. Then conditional on the molecules being within reaction distance, they will not necessarily react, but will do so with a probability independent of V (as other important variables, such as the velocity distributions, are independent of V), thus preserving the inverse dependence on V in the combined probability of being within reaction distance and reacting. The case of time-varying V is a little messy to deal with, so a detailed discussion will be deferred until Section 8.2.3. A fixed volume V will be assumed throughout the rest of this chapter. The case of non-constant temperature and other time-varying environmental factors will not have such a straightforward relationship to the reaction hazard, and so again, throughout this chapter, it will be assumed that temperature, pressure, pH, and all other environmental factors not explicitly described in the reaction network are held constant.

Before going on to explore the theory of discrete stochastic chemical kinetics in detail, it is worth considering other sources of 'noise', uncertainty, randomness, and heterogeneity in biological systems (Wilkinson, 2009). Uncertainty due to the statistical mechanical consideration of the discreteness of chemical kinetic systems is often referred to as *intrinsic noise* (Swain et al., 2002). There are, however, many other factors which can cause biological systems to behave unpredictably, and these factors are often grouped together under the heading *extrinsic noise*. As an example, the rate 'constant' of a reaction may appear to vary randomly over time, due to the effect of processes not explicitly modelled. The apparently random fluctuations of the rate parameter may be modelled as a stochastic process, and will then provide additional noise in the system dynamics. But there are many other ways that unpredictability can arise; for example, due to model simplifications, inadequacies, and spatial effects. Also, when modelling populations of cells, there will most likely be genuine cell–to–cell variations in rate parameters in addition to varying initial conditions, and variation in processes not explicitly modelled. It is difficult to give general advice on the study and modelling of extrinsic noise processes, but we will examine the case of randomly varying initial conditions in Chapter 7 and randomly varying rate parameters in Chapter 8. For the rest of this chapter, we focus on the case of modelling intrinsic noise.

6.3 Mass-action stochastic kinetics

We will consider a system of reactions involving u species $\mathcal{X}_1$, $\mathcal{X}_2, \ldots, \mathcal{X}_u$ and v reactions, $\mathcal{R}_1$, $\mathcal{R}_2, \ldots, \mathcal{R}_v$. Typically (but not always) there will be more reactions than species, $v > u$. We will assume that the qualitative structure of the reaction network can be encoded in the form of a Petri net $N = (P, T, Pre, Post, M)$, where $P = (\mathcal{X}_1, \mathcal{X}_2, \ldots, \mathcal{X}_u)^\mathsf{T}$ and $T = (\mathcal{R}_1, \mathcal{R}_2, \ldots, \mathcal{R}_v)^\mathsf{T}$, as described in Section 2.3. Denote the number of molecules of $\mathcal{X}_i$ at time t by X_{it}, and put $X_t = (X_{1t}, X_{2t}, \ldots, X_{ut})^\mathsf{T}$ for the state of the system at time t. Let R_{it} denote the number reactions of type $\mathcal{R}_i$ in the time window $(0, t]$, and then define $R_t = (R_{1t}, \ldots, R_{vt})^\mathsf{T}$. From the state updating equation discussed in Chapter 2, we have

$$X_t - X_0 = SR_t, \tag{6.6}$$

where

$$S = (Post - Pre)^\mathsf{T}$$

is the $u \times v$ stoichiometry matrix of the reaction network. In addition, each reaction, $\mathcal{R}_i$, will have a *stochastic rate constant*, c_i, and an associated *rate law* (or *hazard* function), $h_i(x, c_i)$, where $x = (x_1, x_2, \ldots, x_u)^\mathsf{T}$ is the current state (or marking) of the system (and so at time t this will be $h_i(X_t, c_i)$). The form of $h_i(x, c_i)$ (and the interpretation of the rate constant c_i) is determined by the order of reaction $\mathcal{R}_i$. In all cases the hazard function has the same interpretation, namely that conditional on the state being x at time t, the probability that an $\mathcal{R}_i$ reaction (or transition) will occur in the time interval $(t, t + dt]$ is given by $h_i(x, c_i)\, dt$. Thus, *in the absence of any other reactions taking place*, the time to such a reaction event *would be* an $Exp(h_i(x, c_i))$ random quantity. Note, however, that since the hazard depends on the state x, and other reactions could change the state, the actual time until an $\mathcal{R}_i$ reaction will typically not be exponential.

6.3.1 Zeroth-order reactions

First consider a reaction of the form

$$\mathcal{R}_i : \quad \emptyset \xrightarrow{c_i} X.$$

Although in practice things are not created from nothing, it can sometimes be useful to model a constant rate of production of a chemical species (or influx from another compartment) via a zeroth-order reaction. In this case, c_i is the hazard of a reaction of this type occurring, and so

$$h_i(x, c_i) = c_i.$$

6.3.2 First-order reactions

Consider the first-order reaction

$$\mathcal{R}_i : \qquad \mathcal{X}_j \xrightarrow{c_i} ?.$$

Here c_i represents the hazard that a particular molecule of $\mathcal{X}_j$ will undergo the reaction. However, there are x_j molecules of $\mathcal{X}_j$, each of which have a hazard of c_i of reacting. This gives a combined hazard of

$$h_i(x, c_i) = c_i x_j$$

for a reaction of this type. Note that first-order reactions of this nature are intended to capture the spontaneous change of a molecule into one or more other molecules, such as radioactive decay, or the spontaneous dissociation of a complex molecule into simpler molecules. They are not intended to model the conversion of one molecule into another in the presence of a catalyst, as this is really a second-order reaction. However, in the presence of a large pool of catalyst that can be considered not to vary in level during the time evolution of the reaction network, a first-order reaction may provide a reasonable approximation. Michaelis–Menten enzyme kinetics will be examined in detail in the next chapter.

6.3.3 Second-order reactions

For second-order reactions of the form

$$\mathcal{R}_i : \qquad \mathcal{X}_j + \mathcal{X}_k \xrightarrow{c_i} ?,$$

c_i represents the hazard that a particular pair of molecules of type $\mathcal{X}_j$ and $\mathcal{X}_k$ will react. But since there are x_j molecules of $\mathcal{X}_j$ and x_k molecules of $\mathcal{X}_k$, there are $x_j x_k$ different pairs of molecules of this type, and so this gives a combined hazard of

$$h_i(x, c_i) = c_i x_j x_k$$

for this type of reaction. There is another type of second-order reaction which needs to be considered:

$$\mathcal{R}_i : \qquad 2\mathcal{X}_j \xrightarrow{c_i} ?.$$

Again c_i represents the hazard of a particular pair of molecules reacting. But here there are only $x_j(x_j - 1)/2$ pairs of molecules of type $\mathcal{X}_j$, and so

$$h_i(x, c_i) = c_i \frac{x_j(x_j - 1)}{2}.$$

Note that this does not match exactly the form of the corresponding deterministic mass-action rate law — this will be further discussed in Section 6.7.

6.3.4 Higher-order reactions

It is straightforward to extend this theory to higher-order reactions. Consideration of the number of available combinations of reacting molecules leads directly to the formula

$$h_i(x, c_i) = c_i \prod_{j=1}^{u} \binom{x_j}{p_{ij}},$$

where p_{ij} is the (i, j)th element of the matrix Pre. In reality, however, most (if not all) reactions that are normally written as a single reaction of order higher than two in fact represent the combined effect of two or more reactions of order one or two. In these cases, it is usually better to model the reactions in detail rather than via high-order stochastic kinetics. Consider, for example, a trimerisation reaction

$$R_i : \qquad 3X \xrightarrow{c_i} X_3.$$

Taken at face value, the rate constant c_i should represent the hazard of triples of molecules of X coming together simultaneously and reacting, leading to a rate law of the form

$$h(x, c_i) = c_i \binom{x}{3} = c_i \frac{x!}{(x-3)!3!} = c_i \frac{x(x-1)(x-2)}{6}.$$

However, in most cases it is likely to be more realistic to model the process as the pair of second-order reactions

$$2X \longrightarrow X_2$$
$$X_2 + X \longrightarrow X_3,$$

and this pair of second-order reactions may have quite different dynamics to the corresponding third-order system.

6.4 The Gillespie algorithm

The discussion in the previous sections shows that the time-evolution of a reaction system can be regarded as a stochastic process. Further, due to the fact that the reaction hazards depend only on the current state of the system (the number of molecules of each type), it is clear that the time-evolution of the state of the reaction system can be regarded as a continuous time Markov process with a discrete state space. Detailed mathematical analysis of such systems is usually intractable, but stochastic simulation of the time-evolution of the system is quite straightforward.

In a given reaction system with v reactions, we know that the hazard for a type i reaction is $h_i(x, c_i)$, so the hazard for a reaction of some type occurring is

$$h_0(x, c) \equiv \sum_{i=1}^{v} h_i(x, c_i).$$

We now follow the discrete event stochastic simulation procedure (Kendall, 1950) discussed in Section 5.4.2 to update the state of the process. It is clear that the time to the next reaction is $Exp(h_0(x, c))$, and also that this reaction will be a random type, picked with probabilities proportional to the $h_i(x, c_i)$, independent of the time to the next event. That is, the reaction type will be i with probability $h_i(x, c_i)/h_0(x, c)$. Using the time to the next event and the event type, the state of the system can be updated, and simulation can continue. In the context of chemical kinetics, this standard discrete event simulation procedure is known as 'the Gillespie algorithm' (or 'Gillespie's direct method'), after Gillespie (1977).[†] The algorithm can be summarised as follows:

The Gillespie algorithm

1. Initialise the system at $t = 0$ with rate constants $c_1, c_2, \ldots, c_v$ and initial numbers of molecules for each species, $x_1, x_2, \ldots, x_u$.

2. For each $i = 1, 2, \ldots, v$, calculate $h_i(x, c_i)$ based on the current state, x.

3. Calculate $h_0(x, c) \equiv \sum_{i=1}^{v} h_i(x, c_i)$, the combined reaction hazard.

4. Simulate time to next event, t', as an $Exp(h_0(x, c))$ random quantity.

5. Put $t := t + t'$.

6. Simulate the reaction index, j, as a discrete random quantity with probabilities $h_i(x, c_i) / h_0(x, c)$, $i = 1, 2, \ldots, v$.

7. Update x according to reaction j. That is, put $x := x + S^{(j)}$, where $S^{(j)}$ denotes the jth column of the stoichiometry matrix S.

8. Output x and t.

9. If $t < T_{max}$, return to step 2.

We will examine different ways of turning this algorithm into R code in the following sections. Note that Step 6 is usually executed via some kind of 'lookup method'. Efficient implementation of this step is crucial to obtaining a simulation algorithm which performs well — see the discussion in Section 4.5 for further details.

6.5 Stochastic Petri nets (SPNs)

A *stochastic Petri net* (SPN) is a Petri net where the state (represented by the number of tokens at each node) changes dynamically and randomly by choosing event firings in a carefully prescribed random manner (Goss and Peccoud, 1998).

If the Petri net is used to describe a chemical reaction network (Section 2.3), where the number of tokens at a node represents the number of molecules of a given type, then the rates of event firings are given by stochastic rate laws, and the Gillespie algorithm can be used to determine the time to the next event firing and which event to fire. So an SPN is simply a convenient mathematical and graphical representation of a stochastic kinetic process.

A very simple R function for simulating the time-evolution of an SPN using the Gillespie algorithm is given in Figure 6.5. To illustrate its use, consider the stochastic kinetic formulation of the Lotka–Volterra system.

[†] This classic paper is still worth reading.

```
gillespie <- function(N, n, ...)
{
        tt = 0
        x = N$M
        S = t(N$Post-N$Pre)
        u = nrow(S)
        v = ncol(S)
        tvec = vector("numeric",n)
        xmat = matrix(ncol=u,nrow=n+1)
        xmat[1,] = x
        for (i in 1:n) {
                h = N$h(x,tt, ...)
                tt = tt+rexp(1,sum(h))
                j = sample(v,1,prob=h)
                x = x+S[,j]
                tvec[i] = tt
                xmat[i+1,] = x
        }
        return(list(t=tvec, x=xmat))
}
```

Figure 6.5 *An R function to implement the Gillespie algorithm for a stochastic Petri net representation of a coupled chemical reaction system.*

Here we will use the usual equations

$$Y_1 \xrightarrow{c_1} 2Y_1$$
$$Y_1 + Y_2 \xrightarrow{c_2} 2Y_2$$
$$Y_2 \xrightarrow{c_3} \emptyset$$

leading to stochastic rate laws

$$h_1(y, c_1) = c_1 y_1$$
$$h_2(y, c_2) = c_2 y_1 y_2$$
$$h_3(y, c_3) = c_3 y_2.$$

The SPN corresponding to this system could be written

$$N = (P, T, Pre, Post, M, h, c), \quad P = (\text{Prey}, \text{Predator})^\mathsf{T},$$

$$T = (\text{Prey reproduction}, \text{Predator–prey interaction}, \text{Predator death})^\mathsf{T},$$

$$Pre = \begin{pmatrix} 1 & 0 \\ 1 & 1 \\ 0 & 1 \end{pmatrix}, \quad Post = \begin{pmatrix} 2 & 0 \\ 0 & 2 \\ 0 & 0 \end{pmatrix}, \quad h(y, c) = (c_1 y_1, c_2 y_1 y_2, c_3 y_2)^\mathsf{T}.$$

It remains only to specify the initial state of the system, M, and the vector of rate

```
N=list()
N$M=c(x1=50,x2=100)
N$Pre=matrix(c(1,0,1,1,0,1),ncol=2,byrow=TRUE)
N$Post=matrix(c(2,0,0,2,0,0),ncol=2,byrow=TRUE)
N$h=function(x,t,th=c(th1=1,th2=0.005,th3=0.6))
{
  with(as.list(c(x,th)),{
          return(c(th1*x1, th2*x1*x2, th3*x2 ))
          })
}

# simulate a realisation of the process and plot it
out = gillespie(N,10000)
op = par(mfrow=c(2,2))
plot(stepfun(out$t,out$x[,1]),pch="")
plot(stepfun(out$t,out$x[,2]),pch="")
plot(out$x,type="l")
par(op)
```

Figure 6.6 *Some R code to set up the LV system as an SPN and then simulate it using the Gillespie algorithm. The state of the system is initialised to 50 prey and 100 predators, and the stochastic rate constants are $c = (1, 0.005, 0.6)^T$.*

constants, c. Some R code that formulates this problem as an SPN and then simulates it using the Gillespie algorithm assuming initial state $M = (50, 100)^T$ and stochastic rate constants $c = (1, 0.005, 0.6)^T$ is given in Figure 6.6. Note that the vector of rate constants is called th (for θ) rather than c, as c has special meaning in R. The output from the Gillespie algorithm consists of a list containing two items. The first item, t, is a vector of event times, and the second item, x, is a matrix whose rows represent the state of the system immediately *before* the corresponding event time. The matrix x therefore has an additional row, corresponding to the state of the system immediately *after* the final event simulated. As shown in the Figure 6.6, this output can be used to construct R 'step function' objects which can be plotted. A single realisation of this process is shown in Figure 6.7 and Figure 6.8. These should be contrasted with the corresponding deterministic kinetics (Figure 6.1 and Figure 6.2). It is clear that although the stochastic solution approximately follows the path of the deterministic phase-space orbits, it is not constrained to follow them slavishly, but rather free to wander to nearby orbits in a stochastic manner.

The SBML-shorthand for this model is given in Figure 6.9, and the full SBML is listed in Appendix A.2 (as well as on this book's website). smfsbSBML is an R package for parsing an SBML model into an R SPN object for simulation and analysis, and is described in Appendix B.3. Other software suitable for taking the SBML as input and simulating the system dynamics will be discussed in Section 6.9.

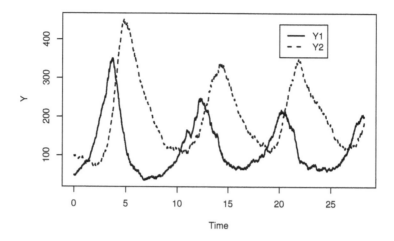

Figure 6.7 *A single realisation of a stochastic LV process. The state of the system is initialised to 50 prey and 100 predators, and the stochastic rate constants are $c = (1, 0.005, 0.6)^T$.*

6.6 Structuring stochastic simulation codes

Although it is sometimes useful to have output corresponding to each event that occurs in the simulation of the reaction network, often this is not desirable. This is because for systems of realistic size and complexity, there will be a very large number of events, and all that is likely to be of interest is the state of the system on a relatively fine regular grid of time points. One approach to this problem is to time-discretise the discrete-event output *post-hoc*. An R function to accomplish this is given in Figure 6.10. This could be used with a command such as **plot** (discretise(out, dt =0.01)), where out is the result of a call to the gillespie function. However, even this solution turns out to be slightly unwieldy in practice. In particular, one cannot know in advance how many reaction events correspond to a particular length of simulation time. For this reason (and others, relating to code efficiency), it is usually better to time-discretise the process as the simulation algorithm progresses. This turns out to be relatively straightforward, but there are various ways to do it. The obvious way is to run a Gillespie algorithm for the full duration required and carefully record the state at pre-specified times. A discretised version of gillespie, called gillespied, is given in Figure 6.11. It can be called with a pre-defined SPN with a command such as **plot** (gillespied(N, T=100, dt=0.01)). Note that the function gillespied also contains a small amount of code to gracefully handle the case of the reaction network going 'extinct,' that is, reaching a point where no more reactions will occur. The previous function (Figure 6.5) would fail in that case, but can be easily modified to return something sensible in this eventuality.

In fact, this approach too has limitations. For example, what if the output is re-

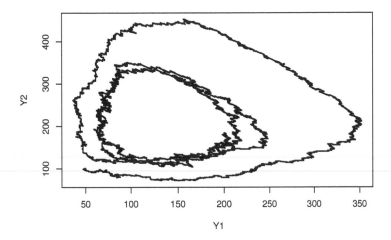

Figure 6.8 *A single realisation of a stochastic LV process in phase-space. The state of the system is initialised to 50 prey and 100 predators, and the stochastic rate constants are $c = (1, 0.005, 0.6)^T$.*

quired on an irregular time grid? It is useful to be able to separate the code for advancing the state of the simulation from code associated with recording the output. There are many ways to achieve this, but most 'clean' methods require some kind of information 'hiding' approach, using ideas from either object-oriented or functional programming. In R it is possible to implement this idea nicely by using *function closures*. Function closures are straightforward to implement in R, due to its being lexically scoped, and having functions as first-class objects. In languages which don't directly support function closures, the same effect can be achieved using objects and methods. The idea is to create a function which advances the state of the process, which can then be used by other functions that do not need to know any details of the model or the simulation algorithm. The idea is most easily explained by example. The function StepGillespie is shown in Figure 6.12. This function accepts an SPN, and returns a function (closure) which can be used to advance the state of the SPN using the Gillespie algorithm. Because R is lexically scoped, the returned function will be able to access the variables relating to the SPN that were in scope at the time the function was created. The returned function has as inputs an initial state and time, together with the amount of time by which the process should be advanced using the Gillespie algorithm. This function can be used in many different ways. For example, if output is required on a regular time grid, a function for achieving this, simTs, is given in Figure 6.13. Some code which uses these two functions in order to simulate a realisation from the Lotka–Volterra model we have been considering is given in Figure 6.14. The advantage of this approach is that it separates the simulation code from code which uses the simulation output. For example, the smfsb

```
@model:3.1.1=LotkaVolterra
 s=item, t=second, v=litre, e=item
@compartments
 Cell
@species
 Cell:Prey=50 s
 Cell:Predator=100 s
@reactions
@r=PreyReproduction
 Prey -> 2Prey
 c1*Prey : c1=1
@r=PredatorPreyInteraction
 Prey+Predator -> 2Predator
 c2*Prey*Predator : c2=0.005
@r=PredatorDeath
 Predator ->
 c3*Predator : c3=0.6
```

Figure 6.9 *SBML-shorthand for the stochastic Lotka–Volterra system.*

R package includes a function simTimes which records the output on an irregular grid of times, and another, simSample, which runs the simulation many times with the same starting conditions and records the distribution of the final state. None of these functions need know anything of the structure of the model or the simulation algorithm. In particular, a different simulation algorithm could be used in place of the one implemented by StepGillespie, and functions such as simTs could still be used. The smfsb R package also includes a function StepEulerSPN which accepts an SPN as input, but returns a function for advancing the state of the SPN using a simple continuous deterministic Euler integration rule.[‡] Other exact and approximate simulation methods will be considered in Chapter 8, and they will all be implemented using this method, and hence can all be used in conjunction with functions such as simTs. Thinking back to the theory of Markov processes developed in Chapter 5, this approach corresponds to developing a method for simulating from the transition kernel of the Markov process, with all other details of the Markov process and how to simulate it hidden from subsequent code.

Interestingly, although Petri nets are often used to model concurrent systems acting locally and in parallel (which is entirely appropriate for a reaction network), the Gillespie algorithm is 'global' in the sense that it acts on the network as a whole, and not 'locally' at the level of reaction nodes. So the Gillespie algorithm does not really feel quite right in this context. However, the Gillespie algorithm is just one possible way of carrying out exact stochastic simulation of the underlying Markov process, and some of the other methods are more local in nature and more naturally lead to

---

[‡] Note that I am adopting a convention where any function returning a function closure starts with an uppercase letter, whereas all other functions start with a lowercase letter. This convention makes it slightly easier to read the code examples.

```
discretise <- function (out, dt = 1, start = 0)
{
    events = length (out$t)
    end = out$t [events]
    len = (end - start)%/%dt + 1
    x = matrix (nrow = len, ncol = ncol (out$x))
    target = 0
    j = 1
    for (i in 1:events) {
        while (out$t [i] >= target) {
            x[j, ] = out$x[i, ]
            j = j + 1
            target = target + dt
        }
    }
    ts (x, start = 0, deltat = dt)
}
```

Figure 6.10 *An R function to discretise the output of* gillespie *onto a regular grid of time points. The result is returned as an R multivariate time series object.*

parallel implementations, which in turn fit better with a Petri net formulation of the problem. We will explore some of these alternative simulation strategies in Chapter 8.

6.7 Rate constant conversion

Much of the literature on biochemical reactions is dominated by a continuous deterministic view of kinetics. Consequently, where rate constants are documented, they are usually deterministic rate constants, k. In order to carry out a stochastic simulation, these constants must be converted in an appropriate way to stochastic rate constants, c, representing molecular reaction hazards. To make this conversion, we need to understand the relationship between the deterministic and stochastic kinetic models.

6.7.1 Concentrations to molecule numbers

The first issue that needs to be addressed is the difference in the representation of the *amount* of any species. In the stochastic model, this is an integer representing the number of molecules of the species, but in the deterministic model, it is usually a concentration, measured in M (moles per litre). In order to carry out the conversion from concentration to numbers of molecules, we also need to know the volume of the container, V, measured in litres.

Then for a concentration of X of $[X]$ M in a volume of V litres, there are clearly

```
gillespied <- function (N, T=100, dt=1, ...)
{
        tt = 0
        n = T%/%dt
        x = N$M
        S = t(N$Post-N$Pre)
        u = nrow(S)
        v = ncol(S)
        xmat = matrix(ncol=u, nrow=n)
        i = 1
        target = 0
        repeat {
                h = N$h(x, tt, ...)
                h0 = sum(h)
                if (h0 < 1e-10)
                        tt = 1e99
                else
                        tt = tt+rexp(1,h0)
                while (tt >= target) {
                        xmat[i,] = x
                        i = i+1
                        target = target+dt
                        if (i > n)
                                return(ts(xmat,start=0,
                                   deltat=dt))
                }
                j = sample(v,1,prob=h)
                x = x+S[,j]
        }
}
```

Figure 6.11 *An R function to implement the Gillespie algorithm for an SPN, recording the state on a regular grid of time points. The result is returned as an R multivariate time series object.*

$[X]V$ moles of X and hence $n_A[X]V$ molecules, where $n_A \simeq 6.023 \times 10^{23}$ is Avogadro's constant (the number of molecules in a mole).

Example

Consider the following example, based loosely on an example from Bower and Bolouri (2000). An *E. coli* cell is a rod-shaped bacterium 2μm long with a diam-

```
StepGillespie <- function(N)
{
        S = t(N$Post-N$Pre)
        v = ncol(S)
        return(
                function(x0, t0, deltat, ...)
                {
                        t = t0
                        x = x0
                        termt = t0+deltat
                        repeat {
                                h = N$h(x,t,...)
                                h0 = sum(h)
                                if (h0 < 1e-10)
                                        t = 1e99
                                else if (h0 > 1e6) {
                                        t = 1e99
                                        warning("Hazard
                                                too big -
                                                terminating
                                                simulation!"
                                                )
                                }
                                else
                                        t = t+rexp(1,h0)
                                if (t >= termt)
                                        return(x)
                                j = sample(v,1,prob=h)
                                x = x+S[,j]
                        }
                }
        )
}
```

Figure 6.12 *An R function which accepts as input an SPN, and returns as output a function (closure) for advancing the state of the SPN using the Gillespie algorithm.*

```
simTs <- function(x0, t0=0, tt=100, dt=0.1, stepFun,
    ...)
{
        n = (tt-t0) %/% dt + 1
        u = length(x0)
        names = names(x0)
        mat = matrix(nrow=n, ncol=u)
        x = x0
        t = t0
        mat[1,] = x
        for (i in 2:n) {
                t = t+dt
                x = stepFun(x,t,dt,...)
                mat[i,] = x
        }
        ts(mat, start=t0, deltat=dt, names=names)
}
```

Figure 6.13 *An R function to simulate a process on a regular time grid using a stepping function such as output by* StepGillespie.

```
N=list()
N$Pre=matrix(c(1,0,1,1,0,1),ncol=2,byrow=TRUE)
N$Post=matrix(c(2,0,0,2,0,0),ncol=2,byrow=TRUE)
N$h=function(x,t,th=c(th1=1,th2=0.005,th3=0.6))
{
  with(as.list(c(x,th)),{
        return(c(th1*x1, th2*x1*x2, th3*x2 ))
        })
}
# create a stepping function
stepLV = StepGillespie(N)
# step the function
print(stepLV(c(x1=50,x2=100),0,1))
# simulate a realisation of the process and plot it
out = simTs(c(x1=50,x2=100),0,100,0.1,stepLV)
plot(out)
```

Figure 6.14 *R code showing how to use the functions* StepGillespie *and* simTs *together in order to simulate a realisation from an SPN.*

eter of 1μm. The cell volume is therefore

$$V = \pi r^2 l$$
$$= \pi(0.5 \times 10^{-6})^2(2 \times 10^{-6})$$
$$= \frac{\pi}{2} \times 10^{-18} \text{ m}^3$$
$$= \frac{\pi}{2} \times 10^{-15} \text{ L}.$$

Now if a chemical species X has a concentration $[X] = 10^{-5}$ M within an *E. coli* cell, the number of molecules is

$$n_A[X]V = 6.023 \times 10^{23} \times 10^{-5} \times \frac{\pi}{2} \times 10^{-15} \simeq 9,461.$$

Once we are happy with converting amounts, we can think about converting rate constants.

6.7.2 Zeroth-order

For the reaction

$$\emptyset \longrightarrow X,$$

the deterministic rate law is k Ms^{-1}, and so for a volume V, X is produced at a rate of $n_A k V$ molecules per second. As the stochastic rate law is just c molecules per second, we have

$$c = n_A V k.$$

6.7.3 First-order

For the reaction

$$X \longrightarrow ?,$$

the deterministic rate law is $k[X]$ Ms^{-1}. As this involves $[X]$, we need to know that for a volume V, a concentration of $[X]$ corresponds to $x = n_A[X]V$ molecules. Now since X decreases at rate $n_A k[X]V = kx$ molecules per second, and the stochastic rate law is cx molecules per second, we have

$$c = k.$$

That is, for first-order reactions, the stochastic and deterministic rate constants are always equal.

6.7.4 Second-order

For the reaction

$$X + Y \longrightarrow ?,$$

the deterministic rate law is $k[X][Y]$ Ms^{-1}. Here, for a volume V, the reaction proceeds at a rate of $n_A k[X][Y]V = kxy/(n_AV)$ molecules per second. Since the stochastic rate law is cxy molecules per second, we have

$$c = \frac{k}{n_A V}.$$

We also need to consider dimerisation-style reactions, of the form

$$2X \longrightarrow ?.$$

Here the deterministic rate law is $k[X]^2$, so the concentration of X decreases at rate $n_A 2k[X]^2 V = 2kx^2/(n_A V)$ molecules per second. Now the stochastic rate law is $cx(x-1)/2$ so that molecules of X are consumed at a rate of $cx(x-1)$ molecules per second. Now these two laws do not match, but for large x, $x(x-1)$ can be approximated by x^2, and so to the extent that the kinetics match, we have

$$c = \frac{2k}{n_A V}.$$

Note the additional factor of two in this case.

6.7.5 Higher-order

It should be fairly clear how to extend this analysis to higher-order reactions, but such reactions are not often used in stochastic kinetic models.

6.8 Kolmogorov's equations and other analytic representations

6.8.1 Kolmogorov's equations and the chemical Master equation

In the stochastic kinetics literature, there is often reference to 'the (chemical) master equation'. This seems to be a slightly overused term, sometimes applied to any set of differential equations whose solution gives the full transition probability kernel for the system dynamics. We have already seen sets of differential equations that determine the time evolution of the transition kernel of a Markov process — the Kolmogorov differential equations (5.8, 5.9). It turns out that the set of differential equations most often labelled as the 'master equation' is just Kolmogorov's forward equation for a stochastic kinetic process.

Proposition 6.1 *Kolmogorov's forward equations (5.9) for an SPN can be written in the form*

$$\frac{d}{dt} p(x_0, t_0, x, t) = \sum_{i=1}^{v} \left[h_i(x - S^{(i)}, c_i) p(x_0, t_0, x - S^{(i)}, t) \right.$$

$$\left. - h_i(x, c_i) p(x_0, t_0, x, t) \right], \quad \forall t_0 \in \mathbb{R}, \ x_0, x \in \mathcal{M},$$

where $\mathcal{M}$ is the countable state space of the process (the set of all possible markings of the SPN). This set of differential equations is often referred to as the chemical master equation.

Proof. We start with the forward equation (5.9) for a move from (x_0, t_0) to (x, t), and then expand $q_{x,x}$ as follows,

$$
\begin{aligned}
\frac{d}{dt}p(x_0, t_0, x, t) &= \sum_{\{x' \in \mathcal{M}\}} q_{x',x}p(x_0, t_0, x', t) \\
&= \left[\sum_{\{x' \in \mathcal{M}|x' \neq x\}} q_{x',x}p(x_0, t_0, x', t) \right] + q_{x,x}p(x_0, t_0, x, t) \\
&= \left[\sum_{\{x' \in \mathcal{M}|x' \neq x\}} q_{x',x}p(x_0, t_0, x', t) \right] - p(x_0, t_0, x, t) \sum_{\{x' \in \mathcal{M}|x' \neq x\}} q_{x,x'} \\
&= \sum_{\{x' \in \mathcal{M}|x' \neq x\}} \left[q_{x',x}p(x_0, t_0, x', t) - q_{x,x'}p(x_0, t_0, x, t) \right].
\end{aligned}
$$

This is just Kolmogorov's forward equation rewritten more appropriately for a Markov process with general countable state space $\mathcal{M}$. The equation involves a sum over all possible transitions, but for an SPN, only a finite number of transition events are possible, corresponding to the v different reaction channels. Considering first the hazard $q_{x,x'}$, we note that starting from x, it is only possible to move to $x + S^{(i)}$, $i = 1, 2, \ldots, v$, and then by definition we have $q_{x,x+S^{(i)}} = h_i(x, c_i)$. Similarly, in order to get to x, the process must have come from one of $x - S^{(i)}$, $i = 1, 2, \ldots, v$, and in this case we have $q_{x-S^{(i)},x} = h_i(x - S^{(i)}, c_i)$. Substituting these into the above equation gives the result.
□

Note that dependence of the transition kernel on the initial state x_0 at time t_0 is often dropped, and the chemical master equation is often written for the probability mass function at time t, $p(x, t)$ directly as

$$
\frac{d}{dt}p(x, t) = \sum_{i=1}^{v} \left[h_i(x - S^{(i)}, c_i)p(x - S^{(i)}, t) - h_i(x, c_i)p(x, t) \right]. \tag{6.7}
$$

A more extensive discussion of the chemical master equation can be found in Van Kampen (1992). We will not have much more to say about it, as the cases where it can be solved exactly and explicitly are very few in number. Such special cases are examined in McQuarrie (1967). The cases that can be solved exactly are interesting for a variety of reasons, including the testing of stochastic simulation algorithms. The derivation of the stationary distribution of the immigration-death process in Section 5.4.3 could be described as a 'master equation approach', and we re-examine that example below. We will do something similar with the analysis of stochastic dimerisation kinetics in Chapter 7. Analytic solutions to the master equation can be found in the case of mass-action stochastic kinetic models involving only zeroth- and first-order reactions (Jahnke and Huisinga, 2007). Second-order reactions are problematic, due to the non-linearities that they introduce. Some progress with numerical methods for the solution of simple systems can be made using techniques from theoretical physics (Walczak et al., 2009; Mugler et al., 2009), and approximate solutions may be obtained by using moment closure approaches (Gillespie, 2009; Milner et al., 2011; Schnoerr et al., 2015), which some authors refer to as the 2MA approach (Ullah and Wolkenhauer, 2009) in the special case of two-moment normal closures. In general, however, a master equation approach to the analysis of stochastic kinetic models of realistic size and complexity will not be possible, and then stochastic simulation will be the only practical approach to gaining insight into the system dynamics.

Note that if an equilibrium distribution, $p(x)$, exists for the system, it will be stationary, and hence will satisfy the system of equations

$$\sum_{i=1}^{v}\left[h_i(x - S^{(i)}, c_i)p(x - S^{(i)}) - h_i(x, c_i)p(x)\right] = 0 \quad \forall x \in \mathcal{M}. \tag{6.8}$$

If the state space is infinite, then this will correspond to an infinite set of equations for the stationary PMF. However, the equations often form a set of recurrence relations for the stationary PMF, which are soluble in certain special cases.

Example: Immigration-death process

We now reconsider the immigration-death process from Chapter 5 as a stochastic kinetic model with reactions

$$\emptyset \xrightarrow{\lambda} X$$
$$X \xrightarrow{\mu} \emptyset,$$

from which we obtain $h_1(x, \lambda) = \lambda$, $h_2(x, \mu) = \mu x$ and $S = (1, -1)$. Substituting these into the chemical master equation (6.7) gives

$$\frac{d}{dt}p(x, t) = \lambda[p(x - 1, t) - p(x, t)] + \mu[(x + 1)p(x + 1, t) - x\,p(x, t)]$$
$$= \lambda p(x - 1, t) - (\lambda + \mu x)p(x, t) + \mu(x + 1)p(x + 1, t), \quad x = 0, 1, 2, \ldots$$

Since this system involves only zeroth- and first-order reactions, it is in fact analytically tractable, and for an initial condition of zero at time zero, has a solution of Poisson form for all t (see the end-of-chapter exercises). However, as before we will focus here on the stationary distribution, which is a solution to

$$\lambda p(x - 1) - (\lambda + \mu x)p(x) + \mu(x + 1)p(x + 1) = 0, \quad x = 0, 1, 2, \ldots.$$

This is clearly the same set of equations we found in Chapter 5, and these can be solved to obtain the solution $X \sim Po(\lambda/\mu)$.

Before leaving the master equation, it is instructive to see that it sheds light on the relationship between the continuous deterministic formulation and the expected value of the stochastic kinetic model. In certain special cases these are the same, and we can see this by using the master equation to derive a set of differential equations for the expected value of the stochastic kinetic model as

$$\frac{\partial}{\partial t} E(X_t) = \frac{\partial}{\partial t} \sum_{x \in \mathcal{M}} x \, p(x, t)$$

$$= \sum_{x \in \mathcal{M}} x \frac{\partial}{\partial t} p(x, t)$$

$$= \sum_{x \in \mathcal{M}} x \sum_{i=1}^{v} \left[h_i(x - S^{(i)}, c_i) p(x - S^{(i)}, t) - h_i(x, c_i) p(x, t) \right]$$

$$= \sum_{i=1}^{v} \left[\sum_{x \in \mathcal{M}} x \, h_i(x - S^{(i)}, c_i) p(x - S^{(i)}, t) - \sum_{x \in \mathcal{M}} x \, h_i(x, c_i) p(x, t) \right]$$

$$= \sum_{i=1}^{v} \left[\sum_{x \in \mathcal{M}} (x + S^{(i)}) \, h_i(x, c_i) p(x, t) - \sum_{x \in \mathcal{M}} x \, h_i(x, c_i) p(x, t) \right]$$

$$= \sum_{i=1}^{v} \left[E\left((X_t + S^{(i)}) h_i(X_t, c_i)\right) - E(X_t h_i(X_t, c_i)) \right]$$

$$= \sum_{i=1}^{v} E\left(S^{(i)} h_i(X_t, c_i)\right)$$

$$= \sum_{i=1}^{v} S^{(i)} E(h_i(X_t, c_i)).$$

Now in general it is not possible to solve this set of differential equations directly, but in the case where all reactions have zero- or first-order mass action rate laws, we can use the linearity of expectation to get $E(h_i(X_t, c_i)) = h_i(E(X_t), c_i)$ giving

$$\frac{\partial}{\partial t} E(X_t) = \sum_{i=1}^{v} S^{(i)} h_i(E(X_t), c_i).$$

Putting $y(t) = E(X_t)$ we get

$$\frac{d}{dt} y(t) = \sum_{i=1}^{v} S^{(i)} h_i(y(t), c_i)$$

$$\Rightarrow \frac{d}{dt} y(t) = S h(y(t), c), \tag{6.9}$$

which is just the ODE system for the deterministic model.§ So when all reactions are zero- and first-order mass-action kinetics, the deterministic solution (6.9) will correctly describe the expected value of the stochastic kinetic model. However, it will not give any insight into variability, and in any case, cannot be used to describe the expectation of any system containing second-order reactions. In general, the expected

---

§ Note that this ODE model uses mass rather than concentration units (and the mass is measured in molecules rather than moles), and uses stochastic rate constants.

value of a model containing second order reactions is not described by the solution of the deterministic approximation (see the end-of-chapter exercises).

6.8.2 The random time-change representation

We have examined the classical mathematical representation of a stochastic kinetic model — the chemical master equation. We have also examined the standard computational approach to generating realisations from the process — the Gillespie algorithm. There is a significant disconnect between these two representations. The chemical master equation does not directly relate to sample paths of the stochastic process, and this is a significant drawback in practice. An alternative mathematical representation of this Markov jump process can be constructed, known as the random time-change representation (Kurtz, 1972), which turns out to be very helpful for mathematical analysis of the system. Now for $i = 1, \ldots, v$, define $N_i(t)$ to be the count functions for v independent unit Poisson processes (so that $N_i(t) \sim Po(t), \forall t$). Then, by time-changing the unit Poisson processes so that their rates match those of the reaction processes, using properties of the time-change discussed in Section 5.4.4, we have

$$R_{it} = N_i \left(\int_0^t h_i(X_\tau, c_i) d\tau \right), \quad i = 1, 2, \ldots, v,$$

recalling that R_{it} denotes the number of type i reaction events occurring up to time t. Then by putting $N((t_1, \ldots, t_v)^\mathsf{T}) = (N_1(t_1), \ldots, N_v(t_v))^\mathsf{T}$, we can write the above equations in vector form as

$$R_t = N \left(\int_0^t h(X_\tau, c) d\tau \right),$$

and then apply the state updating equation (6.6) to get:

Proposition 6.2 (random time-change representation) *The sample path, $\{X_t | t \geq 0\}$ of a stochastic kinetic model satisfies the stochastic integral equation*

$$X_t - X_0 = S \, N \left(\int_0^t h(X_\tau, c) d\tau \right). \tag{6.10}$$

Equation (6.10) is known as the random time-change representation *of the Markov jump process.*

This simple mathematical representation of the process is typically an insoluble stochastic integral equation, but nevertheless lends itself to a range of analyses, and in many situations turns out to be more useful than the chemical master equation, due to its direct connection to the sample paths of the process (Anderson and Kurtz, 2015). In particular, the time-change representation is useful for analysing and deriving exact and approximate algorithms for simulating realisations of the stochastic process. It can be used to justify Gillespie's direct method, but also suggests related algorithms, such as Gillespie's first reaction method (Gillespie, 1976) and the next reaction method of Gibson and Bruck (2000). In addition, it provides a natural way of understanding asymptotic properties, scaling limits, and approximations. See Ball et al. (2006) for applications of this representation to analysis of approximate system dynamics. The time-change representation also aids understanding of some fast hybrid stochastic simulation algorithms, such as those described in Haseltine and Rawlings (2002) or Salis and Kaznessis (2005a), and these will be considered further in Chapter 8.

Example: Immigration-death process

Considering again our immigration-death process example, the random time-change representation of the process is given by

$$X_t = X_0 + (1, -1)N\left(\int_0^t \binom{\lambda}{\mu X_\tau} d\tau\right)$$

$$= X_0 + N_1\left(\int_0^t \lambda d\tau\right) - N_2\left(\int_0^t \mu X_\tau d\tau\right)$$

$$= X_0 + N_1\left(\lambda t\right) - N_2\left(\mu \int_0^t X_\tau d\tau\right).$$

Again this can be solved explicitly to give a Poisson transition kernel, but the solution of discrete stochastic integral equations such as this is beyond the scope of this text. However, consideration of the above formula at time t makes it clear that type 1 reactions occur at rate λ and type 2 reactions occur at rate μX_t.

6.9 Software for simulating stochastic kinetic networks

Although it is very instructive to develop simple algorithms for simulating the dynamic evolution of biochemical networks using a high-level language such as R, this approach will not scale well to large, complex networks with many species and many reaction channels. For such models, it will be desirable to encode them in SBML, and then import them into simulation software designed with such models in mind.[¶] Such 'industrial strength' simulators are often developed in fast compiled languages such as C/C++ or Java, and written carefully to be memory efficient, accurate, and fast. A prototype of such a software library, written in Scala, is described in Appendix B.4. This mimics the structure of the `smfsb` R library, and so it should provide a relatively accessible introduction to writing stochastic simulation software in an efficient programming language.

There are many software systems available for simulating the continuous deterministic kinetics corresponding to an SBML model (many such packages are listed on the SBML.org web page). Increasingly, many simulation systems now support discrete stochastic kinetic model simulation as well. Historically, implementation of good support for correct simulation of stochastic kinetic models was poor, and many software tools contained bugs leading to incorrect simulation behaviour. The situation has improved greatly in recent years, partly due to the development of a standardised SBML test suite for discrete stochastic models (Evans et al., 2008). COPASI

[¶] But note that the `smfsbSBML` R package allows the importing of models into R for simulation with the `smfsb` R package.

(Hoops et al., 2006) is an example of a software library containing stochastic simulation code tested against the test suite. There are many other tools, but look for those which claim to pass the SBML discrete stochastic models test suite or, ideally, test the simulator for correctness and accuracy yourself.

6.10 Exercises

1. Define the auto-regulatory network from Chapter 2 (given as SBML in Appendix A.1) as an SPN in R, and then use the functions `StepGillespie` and `simTs` to simulate its time-course behaviour.

2. Write an R function to simulate the time-course behaviour of the LV system 1,000 times and compute the sample mean of the prey and predator numbers at times $1, 2, \ldots, 20$. Plot them. Overlay the deterministic solution, and verify that the expected trajectory of the LV system is not the deterministic solution.

3. For the Lotka–Volterra system, using the parameters and initial conditions from this chapter, modify one of the simulation functions in order to record the time to predator extinction. Use stochastic simulation to investigate the distribution of the time to extinction. What is the mean time to extinction?

4. Install some SBML-compliant stochastic simulation software. Use it to simulate the auto-regulatory network from Appendix A.1. Check that the results seem consistent with those obtained from Exercise 1.

5. Following the guidance in Appendix B.3, install the `smfsbSBML` library, and use it to import the SBML version of the auto-regulatory network model. Use this for simulation and compare your results with those from the previous exercise.

6. Following the guidance in Appendix B.4, install the `scala-smfsb` Scala library, and run the demo code. Then use it to repeat the previous exercise.

7. Simulate the auto-regulatory model using a deterministic simulator and interpret the results.

8. Consider the dimerisation kinetics example with parameters and initial conditions as given in Figure 6.3. Assume that the reaction is taking place in a compartment with volume V. By converting to a stochastic kinetic model and simulating, discover how small the volume V needs to be in order for stochastic effects to become important and prevalent.

9. Write out explicit forms for the chemical master equation and random time-change representation of the LV system.

10. Use the fact that the immigration-death process is linear to show that the expected value of the immigration-death process at time t for an initial condition $x_0 = 0$ is given by

$$E(X_t) = \frac{\lambda}{\mu}(1 - e^{-\mu t}).$$

By substitution into the chemical master equation, show that the distribution of the process at time t is Poisson with this mean.

6.11 Further reading

Cornish-Bowden (2004) is a classic text on modelling biochemical reactions from a continuous deterministic perspective. For numerical techniques for integrating ODEs, start with a basic numerical analysis text such as Burden and Faires (2010). Kitano (2001) and Bower and Bolouri (2000) give a good overview of systems biology and the role that biochemical network modelling and simulation has to play in it. The latter also includes a couple of chapters on stochastic simulation. To understand the role of Dan Gillespie in making physical scientists aware of the utility of stochastic modelling and simulation techniques, it is worth reading Gillespie (1977, 1992a, and 1992b). For more formal analysis of stochastic biochemical networks, see Anderson and Kurtz (2015). Another important book for physical scientists is Van Kampen (1992). Seminal papers on the modelling of stochastic kinetic effects in the regulation of gene expression include McAdams and Arkin (1997) and Arkin et al. (1998). A key paper examining noise in biological systems at the single cell level was Elowitz et al. (2002). For a review of stochasticity in biological systems, see Wilkinson (2009).

CHAPTER 7

Case studies

7.1 Introduction

This chapter will focus on how the theory developed so far can be used in practice by applying it to a range of illustrative examples. The examples are relatively 'small' compared to the kinds of models that systems biologists are mainly interested in (see McAdams and Arkin (1997) and Arkin et al. (1998) for a couple of good early examples), but smaller models tend to be more effective for elucidating key principles. Although the examples themselves are interesting in their own right, each one will be used to address particular modelling issues that arise in practice. So Section 7.2 will illustrate the conversion of a deterministic model to a stochastic model, and the visualisation, summarisation, and analysis of the output of stochastic simulators. Section 7.3 will discuss conservation laws and dimensionality reduction, Section 7.4 will illustrate sensitivity and uncertainty analysis for stochastic models, and introduce the idea of modelling cell populations. Section 7.5 will examine the analysis of external interventions. Confusingly, external interventions are referred to as discrete *events* in the SBML world, but this has nothing to do with the discrete event simulation methods we have been considering.

7.2 Dimerisation kinetics

Let us consider in greater detail the problem of dimerisation kinetics briefly examined in Section 6.1.4. For this problem we will consider the dimerisation kinetics of a protein, P, at very low concentrations in a bacterial cell. We will begin by considering the usual continuous deterministic kinetics and then go on to examine the corresponding stochastic kinetic behaviour of the system. The forward and backward reactions, respectively, are

$$2P \xrightarrow{k_1} P_2, \quad \text{and} \quad P_2 \xrightarrow{k_2} 2P,$$

where k_1 and k_2 denote the usual deterministic mass-action kinetic rate constants, leading to the ordinary differential equations (6.1) for the time evolution of the system. We will assume an initial concentration of p_0 M (moles per litre) for P at time $t = 0$, and an initial concentration of 0 for P_2. Although knowing the volume of the container (in this case the bacterium) is not strictly necessary for either a deterministic or stochastic analysis, it is required in order to compare the two. So here we assume a volume of V L (litres, or dm^3).

For the particular problem we are interested in, the initial concentration of P is 0.5 μM, giving $p_0 = 5 \times 10^{-7}$, and the volume of the bacterium is $V = 10^{-15}$. It will

```
@model:3.1.1=DimerKineticsDet "Dimerisation Kinetics (
    deterministic)"
 s=mole, t=second, v=litre, e=mole
@compartments
 Cell=1e-15
@species
 Cell:[P]=5e-7
 Cell:[P2]=0
@reactions
@r=Dimerisation
 2P->P2
 Cell*k1*P*P : k1=5e5
@r=Dissociation
 P2->2P
 Cell*k2*P2 : k2=0.2
```

Figure 7.1 *SBML-shorthand for the dimerisation kinetics model (continuous deterministic version).*

be assumed that the values of k_1 and k_2 have been determined from a macroscopic experiment and found to be $k_1 = 5 \times 10^5$, $k_2 = 0.2$. The SBML-shorthand that encodes this model is given in Figure 7.1 (and the full SBML is listed in Appendix A.3). Note the additional factor of Cell in the rate laws. This is interpreted as the volume of the container and is necessary because SBML rate laws are expected to be in units of substance (here moles) per unit time, and not concentration per unit time, which is how continuous deterministic rate laws are traditionally written.

The dynamics associated with this model can be simulated either by integrating the ODEs directly or by using an SBML-compliant simulator to give the dynamics shown in Figure 7.2 (left). However, we know from the stochastic kinetic theory developed in the previous chapter that for reactions involving species at low concentration in small volumes, the continuous deterministic formulation is a poor approximation to the true stochastic kinetic behaviour of the system. We therefore now turn our attention to recasting the above model in the stochastic kinetic framework in order to be able to study it from this perspective.

Algebraically, the initial amount of P is $n_A p_0 V$ molecules, and the stochastic rate constants are obtained as $c_1 = 2k_1/(n_A V)$, $c_2 = k_2$, using the results from Section 6.7. For our particular constants, this gives an initial value of 301 molecules of P (to the nearest integer), zero molecules of P_2, and stochastic rate constants $c_1 = 1.66 \times 10^{-3}$, $c_2 = 0.2$. The SBML-shorthand encoding of the discrete stochastic version of this model is given in Figure 7.3 (and the full SBML is listed in Appendix A.3.2). Note that in principle it should be possible for an SBML-aware software tool to automatically convert the SBML in Appendix A.3.2 into the SBML listed in Appendix A.3.1 (and possibly even the reverse, though this is harder). However, careful study of the two models reveals that this is not quite as trivial as it might first seem, as it will require the tool to have a fairly deep understanding of SBML

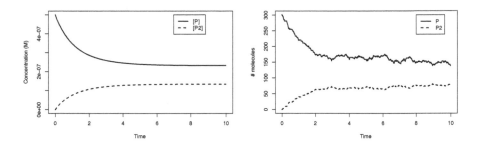

Figure 7.2 *Left: Simulated continuous deterministic dynamics of the dimerisation kinetics model. Right: A simulated realisation of the discrete stochastic dynamics of the dimerisation kinetics model.*

```
@model:3.1.1=DimerKineticsStoch "Dimerisation Kinetics (
    stochastic)"
 s=item, t=second, v=litre, e=item
@compartments
 Cell=1e-15
@species
 Cell:P=301 s
 Cell:P2=0 s
@reactions
@r=Dimerisation
 2P->P2
 c1*P*(P-1)/2 : c1=1.66e-3
@r=Dissociation
 P2->2P
 c2*P2 : c2=0.2
```

Figure 7.3 *SBML-shorthand for the dimerisation kinetics model (discrete stochastic version).*

units and semantics, as well as the relation between deterministic and stochastic rate laws, and the ability to recognise mass-action rate laws (possibly written in slightly different ways). Consequently, at the time of writing, it is typically easier to make the conversion by hand. It is also worth noting that there is nothing particularly discrete or stochastic about the discrete stochastic version of the model. A continuous deterministic simulator with a good understanding of SBML units should be able to correctly simulate the dynamics of the model described in Figure 7.3 to give output similar to that shown in Figure 7.2 (left).

We now have the model encoded in a suitable format for studying the stochastic dynamics. We will see shortly that this model is analytically/numerically tractable. However, we will ignore this fact for the present and instead use stochastic simulation as our primary investigative tool. The dynamics can be simulated using an SBML-

```
Dimer=list()
Dimer$Pre=matrix(c(2,0,0,1),ncol=2,byrow=TRUE)
Dimer$Post=matrix(c(0,1,2,0),ncol=2,byrow=TRUE)
Dimer$h=function(x,t,th=c(th1=1.66e-3,th2=0.2)) {
  with(as.list(c(x,th)),{
        return(c(th1*x1*(x1-1)/2, th2*x2))
        })
    }
```

Figure 7.4 *R code to build an SPN object representing the dimerisation kinetics model.*

aware stochastic simulator, or by reading it into R with the `smfsbSBML` package and using the R functions from Chapter 6. An appropriate SPN object for R can be built with the commands given in Figure 7.4 (and this model is included as a sample 'data set' in the `smfsb` R package).

Regardless of the tool used to conduct the simulation, the results should be the same. A single realisation of the process is given in Figure 7.2 (right). Note again that a different realisation will be obtained each time the simulation is run (provided the random number seed is not fixed). Comparing this to Figure 7.2 (left) it is clear that the qualitative behaviour is somewhat similar, but that the stochastic fluctuations are very pronounced. Again it must be emphasised that these fluctuations are intrinsic to the system and have nothing to do with experimental measurement error (which we have not yet considered at all). Obviously the scales are different in the two cases, but if desired, the stochastic output can easily be mapped onto a concentration scale by dividing through by $n_A V$. A different realisation of the process is shown on this scale in Figure 7.5 (left). These two realisations are not really sufficient to give good insight into the range of behaviour that the model is likely to exhibit. This insight is typically obtained by running the simulation model many times and summarising the output in a sensible way. If we focus now on P (remember that P and P_2 are deterministically related, so it is not necessary to consider both), we begin to understand the range of dynamics it exhibits by running a relatively small number of simulations and overlaying the trajectories (Figure 7.5, right). A more sophisticated approach is to carry out many runs of the simulator and summarise the *distribution* of the level by computing appropriate statistics (such as the sample mean and variance). These can then be used to produce a plot such as Figure 7.6 (left), which is based on 1,000 runs of the simulator. Note that although it is the case here that the results of the deterministic analysis are consistent with the mean of the stochastic kinetic study, this is not true in general (and even here, the mean of the stochastic process is not exactly the deterministic solution, but it provides a reasonable approximation). In general the deterministic analysis provides no useful information about the stochastic kinetic analysis. If the marginal distribution at each time point was normal (which clearly is not the case, but is often a reasonable approximation), we would then expect over 99% of realisations to lie within 3 standard deviations of the mean. It therefore seems reasonable to use the sample mean plus/minus 3 sample standard deviations

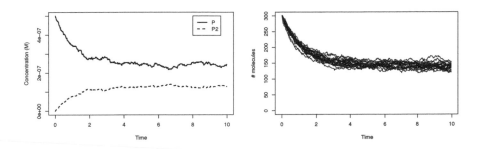

Figure 7.5 *Left: A simulated realisation of the discrete stochastic dynamics of the dimerisation kinetics model plotted on a concentration scale. Right: The trajectories for levels of P from 20 runs overlaid.*

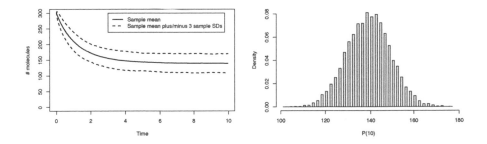

Figure 7.6 *Left: The mean trajectory of P together with some approximate (point-wise) 'confidence bounds' based on 1,000 runs of the simulator. Right: Density histogram of the simulated realisations of P at time $t = 10$ based on 10,000 runs, giving an estimate of the PMF for $P(10)$.*

as a guide to the range of likely values at each time point. For a more detailed understanding of the distribution of values at particular time points, it is possible to study the full set of realisations at a given time. For example, Figure 7.6 (right) shows the empirical PMF for the distribution of P at time $t = 10$. Note the discrete nature of the distribution and the fact that only odd-numbered values are possible (the process was initialised at an odd number of molecules and can only change by a multiple of two). Also note how 'normal' the distribution of realisations appears to be, justifying the use of a 'mean plus/minus 3 SD' approach to summarising the output. Also note that Figure 7.6 (left) suggests that the system seems to have converged to a stationary distribution at time $t = 10$, and hence Figure 7.6 (right) is essentially the equilibrium distribution of the process.*

---

* A demo script which shows how to produce plots similar to those in Figure 7.5 and Figure 7.6 is included as part of the smfsb R package. It can be run by typing **demo** (DimerisationKinetics) after loading the package.

For a problem as simple as this one, it is possible to make a direct analytic attack on the equilibrium probability distribution. In this case, it is made most straightforward by using the deterministic relationship between the number of molecules of P and P_2 to reduce the 2d state-space to a 1d state-space, which can then be analysed in a similar way to the immigration-death model from Chapter 5. So to start, we put

$$P + 2P_2 = n,$$

where n is the number of molecules of P that would be present if they were fully disassociated (so $n = 301$ in our example). Next we can rewrite the rate laws in terms of P_2 only, which we will denote by x,

$$h_1(x, c_1) = c_1 \frac{(n - 2x)(n - 2x - 1)}{2}$$

$$h_2(x, c_2) = c_2 x$$

$$h_0(x, c) = c_1 \frac{(n - 2x)(n - 2x - 1)}{2} + c_2 x.$$

Since reaction 1 corresponds to increasing x by 1 and reaction 2 corresponds to decreasing x by 1, the transition rate matrix Q again has a tri-diagonal form. Note that here the state space (and corresponding Q matrix) is finite, as the size of x cannot exceed $n/2$. Numerical or analytical analysis of this matrix can then yield information regarding the dynamic behaviour of the system. This is essentially what physical scientists would refer to as a 'master equation approach' (although they probably would not approach the problem in exactly this way). However, the actual analysis is somewhat technical and not particularly relevant in the context of this book. It is important to bear in mind that although analytic analysis of simple processes is intellectually attractive and can sometimes give insight into more complex problems, the class of models where analytic approaches are possible is very restricted and does not cover any models of serious interest in the context of systems biology (where we are typically interested in the complex interactions between several intricate mechanisms). Therefore, computationally intensive study based on stochastic simulation and analysis is the only realistic way to gain insight into system dynamics in general.

7.3 Michaelis–Menten enzyme kinetics

Another reaction system worthy of special study is the Michaelis–Menten enzyme kinetic system (Cornish-Bowden, 2004). Here a substrate S is converted to a product P only in the presence of a catalyst E (for enzyme). A plausible model for this is

$$S + E \xrightarrow{k_1} SE$$

$$SE \xrightarrow{k_2} S + E$$

$$SE \xrightarrow{k_3} P + E.$$

If we assume that k_1, k_2, and k_3 are deterministic mass-action kinetic rate constants, then the ODEs governing the deterministic dynamics are

$$\frac{d[S]}{dt} = k_2[SE] - k_1[S][E]$$

$$\frac{d[E]}{dt} = (k_2 + k_3)[SE] - k_1[S][E]$$

$$\frac{d[SE]}{dt} = k_1[S][E] - (k_2 + k_3)[SE]$$

$$\frac{d[P]}{dt} = k_3[SE].$$

This ODE system is most easily constructed from its matrix representation,

$$\frac{d}{dt} \begin{pmatrix} [S] \\ [E] \\ [SE] \\ [P] \end{pmatrix} = \begin{pmatrix} -1 & 1 & 0 \\ -1 & 1 & 1 \\ 1 & -1 & -1 \\ 0 & 0 & 1 \end{pmatrix} \begin{pmatrix} k_1[S][E] \\ k_2[SE] \\ k_3[SE] \end{pmatrix}.$$

For a given set of rate constants and initial conditions, we can integrate this system numerically on a computer.

Again we will assume the setting of low concentrations in small volumes. The compartmental volume is $V = 10^{-15}$L, the initial concentrations of S and E will be 5×10^{-7}M and 2×10^{-7}M, respectively, and the initial concentrations of SE and P will be zero. The model specification is completed with the three rate constants $k_1 = 1 \times 10^6$, $k_2 = 1 \times 10^{-4}$, $k_3 = 0.1$. The SBML-shorthand for this model is given in Figure 7.7. The simulated dynamics for this model are shown in Figure 7.8 (left).

It is clear from this plot that there are conservation laws in this system (it is particularly clear that the sum of $[E]$ and $[SE]$ is constant). Such laws can be used to reduce the dimensionality of the system under consideration. Recalling the Petri net theory from Chapter 2, the reaction matrix has the form

$$A = \begin{pmatrix} -1 & -1 & 1 & 0 \\ 1 & 1 & -1 & 0 \\ 0 & 1 & -1 & 1 \end{pmatrix}.$$

It is clear that this matrix is rank deficient (the first two rows sum to zero), so to find the P-invariants we just need to solve the linear system

$$\begin{pmatrix} 1 & 1 & -1 & 0 \\ 0 & 1 & -1 & 1 \end{pmatrix} y = 0.$$

Straightforward Gaussian elimination leads to the two invariants $y = (0, 1, 1, 0)^{\mathsf{T}}$ and $(1, 0, 1, 1)^{\mathsf{T}}$, corresponding to the conservation laws

$$[E] + [SE] = e_0$$
$$[S] + [SE] + [P] = s_0,$$

where the conservation constants are determined from the initial conditions of the

```
@model:3.1.1=MMKineticsDet "Michaelis-Menten Kinetics (
    deterministic)"
 s=mole, t=second, v=litre, e=mole
@compartments
 Cell=1e-15
@species
 Cell:[S]=5e-7
 Cell:[E]=2e-7
 Cell:[SE]=0
 Cell:[P]=0
@reactions
@r=Binding
 S+E->SE
 Cell*k1*S*E : k1=1e6
@r=Dissociation
 SE->S+E
 Cell*k2*SE : k2=1e-4
@r=Conversion
 SE->P+E
 Cell*k3*SE : k3=0.1
```

Figure 7.7 *SBML-shorthand for the Michaelis–Menten kinetics model (continuous deterministic version).*

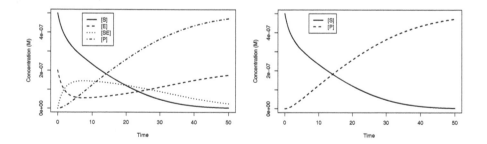

Figure 7.8 *Left: Simulated continuous deterministic dynamics of the Michaelis–Menten kinetics model. Right: Simulated continuous deterministic dynamics of the Michaelis–Menten kinetics model based on the two-dimensional representation.*

system (here, e_0 is the initial value for $[E]$ and s_0 is the initial value for $[S]$). The first conservation law can be used to eliminate $[E]$ from the ODE system, then the second can be used to eliminate $[SE]$, giving

$$\frac{d[S]}{dt} = k_2(s_0 - [S] - [P]) - k_1[S](e_0 - s_0 + [S] + [P])$$
$$\frac{d[P]}{dt} = k_3(s_0 - [S] - [P]).$$

This two-dimensional system of ODEs is exactly equivalent to the original (apparently) four-dimensional system. To confirm this, the simulated dynamics for this system are given in Figure 7.8 (right) (note that the missing components can be easily reconstructed using the conservation laws if required). Such dimension-reduction techniques are particularly important in the continuous-deterministic context. First, mathematical analysis of the system in most cases requires an ODE system of full-rank. Second (of more direct practical relevance), reducing the dimension of the system will improve the speed, accuracy, and general numerical stability of the ODE-integration algorithm.

It is straightforward to convert the Michaelis–Menten system to a discrete stochastic model. The SBML-shorthand for the conversion is given in Figure 7.9, and a single realisation of the process is given in Figure 7.10 (left). Dimensionality reduction techniques can also be used in the context of discrete stochastic models. Again, the Petri net theory can be used to identify the conservation laws of the system, and these can be used to remove E and SE from the model. The SBML-shorthand for the reduced model obtained in this way is given in Figure 7.11, and a single realisation of the process is given in Figure 7.10 (right). It is clear that a software tool could be written to automatically reduce models in this way.[†] This model is somewhat noteworthy in that it is a valid discrete stochastic model with rate laws that are not immediately recognisable as mass-action. Note, however, that it is not good modelling practice to share models that have been reduced in this way. The original version of the model contains more information regarding the true nature of the biochemical network, and sharing the full model allows the end user the choice of whether and how to reduce dimension.

Although dimensionality reduction is clearly applicable in the context of discrete stochastic modelling, it is somewhat less important than in the continuous deterministic case. This is for two main reasons. The first is that the speed improvement obtained by working with the reduced dimension system is not that significant. The second (arguably more fundamental) reason is that exact simulation algorithms such as the Gillespie algorithm are just that — exact. There is therefore no improvement in accuracy or numerical stability to be gained by working with the reduced dimension system. That said, for some of the fast approximate algorithms to be considered in Chapter 8 (notably those that exploit diffusion or ODE approximations), dimensionality reduction is just as important as in the continuous deterministic framework.

[†] Indeed, using the SBML construct of *assignment rules*, it is possible to do this simply by replacing redundant species with appropriate assignment rules based on the conservation laws.

```
@model:3.1.1=MMKineticsStoch "Michaelis-Menten Kinetics (
    stochastic)"
 s=item, t=second, v=litre, e=item
@compartments
 Cell=1e-15
@species
 Cell:S=301 s
 Cell:E=120 s
 Cell:SE=0 s
 Cell:P=0 s
@reactions
@r=Binding
 S+E->SE
 c1*S*E : c1=1.66e-3
@r=Dissociation
 SE->S+E
 c2*SE : c2=1e-4
@r=Conversion
 SE->P+E
 c3*SE : c3=0.1
```

Figure 7.9 *SBML-shorthand for the Michaelis–Menten kinetics model (discrete stochastic version).*

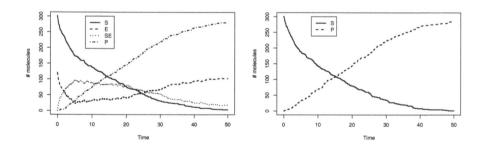

Figure 7.10 *Left: A simulated realisation of the discrete stochastic dynamics of the Michaelis–Menten kinetics model. Right: A simulated realisation of the discrete stochastic dynamics of the reduced-dimension Michaelis–Menten kinetics model.*

7.4 An auto-regulatory genetic network

The dimerisation kinetics and Michaelis–Menten kinetics models are interesting to study from a stochastic viewpoint, but both are somewhat unsatisfactory in the sense that unless the concentrations and volumes involved are really very small, the stochastic fluctuations are not particularly significant. In both cases the continuous deterministic treatment, while clearly an approximation to the truth, actually captures the most

```
@model:3.1.1=RedMMKineticsStoch "Reduced M-M Kinetics (
    stochastic)"
 s=item, t=second, v=litre, e=item
@compartments
 Cell=1e-15
@species
 Cell:S=301 s
 Cell:P=0 s
@reactions
@r=Binding
 S->
 c1*S*(120-301+S+P) : c1=1.66e-3
@r=Dissociation
 ->S
 c2*(301-(S+P)) : c2=1e-4
@r=Conversion
 ->P
 c3*(301-(S+P)) : c3=0.1
```

Figure 7.11 *SBML-shorthand for the reduced dimension Michaelis–Menten kinetics model (discrete stochastic version).*

important aspects of the dynamics reasonably well. If all models of interest to systems biologists were of this nature, it would be quite proper to question whether the additional effort associated with discrete stochastic modelling is worthwhile. However, for any model where there can be only a handful (say, less than ten) of molecules of any of the key reacting species, then stochastic fluctuations can dominate, and the models can (and often do) exhibit behaviour that would be impossible to predict from the associated continuous deterministic analysis. A good example of this is the possibility of the Lotka–Volterra model to go extinct (or explode). Similar things can also happen in the context of molecular cell biology. As a trivial example, random events can trigger apoptosis or other forms of cell death. However, stochastic fluctuations are a normal part of life in the cell, which can have important consequences, and are not just associated with catastrophic events such as cell death.

A good example of a noisy process is gene expression (and its regulation). For this example we will return to the model of prokaryotic gene auto-regulation introduced in Section 1.5.7 and used as the main example throughout Chapter 2. The SBML-shorthand for this model is given in Section 2.6.8, and the full SBML is listed in Appendix A.1. It should be noted that this is an artificial model, with rate constants in arbitrary units, chosen simply to make the model exhibit interesting behaviour. A simulated realisation of this process over a 5,000-second period is shown in Figure 7.12 (left). Only the three key 'outputs' of the model are shown. The discrete bursty stochastic dynamics of the process are clear in this realisation. RNA transcript events are comparatively rare and random in their occurrence. The number of protein monomers oscillates wildly between 10 and 50 molecules, and the number of protein

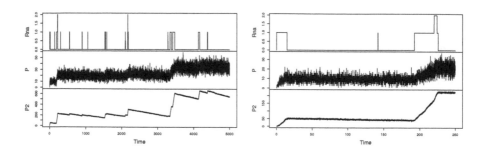

Figure 7.12 *Left: A simulated realisation of the discrete stochastic dynamics of the prokaryotic genetic auto-regulatory network model, for a period of 5,000 seconds. Right: A close-up on the first period of 250 seconds of the left plot.*

dimers jumps abruptly at random times and then gradually decays away. Looking more closely at the first 250 seconds of the same realisation (Figure 7.12, right), it is clear that the jumps in protein dimer levels coincide with the RNA transcript events. This illustrates an important point regarding stochastic variation in complex models: despite the fact that there are a relatively large number of protein dimer molecules, their behaviour is strongly stochastic due to the fact that they are affected by the number of RNA transcripts, and there are very few RNA transcript molecules in the model. Consequently, even if primary interest lies in a species with a relatively large number of molecules, a continuous deterministic model will not adequately capture its behaviour if it is affected by a species which can have a small number of molecules.

Figure 7.13 (left) shows (for the same realisation) the time-evolution of the number of molecules of protein monomers, P, over the first 10 seconds of the simulation. It is clear that even over this short time period the stochastic fluctuations are very significant. In order to understand this variation in more detail, let us now focus on the number of molecules of P at time $t = 10$. By running many simulations of the process it is possible to build up a picture of the probability distribution for the number of molecules, and this is shown in Figure 7.13 (right) (based on 10,000 runs). This clearly shows that there is an almost even chance that there will be no molecules of P at time 10 (and the most likely explanation for this is that there will not yet have been a transcription event). The distribution is clearly far from normal, so a mean plus/minus three SD summary of the distribution is unlikely to be adequate in this case.

Once a model becomes as complex as this one, there is likely to be some uncertainty regarding some quantitative aspects of the model specification. This could be, for example, uncertainty about the initial conditions or the stochastic rate constants adopted. For concreteness, we will suppose here that there is a degree of uncertainty

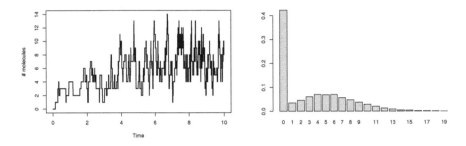

Figure 7.13 *Left: Close-up showing the time-evolution of the number of molecules of P over a 10-second period. Right: Empirical PMF for the number of molecules of P at time $t = 10$ seconds, based on 10,000 runs.*

regarding the value of the gene transcription rate k_2.[‡] A value of $k_2 = 0.01$ was specified in the model, but let us suppose that any value between 0.005 and 0.03 is plausible. It is therefore natural to want to investigate the sensitivity of the model dynamics to this particular specification.

For continuous deterministic models, a sophisticated framework for sensitivity analysis is well established. However, the techniques from this domain do not transfer well to the stochastic modelling paradigm. One of the main motivations for thinking about model sensitivity is a desire to understand uncertainty in the true process dynamics. However, in the context of stochastic modelling, uncertainty about the process dynamics is integral to the whole approach. Consequently, even in the case of complete certainty regarding the model structure, rate laws, rate constants, and initial conditions, the time-evolution of the process is uncertain (or random, or stochastic, depending on choice of terminology). A natural way to incorporate uncertainty regarding model parameters is to adopt a subjective Bayesian interpretation of probability. In the Bayesian paradigm, uncertainty regarding model parameters is not fundamentally different from uncertainty regarding the time evolution of the process due to the stochastic kinetic dynamics. We have already constructed mechanisms for handling uncertainty in the process dynamics by trying to understand the probability distribution of the outcomes, rather than by simply looking at a particular realisation of the process. There is no reason why these probability distributions should not include uncertainty regarding the model parameters in addition to the uncertainty induced by the stochastic kinetics. This is best illustrated by example.

Figure 7.13 (right) shows our uncertainty about the level of P at time $t = 10$ based on a value of $k_2 = 0.01$. Similar plots can be produced based on other plausible values; Figure 7.14 (left) shows the plot corresponding to $k_2 = 0.02$, for example. In order to obtain a summary of our uncertainty regarding the level of P, we need to

[‡] Note that in contradiction to the convention adopted thus far in the book, k_2 is a stochastic rate constant and not a deterministic one. This is because in practice, stochastic modellers often use k rather than c for their rate constants, so it is best not to become over-reliant on this notational cue.

average over our uncertainty for k_2 in an appropriate way. In order to do this properly, we need to specify a probability distribution which reflects our uncertainty in k_2. It was previously stated that all values between 0.005 and 0.03 are plausible. We will now make the much stronger assumption that all values in this range are equally plausible, which leads directly to the probability distribution $U(0.005, 0.03)$. Within the Bayesian framework, this is known as a *prior* probability distribution for a parameter.[§] Having specified the prior probability distribution (in practice, there is likely to be uncertainty regarding several parameters, but it is completely straightforward to assign independent prior probability distributions to each uncertain value), it is then straightforward to incorporate this into the subsequent analysis. Rather than running many simulations with the same parameters, each run begins by first picking uncertain parameters from their prior probability distributions. This has the effect of correctly embedding the parameter uncertainty into the model; all subsequent analysis then proceeds as normal. For example, if interest is in the level of P at time $t = 10$, the values can be recorded at the end of each run to build up a picture of the marginal uncertainty for the level of P. Such a distribution is depicted in Figure 7.14 (right). Unsurprisingly, it looks a bit like a compromise between Figure 7.13 (right) and Figure 7.14 (left).

In practice one is not simply interested in the uncertainty of the level of one particular biochemical species at one particular time, but it is clear that exactly the same approach can be applied to any numerical summary of the simulation output (for example, cell-cycle time, time to cell death, population doubling time, time for RNA expression levels to increase five-fold, etc.). When applied to derived simulation outputs of genuine biological interest (possibly directly experimentally measurable), Bayesian uncertainty analysis provides a powerful framework for model introspection.

7.4.1 Cell population modelling

The discussion in the previous section also provides a framework for beginning to think about modelling (heterogeneous) populations of (independent) cells (Wilkinson, 2009). Each simulated realisation from a model can be thought of as representing the time course trajectory of a single cell within a population. In the case where each cell is started with identical initial conditions and parameters, so that each simulated trajectory is an independent realisation of the same stochastic process, the simulations can be viewed as representing a completely homogeneous, synchronised population of cells. Uncertainty about the initial conditions translates to heterogeneity in the initial conditions across a cell population, perhaps corresponding to an unsynchronised cell population. Uncertainty regarding model parameters can be reinterpreted as a degree of heterogeneity within the cell population, perhaps representing some aspect of extrinsic noise. Of course this approach will only work

---

[§] It is known as a prior distribution because it is possible to use the model and experimental data to update this prior distribution into a *posterior* distribution, which describes the uncertainty regarding the parameter having utilised the information in the data. This is *Bayesian statistics*, and it will be discussed further in the final chapters of this book.

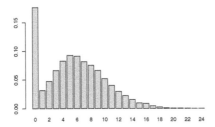 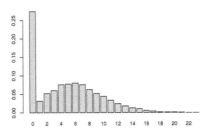

Figure 7.14 *Left: Empirical PMF for the number of molecules of P at time $t = 10$ seconds when k_2 is changed from 0.01 to 0.02, based on 10,000 runs. Right: Empirical PMF for the prior predictive uncertainty regarding the observed value of P at time $t = 10$ based on the prior distribution $k_2 \sim U(0.005, 0.03)$.*

for independently evolving non-dividing cells. Modelling dividing cell populations and cell-to-cell communication requires a more sophisticated approach; see, for example, Lee et al. (2009) and Pahala Gedara et al. (2017).

7.5 The *lac* operon

We now return to the *lac* operon model introduced in Section 1.5.8. Given some appropriate rate constants, we are now in a position to completely specify this model and study its dynamics. Some SBML-shorthand that specifies (a very simplified version of) the model is given in Figure 7.15. The rate constants have been chosen to be biologically plausible (with a time unit of seconds), then fine-tuned to make the model behave sensibly. Again, the model is meant to be illustrative and does not represent a serious effort to accurately model the true dynamics of the *lac* regulation dynamics (or even the actual mechanism, as several simplifying assumptions have been made here as well). Like the auto-regulatory model, from a sensible set of initial conditions, this model will go through a transient phase, then settle down to an equilibrium probability distribution which is not particularly interesting on its own. The most interesting aspect of the *lac* mechanism is the dynamic response to an influx of lactose. Whether or not such an external intervention should be regarded as being part of the model is actually somewhat controversial. However, SBML provides a means for encoding interventions via the event element. SBML events were not discussed in Chapter 2, but the discussion in the SBML specification document is quite readable. Events are also supported in SBML-shorthand — again, see the specification document for further details. The event listed at the end of the shorthand model in Figure 7.15 has the effect of introducing 10,000 molecules of lactose into the cell at time $t = 20,000$.

Conceptually, simulating a model which includes a timed intervention of this kind is quite straightforward. In this particular case, it could be done by running the simulator until time 20,000, then recording the state at this time, adding 10,000 to the

```
@model:3.1.1=lacOperon "lac operon model (stochastic)"
 s=item, t=second, v=litre, e=item
@compartments
 Cell=1e-15
@species
 Cell:Idna=1 s
 Cell:Irna=0 s
 Cell:I=50 s
 Cell:Op=1 s
 Cell:Rnap=100 s
 Cell:Rna=0 s
 Cell:Z=0 s
 Cell:Lactose=20 s
 Cell:ILactose=0 s
 Cell:IOp=0 s
 Cell:RnapOp=0 s
@reactions
@r=InhibitorTranscription
 Idna -> Idna + Irna
 c1*Idna : c1=0.02
@r=InhibitorTranslation
 Irna -> Irna + I
 c2*Irna : c2=0.1
@r=LactoseInhibitorBinding
 I + Lactose -> ILactose
 c3*I*Lactose : c3=0.005
@r=LactoseInhibitorDissociation
 ILactose -> I + Lactose
 c4*ILactose : c4=0.1
@r=InhibitorBinding
 I + Op -> IOp
 c5*I*Op : c5=1
@r=InhibitorDissociation
 IOp -> I + Op
 c6*IOp : c6=0.01
@r=RnapBinding
 Op + Rnap -> RnapOp
 c7*Op*Rnap : c7=0.1
@r=RnapDissociation
 RnapOp -> Op + Rnap
 c8*RnapOp : c8=0.01
@r=Transcription
 RnapOp -> Op + Rnap + Rna
 c9*RnapOp : c9=0.03
@r=Translation
 Rna -> Rna + Z
 c10*Rna : c10=0.1
@r=Conversion
 Lactose + Z -> Z
 c11*Lactose*Z : c11=1e-5
@r=InhibitorRnaDegradation
 Irna ->
 c12*Irna : c12=0.01
@r=InhibitorDegradation
 I ->
 c13*I : c13=0.002
@r=LactoseInhibitorDegradation
 ILactose -> Lactose
 c13*ILactose : c13=0.002
@r=RnaDegradation
 Rna ->
 c14*Rna : c14=0.01
@r=ZDegradation
 Z ->
 c15*Z : c15=0.001
@events
 Intervention = t>=20000 : Lactose=Lactose+10000
```

Figure 7.15 *SBML-shorthand for the lac-operon model (discrete stochastic version).*

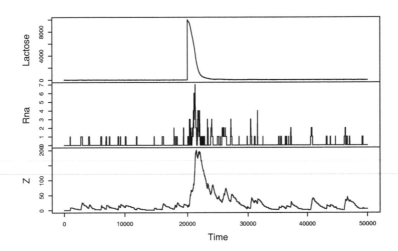

Figure 7.16 *A simulated realisation of the discrete stochastic dynamics of the* lac-*operon model for a period of 50,000 seconds. An intervention is applied at time* $t = 20,000$, *when 10,000 molecules of lactose are added to the cell.*

final level of Lactose, and then restarting the simulator from the new state. In practice, some simulators include built-in support for SBML events, which simplifies the process greatly. A realisation of the process is plotted in Figure 7.16, and shows that under equilibrium conditions the expression level of the *lac-Z* protein is very low. However, in response to the introduction of lactose, the rate of transcription of the *lac* operon is increased, leading to a significant increase in the expression levels of the *lac-Z* protein. The result of this is that the lactose is quickly converted to something else, allowing the cell to gradually return to its equilibrium behaviour.

7.6 Exercises

1. Carry out the simulations described in each section of this chapter and reproduce all of the plots. Of course it will not be possible to *exactly* reproduce the plots showing individual stochastic realisations, but it should be possible to obtain plots showing qualitatively similar behaviour.

2. Use a master equation approach to exactly compute and plot the exact version of Figure 7.6 (right). Also compute the exact equilibrium distribution and compare the two.

3. Identify the conservation laws in the auto-regulatory network and use them to reduce the dimension of the model. Simulate the dynamics of the reduced model to ensure they are consistent with the original.

4. Deduce the continuous deterministic version of the auto-regulatory model (assum-

ing a container volume of 10^{-15}L) and simulate it. Compare it to the stochastic version. How well does it describe the mean behaviour of the stochastic version?

5. In the dimerisation kinetics model, the dissociation rate constant c_2 took the value 0.2. Suppose now that there is uncertainty regarding this value which can be well described by a $U(0.1, 0.4)$ probability distribution. Produce a new version of Figure 7.6 (right) which incorporates this uncertainty.

7.7 Further reading

This chapter provides only the briefest of introductions to stochastic modelling of genetic and biochemical networks, but it should provide sufficient background in order to render the literature in this area more accessible. For a more extensive review, see Wilkinson (2009). To build large and complex stochastic models of interesting biological processes, it is necessary to read around the deterministic modelling literature in addition to the stochastic literature, as the deterministic literature is much more extensive, and can still provide useful information for building stochastic models. Good starting points include Bower and Bolouri (2000), Kitano (2001), and Klipp et al. (2005); also see the references therein. The existing literature on stochastic modelling is of course particularly valuable, and I have found the following articles especially interesting: McAdams and Arkin (1997), Arkin et al. (1998), McAdams and Arkin (1999), Goss and Peccoud (1998), Pinney et al. (2003), Hardy and Robillard (2004), Salis and Kaznessis (2005b), Perkins and Swain (2009), Paszek et al. (2010). Readers with an interest in stochastic modelling of ageing mechanisms may also be interested in Proctor et al. (2005). Information on cell population modelling can be found in Lee et al. (2009). In addition to the modelling literature, it is invariably necessary to trawl the wet-biology literature and on-line databases for information on mechanisms and kinetics — mastering this process is left as an additional exercise for the reader, though some guidance is provided in Klipp et al. (2005).

CHAPTER 8

Beyond the Gillespie algorithm

8.1 Introduction

In Chapter 6 an algorithm for simulating the time-course behaviour of a stochastic kinetic model was introduced. This discrete-event simulation algorithm, usually referred to as the Gillespie algorithm, has the nice properties that it simulates every reaction event and is exact in the sense that it generates exact independent realisations of the underlying stochastic kinetic model. It is also reasonably efficient in terms of computation time among all such algorithms with those properties. However, it should be emphasised that the Gillespie algorithm is just one approach out of many that could be taken to simulating stochastic biochemical dynamics. In this chapter we will look at some other possible approaches, motivated by problems in applying the Gillespie algorithm to large models containing species with large numbers of molecules and fast reactions. In Section 8.2, refinements of the Gillespie algorithm will be considered which still generate exact realisations of the stochastic kinetic process. In Section 8.3, methods based on approximating the process by another that is faster to simulate will be examined. Then in Section 8.4, hybrid algorithms which combine both exact and approximate updating strategies will be considered.

8.2 Exact simulation methods

Let us begin by reconsidering discrete event simulation of a continuous time Markov process with a finite number of states, as described in Section 5.4 (p. 139). We saw how to do this using the *direct method* in Section 5.4.2. There is an alternative algorithm to this, known as the *first event method*, which also generates an exact realisation of the Markov process. It can be stated as follows:

1. Initialise the process at $t = 0$ with initial state i.
2. Call the current state i. For each potential other state k ($k \neq i$), compute a putative time to transition to that state $t_k \sim Exp(q_{ik})$.
3. Let j be the index of the smallest of the t_k.
4. Put $t := t + t_j$.
5. Let the new state be j.
6. Output the time t and state j.
7. If $t < T_{max}$, return to step 2.

It is perhaps not immediately obvious that this algorithm is exactly equivalent to the direct method. In order to see that it is, we need to recall some relevant properties of the exponential distribution, presented in Section 3.9 (p. 87). By Proposition 3.18, the minimum of the putative times has the correct distribution for the time to the first event, and by Proposition 3.19, the index associated with that minimum time clearly has the correct probability mass function (PMF).

Although the first event method is just as correct as the direct method, the direct method is generally to be preferred, as it is more efficient. In particular, the direct method requires just two random numbers to be simulated per event, whereas the first event method requires $r - 1$ (where r is the number of states). The first event method is interesting, however, as it gives us another way of thinking about the simulation of the process, and forms the basis of a very efficient exact simulation algorithm for stochastic kinetic models.

8.2.1 First reaction method

Before looking at the Gibson–Bruck algorithm in detail, it is worth thinking briefly about how the first event method for finite state Markov processes translates to stochastic kinetic models with a potentially infinite state space. We have already seen how the direct method translates into the Gillespie algorithm (Section 6.4). The first event method translates into a variant of the Gillespie algorithm known as the *first reaction method* (Gillespie, 1976), which can be stated as follows:

1. Initialise the starting point of the simulation with $t := 0$, rate constants $c = (c_1, \ldots, c_v)$ and initial state $x = (x_1, \ldots, x_u)$.
2. Calculate the reaction hazards $h_i(x, c_i)$, $i = 1, 2, \ldots, v$.
3. Simulate a putative time to the next type i reaction, $t_i \sim Exp(h_i(x, c_i))$, $i = 1, 2, \ldots, v$.
4. Let j be the index of the smallest t_i.
5. Put $t := t + t_j$.
6. Update the state x according to the reaction with index j. That is, set $x := x + S^{(j)}$.
7. Output t and x.
8. If $t < T_{max}$ return to step 2.

As stated, this algorithm is clearly less efficient than Gillespie's direct method, but with a few clever tricks it can be turned into a very efficient algorithm. An R function to implement the first reaction method for an SPN is given in Figure 8.1.

8.2.2 The Gibson–Bruck algorithm

The so-called *next reaction method* (also known as the Gibson–Bruck algorithm) is a modification of the first reaction method which makes it much more efficient. A first attempt at presenting the basics of the Gibson–Bruck algorithm follows:

1. Initialise $t := 0$, c and x, and additionally calculate all of the initial reaction hazards $h_i(x, c_i)$, $i = 1, \ldots, v$. Use these hazards to simulate putative first reaction times $t_i \sim Exp(h_i(x, c_i))$.

```
StepFRM <- function (N)
{
        S = t(N$Post-N$Pre)
        v = ncol(S)
        return (
                function (x0, t0, deltat, ...)
                {
                        x = x0
                        t = t0
                        termt = t0+deltat
                        repeat {
                                h = N$h(x, t, ...)
                                pu = rexp (v,h)
                                j = which.min (pu)
                                t = t+pu[j]
                                if (t >= termt)
                                        return (x)
                                x = x+S[,j]
                        }
                }
        )
}
```

Figure 8.1 *An R function to implement the first reaction method for a stochastic Petri net representation of a coupled chemical reaction system. It is to be used in the same way as the* StepGillespie *function from Figure 6.12.*

2. Let j be the index of the smallest t_i.

3. Set $t := t_j$.

4. Update x according to reaction with index j.

5. Update $h_j(x, c_j)$ according to the new state x and simulate a new putative time $t_j := t + Exp(h_j(x, c_j))$.

6. For each reaction $i(\neq j)$ whose hazard is changed by reaction j:

   (a) Update $h_i' = h_i(x, c_i)$ (but temporarily keep the old h_i).
   (b) Set $t_i := t + (h_i/h_i')(t_i - t)$.
   (c) Forget the old h_i.

7. If $t < T_{max}$, return to step 2.

There are several things to note about this algorithm. The first is that it has moved from working with 'relative' times (times from now until the next event) to 'absolute' times (the time of the next event). The reason for doing this is that it saves generating new times for all of the reactions that are not affected by the reaction that has just taken place (thanks to the memoryless property, Proposition 3.16, this is okay). The second thing to note is that the times that are affected by the most recent reaction are 're-used' by appropriately rescaling the old variable (conditional on it being greater than t). Again, a combination of the memoryless property and the rescaling property (Proposition 3.20) ensures that this is okay.

The next thing worth noting is that it is assumed that the algorithm 'knows' which hazards are affected by each reaction. Gibson and Bruck (2000) suggest that this is done by creating a 'dependency graph' for the system. The dependency graph has nodes corresponding to each reaction in the system. A directed arc joins node i to node j if a reaction event of type i induces a change of state that affects the hazard for the reaction of type j. These can be determined (automatically) from the forms of the associated reactions. Using this graph, if a reaction of type i occurs, the set of all children of node i in the graph gives the set of hazards that needs to be updated.

An interesting alternative to the dependency graph is to work directly on the Petri net representation of the system. Then, for a given reaction node, the set of all 'neighbours' (species nodes connected to that reaction node), $\mathcal{X}$ is the set of all species that can be altered. Then the set of all reaction nodes that are 'children' of a node in $\mathcal{X}$ is the set of all reaction nodes whose hazards may need updating. This approach is slightly conservative in that the resulting set of reaction nodes is a superset of the set which absolutely must be updated, but nevertheless represents a satisfactory alternative.

This algorithm is now 'local' in the sense that all computations (bar one) involve only adjacent nodes on the associated Petri net representation of the problem. The only remaining 'global' computation is the location of the index of the smallest reaction time. Gibson and Bruck's clever solution to this problem is to keep all reaction times (and their associated indices) in an 'indexed priority queue'. This is another graph, allowing searches and changes to be made using only fast and local operations; see the original paper for further details of exactly how this is done. The advantage of having local operations on the associated Petri net is that the algorithm becomes straightforward to implement in an event-driven, object-oriented programming style, consistent with the ethos behind the Petri net approach. Further, such an implementation could be multi-threaded on an SMP or hyper-threading machine, and would also lend itself to a full message-passing implementation on a parallel computing cluster. For further information on parallel stochastic simulation, see Wilkinson (2005).

This algorithm is more efficient than Gillespie's direct method in the sense that only one new random number needs to be simulated for each reaction event which takes place, as opposed to the two that are required for the Gillespie algorithm. Note however, that selective recalculation of the hazards, $h_i(x, c_i)$ (and the cumulative hazard $h_0(x, c)$), is also possible (and highly desirable) for the Gillespie algorithm, and could speed up that algorithm enormously for large systems. Given that the Gillespie algorithm otherwise requires fewer operations than the next reaction method, and does not rely on the ability to efficiently store putative reaction times in an indexed priority queue, the relative efficiency of a cleverly implemented direct method and the next reaction method is likely to depend on the precise structure of the model and the speed of the random number generator used.

8.2.3 Time-varying volume

A problem that has so far been overlooked is that of reaction hazards which vary continuously over time. The most common context for this to arise in a practical modelling situation is when a growing cell (or cellular compartment) has its volume modelled as a continuous deterministic function of time. For example, let us suppose that the container volume at time t, $V(t)$, is modelled as

$$V(t) = v_0 + \alpha t, \quad t \geq 0, \tag{8.1}$$

for some constant $\alpha > 0$. If the model contains any second-order reactions, the hazards of these should be inversely proportional to $V(t)$ (Section 6.2). The hazards of first-order reactions are independent of volume and hence unaffected. What to do about any zero-order reactions is somewhat unclear. As zero-order reactions are typically used to model 'production' or 'influx' in a simple-minded way, it is conceivable that at least some zero-order reactions should have volume dependence. Certain production rates might reasonably be considered to be directly proportional to $V(t)$, while influx equations might have hazards proportional to $V(t)^{2/3}$ (as surface area increases more slowly than volume). In general, zero-order reactions should be considered on a case-by-case basis.

In order to keep the presentation as straightforward as possible, we will just consider modifying the (inefficient) first reaction method to take account of time-varying reaction hazards.* It should be reasonably clear that since the only steps involving the hazards are steps 2 and 3, only steps 2 and 3 need modification. Since the hazard is time varying, we should now write it $h_i(x, t, c_i)$, $i = 1, 2, \ldots, v$. Now we could simply run the algorithm using these time-depended hazards but otherwise unmodified. Unfortunately this will lead to an algorithm that is only approximately correct, as we would be essentially assuming that the hazards remain constant between each reaction event, which is not actually true. At this point it is helpful to recall the inhomogeneous Poisson process (Section 5.4.4), as this is exactly what is needed for dealing with non-constant hazards. Proposition 5.4 and the subsequent discussion tell us exactly how to simulate the time of the next event of an inhomogeneous Poisson process, but note that the lower limit of the integral defining the cumulative hazard must be the current simulation time, and not zero.

For concreteness consider a reaction with hazard $h(t) = a/V(t)$, where a will be a function of x and c_i, but constant with respect to t. Suppose further that $V(t)$ is given by (8.1), the current simulation time is t_0, and we wish to simulate the time t' of the next reaction event. We begin by computing the cumulative hazard

$$H(t) = \int_{t_0}^{t} h(t)dt = \int_{t_0}^{t} \frac{a\,dt}{v_0 + \alpha t} = \frac{a}{\alpha} \log\left(\frac{v_0 + \alpha t}{v_0 + \alpha t_0}\right), \quad t \geq t_0,$$

and then compute the cumulative distribution function (CDF) of the time of the next

---

* The next reaction method (Gibson–Bruck algorithm) is also quite straightforward to modify. The direct method (Gillespie algorithm) is actually a bit awkward to modify, and so the desire to work with time-varying hazards is one reason why some people prefer to use an algorithm in the Gibson and Bruck style. These issues are discussed in some detail in Gibson and Bruck (2000).

event as

$$F(t) = 1 - \exp\{-H(t)\} = 1 - \left(\frac{v_0 + \alpha t}{v_0 + \alpha t_0}\right)^{-a/\alpha}, \quad t \geq t_0.$$

Once we have the CDF we can simulate $u \sim U(0,1)$ and solve $u = F(t')$ for t', the time of the next event, to obtain

$$t' = \frac{1}{\alpha}\left[(v_0 + \alpha t_0)u^{-\alpha/a} - v_0\right].$$

Returning to the problem of modifying the first reaction method, one simply simulates putative times to the next event using the above strategy for any reactions with time varying rates, and the rest of the algorithm remains untouched.

This provides an example of coupling a discrete stochastic process with a variable that changes continuously in time. Essentially the same strategy is used in several of the hybrid algorithms to be considered in Section 8.4, where some variables are treated as discrete and others as varying continuously in time. A good understanding of the above technique is a necessary prerequisite for understanding hybrid simulation algorithms.

At this point it is also worth referring back to Section 5.4.5 on exact sampling of jump times, as such methods can also be used here, and in some cases can be easier to automate than the algebraic techniques discussed above. To see how the method can be used in the context of the above example, for $t > t_0$, $h(t_0)$ is an upper bound on $h(t)$, and so putative waiting times t' can be sampled from an $Exp(h(t_0))$ distribution. The waiting time corresponds to a putative event time, $\tilde{t}$. Such a putative time should be accepted as an event time with probability $h(\tilde{t})/h(t_0)$. If rejected, an additional waiting time t' should be simulated, and added to the current putative time to get the next putative event time, and the procedure should continue until an event time is accepted. Note that this procedure requires no algebraic analysis of the hazard functions, other than the initial identification of an upper bound on the hazard.

8.3 Approximate simulation strategies

8.3.1 Time discretisation

Gibson and Bruck's next reaction method is regarded by some to represent the state-of-the-art for *exact* simulation of a stochastic kinetic model. However, if one is prepared to sacrifice the exactness of the simulation procedure, there is a potential for huge speed-up at the expense of a little accuracy. These fast approximate methods are all based on a time discretisation of the Markov process.

The essential idea is that the time axis is divided into (small) discrete chunks, and the underlying kinetics are approximated so that advancement of the state from the start of one chunk to another can be made in one go. Most of the methods work on the assumption that the time intervals have been chosen to be sufficiently small that the reaction hazards can be assumed (roughly) constant over the interval. We know that a point process with constant hazard is a (homogeneous) Poisson process (Section 3.6.5). Based on the definition of the Poisson process, we assume that the number of reactions (of a given type) occurring in a short time interval has a Poisson distribution (independently of other reaction types). We can then simulate Poisson numbers of reaction events and update the system accordingly.

For a fixed (small) time step Δt, we can present an approximate simulation algorithm as follows (we use the matrix notation from Section 2.3.2):

1. Initialise the problem with time $t := 0$, rate constants c, state x, and stoichiometry matrix S.

2. Calculate $h_i(x, c_i)$, for $i = 1, \ldots, v$, and simulate the v-dimensional reaction vector r, with ith entry a $Po(h_i(x, c_i)\Delta t)$ random quantity.

3. Update the state according to $x := x + Sr$.

4. Update $t := t + \Delta t$.

5. Output t and x.

6. If $t < T_{max}$ return to step 2.

We could call this the *Poisson timestep method*. Note that step 3 should (ideally) be accomplished using a sparse matrix update. An R function to implement the Poisson timestep method is given in Figure 8.2. There is a slight flaw in this function, due to the way that the final time step typically 'over-shoots' the required target time. The degree of over-shooting will be insignificant if the time step of the algorithm is small relative to the time step by which the process is to be advanced. However, this is easily corrected by ensuring that the final time step is reduced to exactly 'hit' the target time — see the end of chapter exercises.

The main difficulty with the above method is that of choosing an appropriate timestep Δt so that the method is fast but reasonably accurate. Clearly the smaller Δt, the more accurate, and the larger Δt, the faster. Another problem is that although one particular Δt may be good enough for one part of a simulation, it may not be appropriate for another. This motivates the idea of stepping ahead a variable amount of time τ, based on the rate constants, c and the current state of the system, x. This is the idea behind Gillespie's τ-leap algorithm.

8.3.2 Gillespie's τ-leap method

The τ-leap method (Gillespie, 2001) is an adaptation of the Poisson timestep method to allow stepping ahead in time by a variable amount τ, where at each timestep τ is chosen in an appropriate way in order to ensure a sensible trade-off between accuracy and speed. This is achieved by making τ as large (and hence fast) as possible while still satisfying some constraint designed to ensure accuracy. In this context, the accuracy is determined by the extent to which the assumption of constant hazard over the interval is appropriate. Clearly, whenever any reaction occurs, some of the reaction hazards change, and so an assessment needs to be made of the magnitude of change of the hazards $h_i(x, c_i)$. Essentially, the idea is to choose τ so that the (proportional) change in all of the $h_i(x, c_i)$ is small.

The simplest way to check that a chosen τ is satisfactory is to apply a *post-leap* check. That is, after a leap of τ, check that $|h_i(x', c_i) - h_i(x, c_i)|$ is sufficiently small for each i (where x and x' represent the state of the system before and after the leap).

```
StepPTS <- function (N,dt=0.01)
{
        S = t(N$Post-N$Pre)
        v = ncol(S)
        return (
                function (x0,t0,deltat,...)
                {
                        x = x0
                        t = t0
                        termt = t0+deltat
                        repeat {
                                h = N$h(x, t, ...)
                                r = rpois (v,h*dt)
                                x = x+as.vector (S %*% r)
                                t = t+dt
                                if (t > termt)
                                        return (x)
                        }
                }
        )
}
```

Figure 8.2 *An* R *function to implement the Poisson timestep method for a stochastic Petri net representation of a coupled chemical reaction system. It is to be used in the same way as the* StepGillespie *function from Figure 6.12. Note that the variable* dt *corresponds to the variable* Δt *from the text. The variable* deltat *corresponds to the amount of time by which the advancement of the process is required, which could be much bigger than the time step of the algorithm,* Δt.

If any of the differences are too large, try again with a smaller value of τ. One of the problems with this method is that it biases the system away from large yet legitimate state changes.

A *pre-leap* check seems more promising. Here we can calculate the expected new state as $E(x') = x + E(r) A$, where the ith element of $E(r)$ is just $h_i(x, c_i)\tau$. We can then calculate the change in hazard at this 'expected' new state and see if this is acceptably small (it should be noted that this is *not* necessarily the expected change in hazard, due to the potential non-linearity of $h_i(x, c_i)$). It is suggested that the magnitude of acceptable change should be a fraction of the cumulative hazard $h_0(x, c)$, i.e.,

$$|h_i(x', c_i) - h_i(x, c_i)| \leq \varepsilon h_0(x, c), \quad \forall i.$$

Gillespie provides an approximate method for calculating the largest τ satisfying this property (Gillespie, 2001). Note that if the resulting τ is as small (or almost as small) as the expected time leap associated with an exact single reaction update, then it is preferable to do just that. Since the time to the first event is $Exp(h_0(x, c))$, which has expectation $1/h_0(x, c)$, one should always prefer an exact update if the suggested τ is less than (say) $2/h_0(x, c)$.

Gillespie and Petzold (2003) consider refinements of this basic τ selection algorithm that improve its behaviour somewhat. However, in my opinion, the 'pure' τ-leap method is always likely to be somewhat unsatisfactory in the context of biochemical networks with very small numbers of molecules of some species (say, zero or one copy of an activated gene). On the other hand, a hybrid algorithm known as the *maximal timestep method* uses the τ-leap method for some variables and not others. This algorithm, which seems quite promising, will be briefly described in Section 8.4.2.

8.3.3 Diffusion approximation (chemical Langevin equation)

Another way of speeding up simulation is to simulate from the diffusion approximation to the true process (Section 5.5). A formal discussion of this procedure is beyond the scope of this book. However, we shall here first derive the form of the diffusion approximation using an intuitive procedure (which can be formalised with a little effort; see, for example Gillespie (2000)). We will then give an alternative derivation, exploiting the random time-change representation of the stochastic kinetic model (6.10).

It is clear from the discussion of the Poisson timestep method that the change in state of the process, dX_t, in an infinitesimally small time interval dt is $S\, dR_t$, where dR_t is a v-vector whose ith element is a $Po(h_i(X_t, c_i)dt)$ random quantity. Matching the mean and variance, we put

$$dR_t \simeq h(x, c)dt + \mathrm{diag}\left\{ \sqrt{h(x, c)} \right\} dW_t,$$

where $h(x, c) = (h_1(x, c_1), \dots, h_v(x, c_v))^\mathsf{T}$, dW_t is the increment of a v-d Wiener process, and for a p-d vector v, $\mathrm{diag}\{v\}$ denotes the $p \times p$ matrix with the elements of v along the leading diagonal and zeros elsewhere (as discussed in Chapter 4). We now have the diffusion approximation

$$dX_t = S\, dR_t$$
$$= S\left(h(X_t, c)dt + \mathrm{diag}\left\{ \sqrt{h(X_t, c)} \right\} dW_t \right)$$
$$\Rightarrow dX_t = Sh(X_t, c)dt + S\, \mathrm{diag}\left\{ \sqrt{h(X_t, c)} \right\} dW_t. \tag{8.2}$$

Equation (8.2) is one way of writing the *chemical Langevin equation* (CLE) for a stochastic kinetic model.[†] One slightly inconvenient feature of this form of the equation is that the dimension of X_t (u) is different from the dimension of the driving process W_t (v). Since we will typically have $v > u$, there will be unnecessary redundancy in the formulation associated with this representation. However, using some straightforward multivariate statistics (not covered in Chapter 3), it is easy

---

[†] Note that setting the term in dW_t to zero gives the ODE system corresponding to the continuous deterministic approximation.

to see that the variance–covariance matrix for dX_t is $S \, \text{diag}\{h(X_t, c)\} \, S^{\mathsf{T}} \, dt$ (or $A^{\mathsf{T}} \, \text{diag}\{h(X_t, c)\} \, A \, dt$). So, (8.2) can be rewritten

$$dX_t = Sh(X_t, c)dt + \sqrt{S \, \text{diag}\{h(X_t, c)\} \, S^{\mathsf{T}}} \, dW_t, \qquad (8.3)$$

where dW_t now denotes the increment of a u-d Wiener process, and we use a common convention in statistics for the square root of a matrix.[‡] In some ways (8.3) represents a more efficient description of the CLE. However, there can be computational issues associated with calculating the square root of the diffusion matrix, particularly when S is rank degenerate (as is typically the case, due to conservation laws in the system), so both (8.2) and (8.3) turn out to be useful representations of the CLE, depending on the precise context of the problem, and both provide the basis for approximate simulation algorithms. An R function to integrate the CLE using the Euler method is given in Figure 8.3 (which suffers from the same 'over-shooting' problem as the PTS function). This method can work extremely well if there are more than (say) ten molecules of each reacting species throughout the course of the simulation. However, if the model contains species with a very small number of molecules, simulation based on a pure Langevin approximation is likely to be unsatisfactory. Again, however, a hybrid algorithm can be constructed that uses discrete updating for the low copy-number species and a Langevin approximation for the rest; such an algorithm will be discussed in Section 8.4.3.

We will now give an alternative derivation based on the random time-change representation of the process (6.10). Start with the time-change representation

$$X_t - X_0 = S \, N \left(\int_0^t h(X_\tau, c) d\tau \right)$$

and make the approximation $N_i(t) \simeq t + W_i(t)$, where $W_i(t)$ is an independent Wiener process for each i. This approximation is just the Gaussian approximation to the Poisson distribution extended to the Poisson process. Understanding the extent to which this approximation is adequate is key to understanding when the CLE is a good model. A careful study is beyond the scope of this text, but as for the case of the Poisson distribution, the approximation improves as the mean increases. Roughly speaking, this means that the approximation is generally adequate providing that there are a reasonable number of molecules of each reacting species. For this construction we just accept the approximation as valid and examine its consequences. Substituting in

---

[‡] Here, for a $p \times p$ matrix M (typically symmetric), $\sqrt{M}$ (or $M^{1/2}$) denotes *any* $p \times p$ matrix N satisfying $NN^{\mathsf{T}} = M$. Common choices for N include the symmetric square root matrix and the Cholesky factor; see Golub and Van Loan (1996) for computational details.

and using knowledge of the time-change of a Wiener process (Section 5.5.2) gives:

$$X_t - X_0 = S N \left(\int_0^t h(X_\tau, c) d\tau \right)$$

$$\simeq S \left[\int_0^t h(X_\tau, c) d\tau + W \left(\int_0^t h(X_\tau, c) d\tau \right) \right]$$

$$\Rightarrow dX_t = Sh(X_t, c) \, dt + S \operatorname{diag} \left\{ \sqrt{h_i(X_t, c)} \right\} dW_t,$$

where W_t is the same dimension as $h(\cdot, \cdot)$, but again, since

$$\operatorname{Var}(dX_t) = S \operatorname{diag}\{h_i(X_t, c)\} S^\mathsf{T} \, dt$$

we can re-write as

$$dX_t = Sh(X_t, c) \, dt + \sqrt{S \operatorname{diag}\{h(X_t, c)\} S^\mathsf{T}} \, dW_t,$$

where W_t is now the same dimension as X_t. This derivation is quite rigorous conditional on the key approximating step. More careful derivations attempt to justify the approximation by constructing an appropriate asymptotic for which the CLE is the exact limit; see Ball et al. (2006) for further discussion.

Having used relatively informal methods to derive the CLE, it is worth taking time to understand in more detail its relationship with the true discrete stochastic kinetic model (represented by a Markov jump process). In Section 6.8 we derived the chemical master equation as Kolmogorov's forward equation for the Markov jump process. It is also possible to develop a corresponding forward equation for the chemical Langevin equation (8.3). We know from Chapter 5 that the forward equation for a diffusion process is known as the *Fokker–Planck equation* (5.21). Starting with this,

$$\frac{\partial p}{\partial t} = -\nabla \cdot [\mu(x)p] + \frac{1}{2}\nabla \cdot (\nabla \cdot [\Sigma(x)p]),$$

and substituting in $\mu(x) = Sh(x, c)$ and $\Sigma(x) = S \operatorname{diag}\{h(x, c)\} S^\mathsf{T}$ gives the Fokker–Planck equation for the CLE as

$$\frac{\partial p}{\partial t} = -\nabla \cdot [Sh(x, c)p] + \frac{1}{2}\nabla \cdot \left(\nabla \cdot [S \operatorname{diag}\{h(x, c)\} S^\mathsf{T} p] \right).$$

This is the partial differential equation whose solution gives the density of the system state at time t, $p(x, t)$. This Fokker–Planck equation can be shown to correspond to a second-order approximation to the chemical master equation. Thus the CLE in some sense represents the 'best' SDE approximation to the true stochastic kinetic model. See Gillespie (2000) for further details. Note that if the CLE has an equilibrium density, $p(x)$, it will satisfy

$$\nabla \cdot [Sh(x, c)p] = \frac{1}{2}\nabla \cdot \left(\nabla \cdot [S \operatorname{diag}\{h(x, c)\} S^\mathsf{T} p] \right).$$

```
StepCLE <- function (N, dt=0.01)
{
        S = t (N$Post-N$Pre)
        v = ncol (S)
        sdt = sqrt (dt)
        return (
                function (x0, t0, deltat,...)
                {
                        x = x0
                        t = t0
                        termt = t0+deltat
                        repeat {
                                h = N$h (x, t, ...)
                                dw = rnorm (v, 0, sdt)
                                dx = S %*% (h*dt + sqrt (
                                    h) *dw)
                                x = x+as.vector (dx)
                                t = t+dt
                                if (t > termt)
                                        return (x)
                        }
                }
        )
}
```

Figure 8.3 *An R function to integrate the CLE using an Euler method for a stochastic Petri net representation of a coupled chemical reaction system. It is to be used in the same way as the* StepPTS *function from Figure 8.2.*

Examples

First consider the birth–death process discussed in Chapter 1:

$$\mathcal{X} \xrightarrow{\lambda} 2\mathcal{X}$$
$$\mathcal{X} \xrightarrow{\mu} \emptyset.$$

Since here $S = (1, -1)$ and $h(x, c) = (\lambda x, \mu x)^{\mathsf{T}}$, this model has random time-change representation

$$X_t = X_0 + N_1 \left(\int_0^t \lambda X_s \, ds \right) - N_2 \left(\int_0^t \mu X_s \, ds \right)$$

and corresponding CLE:

$$dX_t = (\lambda - \mu)X_t \, dt + \sqrt{(\lambda + \mu)X_t} \, dW_t.$$

This SDE is a special (degenerate) case of the *CIR model* discussed in Chapter 5.

Next, consider a reversible isomerisation process

$$\mathcal{X} \xrightarrow{\alpha} \mathcal{Y}$$
$$\mathcal{Y} \xrightarrow{\beta} \mathcal{X}.$$

This model has an obvious conservation law, $X_t + Y_t = c$, which can be used to eliminate Y_t. Having done so, we can deduce the random time-change representation

$$X_t = X_0 - N_1 \left(\int_0^t \alpha X_s \, ds \right) + N_2 \left(\int_0^t \beta(c - X_s) \, ds \right)$$

and corresponding CLE

$$dX_t = [\beta c - (\alpha + \beta)X_t] \, dt + \sqrt{\beta c + (\alpha - \beta)X_t} \, dW_t.$$

Note how the noise term does not tend to zero as X_t tends to zero. This suggests that this CLE is not bounded below by zero, unlike the true Markov jump process described by the random time-change representation. See Szpruch and Higham (2010) for further discussion of this issue.

It is therefore clear that the CLE does not always preserve the non-negativity of the discrete stochastic kinetic model that it is approximating. However, it does preserve the noise associated with the process, and other important features, such as conservation laws, unlike some simpler approximate stochastic models. It is easy to see that the CLE preserves conservation laws, since they correspond to P-invariants α such that $\alpha^\mathsf{T} S = 0$, from which it is clear that $\alpha^\mathsf{T} dX_t = 0$.

8.3.4 Modelling extrinsic noise

We will now re-examine the immigration-death model

$$\emptyset \xrightarrow{\lambda} \mathcal{X}$$
$$\mathcal{X} \xrightarrow{\mu} \emptyset,$$

considered in Chapter 5. We have already seen that the CLE for this model is

$$dX_t = (\lambda - \mu X_t) \, dt + \sqrt{\lambda + \mu X_t} \, dW_t,$$

and that this captures the *intrinsic* noise associated with this process reasonably well. We also know that if we ignore the intrinsic noise, we get the deterministic approximation

$$\frac{dX_t}{dt} = (\lambda - \mu X_t).$$

We have also seen that in the case of this simple linear process, the deterministic mean correctly describes the expected value of the associated stochastic process. We will now consider how to inject some *extrinsic noise* into the CLE (and the deterministic approximation). Recall that extrinsic noise is a catch-all term for any stochasticity in the system which does not arise directly from the discreteness of the explicitly modelled reaction processes. Here we focus on one possible aspect — a randomly time-varying degradation rate. We know from the theory of stochastic chemical kinetics that if our model description is complete, then the degradation rate should be of the form μX_t, where μ is a constant. Unfortunately, models are rarely complete, and this lack of complete knowledge of the full system state often manifests itself as additional stochasticity — extrinsic noise. Here, we can imagine that the degradation process involves other chemical species that we have not modelled. The levels of these chemical species will vary (randomly) over time, just as the modelled species do, and this will result in the rate of degradation varying over time. For extrinsic noise processes we do not have an underlying kinetic theory which tells us how to model the variation, so in practice simple models are often used. It is often convenient to model time-varying parameters using stationary stochastic processes. The CIR process can be a convenient choice for variables which must remain non-negative. Here we will replace the degradation rate parameter μ with the CIR process Y_t governed by the SDE

$$dY_t = \alpha(\mu - Y_t)dt + \sigma\sqrt{Y_t}dB_t.$$

Recall that this process has stationary mean μ and stationary variance $\sigma^2\mu/[2\alpha]$. We can then couple this to the CLE to give the pair of SDEs

$$dX_t = (\lambda - Y_tX_t)\,dt + \sqrt{\lambda + Y_tX_t}\,dW_t,$$
$$dY_t = \alpha(\mu - Y_t)dt + \sigma\sqrt{Y_t}dB_t.$$

Obviously, we can re-write this pair of SDEs as a bivariate diffusion process

$$\begin{pmatrix} dX_t \\ dY_t \end{pmatrix} = \begin{pmatrix} \lambda - Y_tX_t \\ \alpha(\mu - Y_t) \end{pmatrix}dt + \begin{pmatrix} \sqrt{\lambda + Y_tX_t} & 0 \\ 0 & \sigma\sqrt{Y_t} \end{pmatrix}\begin{pmatrix} dW_t \\ dB_t \end{pmatrix}. \quad (8.4)$$

For this process, the sample paths of X_t will have noise arising from both intrinsic and extrinsic sources.

If it is felt that intrinsic noise is insignificant relative to extrinsic noise, then the extrinsic noise process can be coupled directly to the deterministic model to give the pair of SDEs

$$dX_t = (\lambda - Y_tX_t)\,dt$$
$$dY_t = \alpha(\mu - Y_t)dt + \sigma\sqrt{Y_t}dB_t,$$

corresponding to the bivariate diffusion process

$$\begin{pmatrix} dX_t \\ dY_t \end{pmatrix} = \begin{pmatrix} \lambda - Y_tX_t \\ \alpha(\mu - Y_t) \end{pmatrix}dt + \begin{pmatrix} 0 & 0 \\ 0 & \sigma\sqrt{Y_t} \end{pmatrix}\begin{pmatrix} dW_t \\ dB_t \end{pmatrix}.$$

This process is clearly degenerate in that there is a rank-deficiency in the diffusion matrix which does not correspond to a conservation law of the system. Such processes are sometimes known as *hypo-elliptic diffusions*. One consequence of this is smoothness in the solution. We know that trajectories of Brownian motion are continuous everywhere but nowhere differentiable. This property is transferred to the extrinsic noise component, Y_t. However, since X_t is essentially the integral of this continuous sample path, trajectories of X_t are not only continuous but also (once) differentiable.

We can simulate the sample paths of arbitrary multivariate diffusion processes using an Euler–Maruyama method such as that shown in Figure 8.4. An example

```
StepSDE <- function (drift, diffusion, dt = 0.01)
{
    sdt = sqrt(dt)
    return (function(x0, t0, deltat, ...) {
        x = x0
        t = t0
        termt = t0 + deltat
        v = length(x)
        repeat {
            dw = rnorm(v,0,sdt)
            x = x + drift(x, t, ...)*dt + as.vector(
                diffusion(x, t,...)%*%dw)
            t = t + dt
            if (t > termt)
                return(x)
        }
    })
}
```

Figure 8.4 *An R function to integrate a multivariate diffusion process using a simple Euler–Maruyama method. The function returns a function closure for simulating from the transition kernel of the SDE, which can then be passed into functions such as* simTs. *An example of calling* StepSDE *and utilising the resulting function closure is given in Figure 8.5.*

demonstrating the use of this function for the immigration death process incorporating both intrinsic and extrinsic noise (8.4) is given in Figure 8.5. Sample paths from the immigration death process with different combinations of intrinsic and extrinsic noise are shown in Figure 8.6.[§]

8.3.5 The linear noise approximation

The linear noise approximation (LNA) is an approximation to the CLE which generally possesses a greater degree of numerical and analytic tractability than the CLE. Again, rigorous derivation and analysis are beyond the scope of this text, but we give an informal derivation here. Start with the CLE in the form of (8.2), and replace the hazard function $h(X_t, c)$ with the rescaled form $vf(X_t/v)$ where v is the volume of the container in which the reactions are taking place, and we drop dependence on the rate constants, c. If the rate laws are of mass-action kinetic form, this ensures that the re-scaled rate function, $f(\cdot)$, is independent of the volume of the container, and can therefore be regarded as constant when considering large volume limits. The CLE becomes

$$dX_t = vSf(X_t/v)dt + \sqrt{v}S \operatorname{diag}\left\{\sqrt{f(X_t/v)}\right\} dW_t.$$

[§] Note that the R code used to generate this figure is included as a demo script in the smfsb R package.

```
theta=c(lambda=2,alpha=1,mu=0.1,sigma=0.2)
myDrift <- function(x,t,th=theta)
     {
            with(as.list(c(x,th)),{
                   c( lambda - x*y ,
                          alpha*(mu-y) )
            })
     }
myDiffusion <- function(x,t,th=theta)
     {
            with(as.list(c(x,th)),{
                   matrix(c( sqrt(lambda + x*y) , 0,
                          0, sigma*sqrt(y) ),ncol=2,
                          nrow=2,byrow=TRUE)
            })
     }
stepProc=StepSDE(myDrift,myDiffusion,dt=0.001)
plot(simTs(c(x=1,y=0.1),0,30,0.01,stepProc))
```

Figure 8.5 *Example showing how to use the function* StepSDE *for the SDE given in (8.4).*

If we think about how we expect the system to scale with container volume, inspired by the central limit theorem, and Poisson process variation, we make the substitution

$$X_t = vx_t + \sqrt{v}Y_t, \tag{8.5}$$

where x_t is a deterministic process, and Y_t is a residual stochastic process independent of the system size. Substituting into our CLE, we get

$$dx_t + \frac{1}{\sqrt{v}}dY_t = Sf(x_t + Y_t/\sqrt{v})dt + \frac{1}{\sqrt{v}}S\operatorname{diag}\left\{\sqrt{f(x_t + Y_t/\sqrt{v})}\right\}dW_t.$$

We now linearise the rate function using

$$f(x_t + Y_t/\sqrt{v}) \simeq f(x_t) + \frac{1}{\sqrt{v}}\frac{\partial f}{\partial x}(x_t)Y_t,$$

where $\frac{\partial f}{\partial x}(x_t)$ is the Jacobian matrix evaluated at x_t, and substitute in to get

$$dx_t + \frac{1}{\sqrt{v}}dY_t = S\left[f(x_t) + \frac{1}{\sqrt{v}}\frac{\partial f}{\partial x}(x_t)Y_t\right]dt$$

$$+ \frac{1}{\sqrt{v}}S\operatorname{diag}\left\{\sqrt{f(x_t) + \frac{1}{\sqrt{v}}\frac{\partial f}{\partial x}(x_t)Y_t}\right\}dW_t.$$

We get the large volume limit by considering terms of $O(1)$ to get the deterministic reaction rate equations

$$dx_t = Sf(x_t)dt,$$

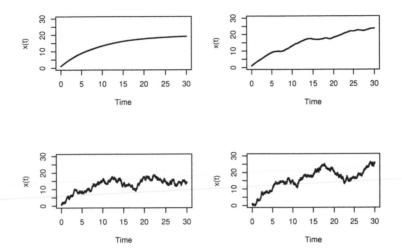

Figure 8.6 *Figure showing realisations of the SDE models for the immigration-death process discussed in Section 8.3.4 incorporating different combinations of intrinsic and extrinsic noise. The plots on the top row do not incorporate intrinsic noise, whereas the plots on the bottom row do. Similarly, the plots on the left-hand side do not incorporate extrinsic noise, whereas the plots on the right-hand side do. Note that all plots exhibit trajectories with continuous sample paths. The trajectories shown in the plots on the top row are also (once) differentiable (the trajectory in the top left plot is infinitely differentiable).*

more usually written as

$$\frac{dx_t}{dt} = Sf(x_t). \tag{8.6}$$

This is the precise sense in which the deterministic RREs are the large volume asymptotic limit of the CLE. We can pick up the residual process Y_t by considering terms of $O(1/\sqrt{v})$ to get

$$dY_t = S\frac{\partial f}{\partial x}(x_t)Y_t dt + S \operatorname{diag}\left\{\sqrt{f(x_t)}\right\} dW_t.$$

This SDE, like the one we started from, (8.2), is driven by noise of different dimension to the state of the process. But again, by considering the infinitesimal variance, we can re-write in more standard form as

$$dY_t = S\frac{\partial f}{\partial x}(x_t)Y_t dt + \sqrt{S \operatorname{diag}\{f(x_t)\} S^{\mathsf{T}}}dW_t. \tag{8.7}$$

Equations (8.5), (8.6), and (8.7) together form the *linear noise approximation* of the CLE (and the original discrete Markov jump process model, although it is worth noting that some authors use the term LNA to refer specifically to (8.7)). There are several things to note about the form of the residual process (8.7). To begin with, suppose that x_t is constant over time, representing an equilibrium of the deterministic component (8.6). In this case the drift is linear in Y_t and the diffusion matrix is constant, giving rise to a (multivariate) Ornstein–Uhlenbeck process (5.20). This process is tractable, and the increments are (multivariate) Gaussian. However, it must also be noted that away from equilibrium, the drift vector and diffusion matrix will be time varying (in a non-linear way), and so the residual process will not be of simple Ornstein–Uhlenbeck form. However, the process is still tractable and the increments are still Gaussian. Solving the LNA equations is beyond the scope of this text, but the residual process turns out to have an explicit solution as an Itô stochastic integral

$$Y_t = \Phi(t) \left[y_0 + \int_0^t \Phi(s)^{-1} \Psi(s) dW_s \right],$$

where $\Psi(t) = S \operatorname{diag}\left\{ \sqrt{f(x_t)} \right\}$, and $\Phi(t)$ is the solution to

$$\frac{d\Phi}{ds} = A(s)\Phi,$$

where $A(t) = S \frac{\partial f}{\partial x}(x_t)$ and $\Phi(0) = I$. From this it is possible to conclude that the increments must be Gaussian, and the distribution at time t is given by

$$(Y_t | Y_0 = y_0) \sim N(\mu, \Xi),$$

where $\mu = \Phi(t) y_0$ and

$$\Xi = \Phi(t) \int_0^t [\Phi^{-1}(s)\Psi(s)][\Phi^{-1}(s)\Psi(s)]^{\mathsf{T}} ds \, \Phi(t)^{\mathsf{T}}.$$

The corresponding approximating distribution for the original process, X_t, is then clearly given by

$$(X_t | X_0) \sim N\left(v x_t + \sqrt{v}\mu, v\Xi \right).$$

For further information on the theory associated with the linear noise approximation, see Gardiner (2004), Van Kampen (1992), Elf and Ehrenberg (2003), Wallace et al. (2012), and Komorowski et al. (2009).

Example

Let us reconsider the immigration–death process with stoichiometry matrix $S = (1, -1)$ and hazard function $h(x) = (\lambda v, \mu x)^{\mathsf{T}}$, where we have changed the zeroth order immigration parameter so that it scales 'correctly' with volume. The rescaled propensity function therefore takes the form $f(x) = (\lambda, \mu x)^{\mathsf{T}}$, which we see is independent of the volume, v. Substituting into the linear noise approximation, we find the deterministic limit

$$\frac{dx_t}{dt} = \lambda - \mu x_t$$

and residual process

$$dY_t = -\mu Y_t dt + \sqrt{\lambda + \mu x_t} dW_t.$$

We see that at equilibrium, $x_t = \lambda/\mu$, the residual process is the univariate Ornstein–Uhlenbeck process

$$dY_t = -\mu Y_t dt + \sqrt{2\lambda} dW_t,$$

which could be solved explicitly using the methods developed in Chapter 5. In particular, using the formula for the stationary variance of the OU process, we see that the asymptotic variance of the residual process is λ/μ. Back on the original scale of the process, this corresponds to fluctuations with an asymptotic variance of $v\lambda/\mu$, as we would expect.

We can now use the results of this section in order to solve exactly for the linear noise approximation subject to the initial condition $X_0 = x_0 = Y_0 = 0$. The ODE for the deterministic component, x_t, may be easily solved using separation of variables, in order to obtain the solution

$$x_t = \frac{\lambda}{\mu}(1 - e^{-\mu t}).$$

The equation

$$\frac{d\Phi}{ds} = -\mu\Phi$$

may be solved similarly to get

$$\Phi(s) = e^{-\mu s}.$$

From this we know that the mean of the residual noise term is $Y_0 e^{-\mu t}$, which is zero for our initial condition. The variance can be found by evaluating

$$\Xi = e^{-2\mu t} \int_0^t \left[e^{\mu s} \sqrt{\lambda + \mu x_s} \right]^2 ds$$

$$= e^{-2\mu t} \int_0^t e^{2\mu s}(\lambda + \mu x_s) ds$$

$$= \lambda e^{-2\mu t} \int_0^t e^{2\mu s}(2 - e^{-\mu s}) ds$$

$$= \frac{\lambda}{\mu}(1 - e^{-\mu t}),$$

after some algebra. Thus the residual noise component has marginal distribution

$$Y_t \sim N\left(0, \frac{\lambda}{\mu}(1 - e^{-\mu t})\right),$$

and the approximation to the marginal distribution of the original process is

$$X_t \sim N\left(\frac{v\lambda}{\mu}(1 - e^{-\mu t}), \frac{v\lambda}{\mu}(1 - e^{-\mu t})\right),$$

which makes sense, since we know that the true distribution is

$$X_t \sim Po\left(\frac{v\lambda}{\mu}(1 - e^{-\mu t})\right).$$

8.4 Hybrid simulation strategies

Hybrid algorithms aim to bridge the gap between the exact algorithms considered in Section 8.2 and the approximate algorithms considered in Section 8.3. The essential idea is to partition model species into two (or possibly more) groups that can be treated in a similar way. The algorithms considered recently in the literature tend to partition the species into two groups; a group of low-copy number species that need to be treated as discrete and updated with an 'exact' method, and a group of species that can be treated with an approximate algorithm of some description. We will denote these two groups X_D and X_A. It then turns out that it is necessary to partition the set of reactions similarly into two groups, R_D and R_A (sometimes referred to as 'stochastic partitioning'). This is typically done by assigning to R_D any reaction that can change the level of any species assigned to X_D (though there are several variations of this idea). Also note that the reactions in R_D are often referred to as 'slow', while those in R_A are 'fast'. See Ball et al. (2006) for a more sophisticated approach to understanding the separation of time scales. Also see Anderson and Kurtz (2015) for more detailed analysis of stochastic biochemical systems, and Cotter et al. (2011) for a more recent approach to multiscale simulation.

8.4.1 Discrete/ODE models

We begin by looking at a method of combining discrete event simulation with conventional ODE models, based loosely on the algorithm of Kiehl et al. (2004), to which the reader is referred for further details. Also see Alfonsi et al. (2005) for a more formal discussion of this approach to hybrid simulation. These methods are appealing to conventional mathematical modellers as they allow the approximate part of the system to be modelled with the usual continuous deterministic techniques that they are most familiar with.

The algorithm begins with a partitioning of species and reactions, and the fixing of a step size for the numerical integration algorithm for the ODE system. Note that it is possible to use any ODE solver in conjunction with this algorithm (Euler, Runge–Kutta, etc.). The basic idea is that an ODE integration step will take place on the assumption that no discrete reaction takes place. Once this has been done, time-varying hazards for the discrete reaction hazards are determined (if an Euler method is being used, the hazards will be linear in time, and if a Runge–Kutta method is used, they will be a higher-order polynomial). It is then possible to use the methods discussed in Section 5.4.4 and Section 8.2.3 to check if a discrete event occurred. If not, the timestep is done. If there is, the time of the event should be identified, the discrete update should take place, and the continuous variables should be updated over this new shorter timestep. There are various possible modifications of this procedure. In particular, the algorithm is greatly simplified if it is deemed reasonable to assume that the hazards for the discrete regimes are approximately constant over the time interval of interest (though this obviously introduces additional error). Another possibility is to reduce the size of the numerical integration step dynamically in order to reduce the number of integration steps that contain a discrete reaction event. The procedure is summarised in the algorithm below:

1. Initialise the system and set $t := 0$.

2. Calculate the discrete reaction hazards at time t.

3. On the basis of these, decide on an appropriate timestep for the ODE integration, Δt.

4. Use an ODE integration step to compute the trajectory of the continuous variables over the interval $[t, t + \Delta t]$.

5. On the basis of these, compute the corresponding time-varying discrete reaction hazards, and decide if a discrete reaction event has taken place.

6. If no discrete reaction has taken place, put $t := t + \Delta t$, and update the continuous variables to the values appropriate for this new time.

7. If a discrete reaction has occurred, find the time, t_1 and type of the (first) reaction. Then put $t := t_1$, update the continuous values to those appropriate for time t_1, and update the discrete variables according the reaction type that has occurred.

8. If $t < T_{max}$, return to step 2.

The algorithm has been presented assuming that Gillespie's direct method will be used for the discrete updating, but is easily adapted to other discrete updating schemes, such as the next reaction method. There are a number of techniques that could be used for determining an appropriate value of Δt. In Kiehl et al. (2004) it is assumed that an appropriate timestep for the ODE solver can be determined a priori as δt. Then an appropriate 'discrete step size' $\delta \tau$ can be chosen, possibly as the expected time until the first discrete reaction, giving $\delta t = 1/h_0(x, c)$. Then $\Delta t = \min\{\delta t, \delta \tau\}$. It may be beneficial to make $\delta \tau$ a bit smaller than the expected time until the first discrete reaction, in order to reduce the number of time intervals containing a discrete event.

This algorithm is fairly straightforward to understand and implement and has a similar structure to the other hybrid algorithms we will consider. It therefore provides a useful starting point for thinking about hybrid simulation strategies. The main problem with the algorithm is that the ODE approximation is too crude in the context of stochastic modelling of genetic and biochemical networks. As Kiehl et al. (2004) fully acknowledge, the algorithm has the effect of suppressing the intrinsic variation of variables assigned to the continuous regime, and this can have important consequences when the study of stochastic dynamics is of direct interest. This therefore motivates the study of hybrid algorithms which better preserve the stochastic variation of the system dynamics.

8.4.2 Maximal timestep algorithm

The *maximal timestep method* is a hybrid algorithm proposed by Puchalka and Kierzek (2004), one that combines an exact updating procedure for the low concentration species with a τ-leaping approximate updating algorithm for the other species. Puchalka and Kierzek (2004) chose to use the next reaction method for the exact updating, but it would be straightforward to reformulate the algorithm using Gillespie's direct method. A greatly simplified version of the full algorithm could be described as follows:

1. Initialise the system and set $t := 0$.

2. Calculate the fast reaction hazards and use these to select an appropriate τ (for the τ-leap updating scheme).

3. Assuming a constant hazard for the slow reactions, decide if a slow reaction has taken place in $[t, t + \tau]$.

4. If no slow reaction has taken place, perform a τ-leap update on the fast reactions to $t := t + \tau$.

5. If a slow reaction has taken place, identify the time, t_1, and type of the (first) reaction. Then put $t := t_1$, perform a τ-leap update of the fast reactions to t_1, and update the slow variables according to the slow reaction that has occurred.

6. If $t < T_{max}$, return to step 2.

This algorithm has a similar structure to the one considered earlier, but in the context of τ-leaping, it is more difficult to take account of the time-varying nature of the reaction hazards, so it is just hoped that the timestep is small enough that a constant hazard assumption is valid. The full algorithm presented in Puchalka and Kierzek (2004) is considerably more complex than the simple overview given above, but most of the additional details are associated with dynamic repartitioning of re-actions into fast and slow categories. The reader is referred to the original paper for further information. Note that these authors also make use of the quasi-steady-state assumption of Rao and Arkin (2003) in order to utilise reactions with rate laws that do not fit into the mass-action class.

The maximal timestep algorithm is quite appealing in that it is reasonably clear that provided a sufficiently small τ is used and sufficiently strict criteria are imposed for admission to the class of fast reactions, then this algorithm is capable of capturing the true stochastic kinetic dynamics arbitrarily well. However, it is not clear how well such an algorithm will scale-up to the very large complex systems that it is intended to tackle.

8.4.3 Discrete/Langevin methods

The final hybrid algorithm we will consider here is based on combining a discrete updating scheme for slow reactions with a diffusion (or Langevin) approach to up-dating the fast reactions. The presentation will be loosely based on that of Salis and Kaznessis (2005a), which is itself a refinement of the basic strategy described by Haseltine and Rawlings (2002). Salis and Kaznessis (2005a) also recommend an au-tomatic partitioning and dynamic repartitioning of fast and slow reactions. Based on a timestep of Δt for the updating of the fast reactions, they suggest that reaction R_j should only be classified as fast if both

$$h_j(x, c_j)\Delta t \geq \lambda$$

and

$$x \geq \varepsilon |S^{(j)}|$$

are true for the current system state x, and suitably chosen λ and ε. The authors suggest using $\lambda = 10$ and $\varepsilon = 100$. Obviously then, as the system state x evolves over time, the set of reactions classified as fast will also change.

While Haseltine and Rawlings (2002) use a direct method for updating the slow reactions, Salis and Kaznessis (2005a) use a next reaction method, and hence refer to their combined scheme as the *next reaction hybrid* (NRH) algorithm. Again, there is nothing particularly fundamental as to the choice of slow updating scheme.

A greatly simplified version of the basic algorithm structure (which is sufficiently general to apply to both the Haseltine and Rawlings (2002) and Salis and Kaznessis (2005a) algorithms) could be stated as follows; see the original papers for further details.

1. Initialise the system and set $t := 0$.

2. Calculate the hazards for the fast reactions.

3. Numerically integrate the CLE for the fast reactions from t to $t + \Delta t$ to obtain a 'sample path' for the continuous variables over the interval $[t, t + \Delta t]$.¶

4. Using the time-dependent hazards for the slow reactions, decide whether or not a slow reaction has happened in $[t, t + \Delta t]$.

5. If no slow reaction has occurred, set $t := t + \Delta t$ and update the continuous variables to their proposed values at this time.

6. If one slow reaction has occurred, identify the time, t_1 and type, set $t := t_1$, and update the system to t_1 (using the type of the discrete reaction and the appropriate continuous state).

7. If more than one slow reaction has occurred, reduce Δt and return to step 3.

8. If $t < T_{max}$, return to step 2.

Here, steps 6 and 7 could (in principle) be replaced with a single combined: 6 and 7. Identify the time and type of the first reaction and update accordingly.

The reason the algorithm is presented this way is that it is slow and difficult to identify the times of slow reactions using numerical integration and solution methods, because the reaction hazards correspond to sample paths of unknown stochastic processes. For this reason, Salis and Kaznessis (2005a) use approximate numerical techniques to estimate the reaction times that are more accurate if the time is first narrowed down to a small interval where it is known that only this one reaction occurs. Salis and Kaznessis (2005a) also present another algorithm (the ANRH algorithm), which is faster but less accurate than the NRH, as it allows multiple slow reactions to fire without affecting the fast reactions; again, see the paper for further details.

The NRH and related algorithms that couple a Markov jump process with a Langevin approximation are (in principle) very attractive, as they are more accurate than the ODE hybrid algorithms and are likely to scale-up to very large models much better than the maximal timestep algorithm.

¶ Of course the true sample path is a diffusion process, which cannot be obtained in its entirety, and this is the main complication for algorithms of this type; see the text immediately following the algorithm for further details.

8.4.4 Discussion

There is a clear and pressing need for accurate stochastic simulation algorithms that are much faster than the exact algorithms. Although the 'pure' approximate algorithms are elegant and appealing, they fail to adequately cope with the common problem of multiple reactions happening on differing time scales. This then motivates the development of hybrid algorithms that can address exactly this issue using a combination of exact and approximate techniques. Unfortunately, all three of the hybrid algorithms presented (and there are other, related algorithms that have not been explicitly covered here) are somewhat 'crude' in terms of their statistical sophistication. The ODE model is unsatisfactory as the jump in scale from discrete stochastic to continuous deterministic is too great. The maximal timestep method is hard to couple without assuming constant hazard over the integration timestep, which makes it heavily reliant on a small integration step size for adequate accuracy, and in any case it is not clear how well it will cope with vastly differing time scales.

Of the approaches presented, the discrete/CLE partitioning adopted by Haseltine and Rawlings (2002) seems to have the greatest potential, and the refinements proposed by Salis and Kaznessis (2005a) lead to an algorithm which should perform reasonably well in practice. However, there are many improvements that could be made to their algorithm to improve both its accuracy and its computational efficiency. Obviously, as noted by Salis and Kaznessis (2005a), better algorithms than the Euler–Maruyama method could be used for numerically integrating the fast reactions. At a more fundamental level, however, improvements could be made to the way that the discrete and continuous regimes are coupled. Recalling the discussion from Section 5.4.5, it is possible to exactly sample the time at which the first reaction takes place, and this should lead to algorithms that are both more efficient and more accurate. Once this has been done, the only sources of error will be the error in adopting the CLE approximation of the fast reactions and the error in the numerical integration step. There is not much that can be done about the error in the CLE approximation within this framework (other than ensuring that only appropriate reactions are designated 'fast'), but it *might* be possible to eliminate the discretisation error completely. Using techniques for exact sampling of non-linear diffusions (Beskos and Roberts, 2005), it could turn out to be possible to directly and efficiently simulate the time to the first slow reaction and the state of the fast system at that time (and any other finite number of intermediate times) exactly, without any time discretisation error at all. We would then be in a situation more comparable with that of the exact discrete algorithms, where the simulated realisations are exact, conditional on the model. However, it is far from clear that the techniques currently being explored for retrospective exact sampling of diffusions will be applicable to non-linear multivariate diffusion processes as general as the CLE. See Beskos et al. (2006) for further information regarding this approach. Other interesting recent papers on fast, accurate stochastic simulation include Ball et al. (2006), E et al. (2007), Samant et al. (2007), Anderson (2007), Crudu et al. (2009), Singh and Hespanha (2010), Higham et al. (2011), and Cotter et al. (2011).

It is worth emphasising that the literature on stochastic simulation is vast, and that there has not been any attempt here to give a comprehensive survey. An interesting area that has not yet been covered is that of spatial modelling of stochastic kinetic systems, and that is the topic of the next chapter.

8.5 Exercises

1. Modify the function `StepPTS` so the the final integration time step exactly hits the required target time. Apply similar modifications to `StepCLE` and `StepSDE`.

2. Read the original source papers for the various different algorithms discussed but not implemented in this chapter, and work through some simple examples by hand on paper to see how they really work.

3. Code up some of the algorithms and compare them against one another for accuracy on some of the example models from Chapter 7. This can be done by comparing sample means, standard deviations, and full empirical probability distributions at a collection of time points, using data from many independent simulation runs; see Salis and Kaznessis (2005a) for further details.

4. After tuning the different algorithms so that they have reasonable accuracy, compare them for speed. Note that the algorithms discussed in this chapter are designed to be efficient for large models with large numbers of species, reactions, and molecules, so it should not be too surprising if the algorithms turn out to be slower than Gillespie's direct method on small models.

5. Download some software packages that implement stochastic simulation using algorithms other than the direct method, and compare them in terms of flexibility, accuracy, and performance. Links to some appropriate software are given on this book's website.

8.6 Further reading

The main source papers for this chapter are Gibson and Bruck (2000), Gillespie (2001), Gillespie and Petzold (2003), Gillespie (2000), Kiehl et al. (2004), Puchalka and Kierzek (2004), Rao and Arkin (2003), Haseltine and Rawlings (2002), and Salis and Kaznessis (2005a). All are required reading for a complete understanding of how the various algorithms work. Ball et al. (2006) provides a nice introduction to mathematical frameworks useful for understanding the properties of fast and approximate simulation methods.

CHAPTER 9

Spatially extended systems

9.1 Introduction

So far we have been assuming that the (bio)chemical reactions we have been modelling have been 'well-mixed', that is, we have ignored any spatial aspects that might influence reaction hazards. Given that biochemical reactions typically take place in very small volumes, such as cells, this seems like a reasonable first approximation. Nevertheless, studies confirm that spatial effects *can* be important in many cases, even in small cellular volumes. There are several reasons for this, but not least that the cell is a very crowded place, so bio-molecules are not able to diffuse as rapidly as they would in pure water, for example.

To incorporate space into the reaction system, we must somehow model the fact that reactive molecules have a position in space, do not diffuse infinitely quickly to other parts of the space, and can only react with molecules nearby in space. When we model reactions and molecular diffusion together, we describe this as a *reaction–diffusion system*. As for the well-mixed case, one can ignore the effects of discreteness and stochasticity to obtain deterministic reaction–diffusion models. These are often described using partial differential equations (PDEs). However, in keeping with the rest of this text, we wish to explicitly keep these stochastic elements in both our mathematical and computational descriptions.

There are two commonly used approaches to developing discrete stochastic descriptions of reaction–diffusion systems. In the first, space is subdivided into some kind of grid, with grid elements typically referred to as *voxels* (a generalisation of *pixels*). It is assumed that within each voxel any reactions are well-mixed, but that only molecules in the same voxel can react. This must be combined with a model allowing molecules to diffuse from one voxel to an adjacent voxel, and these diffusion events can also be described using 'reaction' notation. It is then possible to write down the Kolmogorov forward equation describing the time–evolution of the joint probability distribution of the state of the system, and this is known as the *reaction–diffusion master equation (RDME)*. Then one can simulate the time course behaviour using an algorithm such as the *next subvolume method* (Elf and Ehrenberg, 2004), such as is implemented in the software MesoRD (Hattne et al., 2005); see also Ander et al. (2004) and Lemerle et al. (2005) for a related approach.

Such algorithms will not be particularly efficient if the number of molecules being considered is significantly smaller than the number of subvolumes. In this case, it may be preferable to dispense with the grid and explicitly model the position of every molecule in the system. This is the second commonly used approach to modelling discrete stochastic reaction–diffusion systems, and this general class of methods are typically referred to as *single-particle* or *individual-based* models. Here some kind of model is required for deciding when pairs of molecules should react (e.g., when they get 'sufficiently close'). Furthermore, although this modelling approach seems quite natural for bi-molecular reactions, it is less natural for first-order (or any order other than second) reactions. Again, some convention must be developed, and some authors favour the introduction of phantom *pseudo-molecules* which can be used to turn first-order reactions into second-order reactions so that they don't need to be treated differently. Efficient and accurate numerical simulation of these models is not trivial, and all current implementations rely on some approximations, but there are some implementations freely available. Examples include StochSim (Le Novère and Shimizu, 2001) or Smoldyn; see, e.g., Andrews et al. (2010). Note that single particle methods are not particularly attractive in the context of regular systems biology models such as those considered in Chapter 7, as they will typically be more computationally intensive than the Gillespie algorithm, as well as inexact. However, algorithms that represent each molecule as a single software object can sometimes be useful in spatial settings, and also in the context of modelling complex chemical species where the potential number of distinct molecular species is vast, yet only a small number are expected to be present during the course of a given simulation run; see Hlavacek et al. (2006) for further details relating to complex species modelling.

In this text we will follow the first approach of discretising space to obtain a reaction–diffusion master equation model, as this builds more naturally on the material we have developed thus far. We will begin by considering the unrealistic setting of one-dimensional space, as the notation is much simpler and the computational results are obtained more quickly and are more straightforward to visualise. We will then generalise to two-dimensional models, from which the generalisation to three-dimensional space will be clear.

9.2 One-dimensional reaction–diffusion systems

We begin by considering reaction–diffusion systems in 1-d. Here we assume that every molecule in our system has a position in space which can be described using a single real number. In practice we tend to assume that the position, x, of each molecule must lie in some contiguous closed and bounded subset of $\mathbb{R}$, that is, $x \in [a, b]$ for some $a, b \in \mathbb{R}$. We assume this bounded region has a 'volume' V (which here is just the length $V = b - a$). For our discretisation approach we divide our interval into N equal sized 'sub-volumes' (here, just sub-intervals), each of which has 'volume' $V_s = V/N$ (which here is just a length). We assume that V_s is small enough so that molecular reactions taking place *within* any of the sub-volumes can be regarded as 'well–mixed', and hence can be described using the techniques we have developed in the previous chapters. The sub-volume size V_s is sometimes referred to as the *meso-scopic scale* of the system. Before proceeding further, it is worth emphasising that, for reasons to be elucidated, we are *not* contemplating any kind of limiting calculus-style approach where we let N get large and V_s get arbitrarily small. It turns out that the model under consideration breaks down in such a limit. We will look later at an approximate model which does have a sensible continuum limit, but this one does not, so the choice of meso-scopic scale V_s is an important modelling assumption. Essentially, V_s must be small enough so that the well–mixed assumption is reasonable within each subvolume, but must be such that the diameter of the volume is much larger than the typical reaction distance between the species of interest. Otherwise the assumption that molecules in different subvolumes can't react becomes implausible.

We assume at this stage that we have N independent compartments where reactions proceed according to a typical discrete stochastic kinetic model. The compartments must now be coupled by allowing molecules to *diffuse* from one compartment to an adjacent compartment at an appropriate rate. Consistent with typical stochastic kinetic modelling assumptions, we assume that each molecule has a constant hazard d of moving into any adjacent compartment. In the 1-d case, most compartments have two adjacent compartments (one on the 'left' and one on the 'right'), and so the hazard of any particular molecule moving out of such a compartment will be $2d$. Note that we assume that the hazard is the same for any molecule of a particular species, but we allow molecules of different species to have different diffusion rates (e.g., big molecules might diffuse more slowly than small molecules). Where it is helpful to include this in the notation, we will include it, so for example, species $\mathcal{X}_j$ could have diffusion coefficient d_j.

We are now in a position to enumerate all of the reactions associated with this 1-d system. As usual, we assume that we are modelling u species, $\mathcal{X}_1, \ldots, \mathcal{X}_u$ and v reactions $\mathcal{R}_1, \ldots, \mathcal{R}_v$. Since reactions can only occur within a subvolume, we treat the species and reactions in each subvolume as independent. We therefore introduce a superscript to denote the volume of interest, in order to get a total of uN different species, $\mathcal{X}_j^k$, $j = 1, \ldots, u$, $k = 1, \ldots, N$, and a total of vN different chemical reactions $\mathcal{R}_i^k$, $i = 1, \ldots, v$, $k = 1, \ldots, N$. The reactions associated with compartment k involve only the species in compartment k. We also need reactions corresponding to the diffusion events. There are $u(N-1)$ reactions corresponding to 'diffuse right':

$$\mathcal{D}_j^{k+} : \quad \mathcal{X}_j^k \xrightarrow{d_j} \mathcal{X}_j^{k+1}, \quad j = 1, \ldots, u, \quad k = 1, \ldots, N-1,$$

and another $u(N-1)$ reactions corresponding to 'diffuse left':

$$\mathcal{D}_j^{k-} : \quad \mathcal{X}_j^k \xrightarrow{d_j} \mathcal{X}_j^{k-1}, \quad j = 1, \ldots, u, \quad k = 2, \ldots, N.$$

Our 1-d reaction–diffusion system is therefore just a regular discrete stochastic kinetic model having uN species and $vN + 2u(N-1)$ reactions. We can therefore simulate exact realisations from it using the Gillespie algorithm. In principle this requires no new algorithmic developments, but in practice, convenient and efficient simulation requires a few tricks.

The most important optimisation is to selectively recompute hazards after each reaction event. In particular, a chemical reaction event occurring in subvolume k can only lead to changes in hazards for reactions associated with that volume, that is, reactions $\mathcal{D}_j^{k+}, \mathcal{D}_j^{k-}$, $j = 1, \ldots, u$ and $\mathcal{R}_i^k$, $i = 1, \ldots, v$. Thus, a maximum of $v + 2u$ hazards can change out of a potential $vN + 2u(N-1)$. Similarly, for a diffusion reaction, only hazards in the two adjacent affected subvolumes can change, which corresponds to the hazards of $2(v + 2u)$ reactions. For a large number of subvolumes, N, this selective recalculation of hazards will represent a significant computational saving.

Other potential optimisations follow from grouping together reactions into groups, then picking a group according to the combined hazard associated with that group, before subsequently picking a reaction within that group. More than one level of grouping is possible, and there are different ways to group the reactions. One way to do this is to begin by grouping together all reactions associated with a given sub-volume (both chemical and diffusion reactions). Then by computing the combined hazard associated with each subvolume, we can randomly pick the subvolume containing the next reaction event. Conditional on this, we can then decide whether the event is a reaction or a diffusion. Conditional on this, we can then decide which reaction or diffusion event occurred. By combining this approach with ideas from the next reaction method discussed in Section 8.2.2, and carefully implementing the algorithm with appropriate data structures, it is possible to get an algorithm which scales well with the number of subvolumes, N. This algorithm, known as the *next subvolume method*, is described in Elf and Ehrenberg (2004) and implemented in the software MesoRD (Hattne et al., 2005).

However, there are other ways to group the reactions which can sometimes be more convenient in vector-oriented languages such as R. Here we will consider first grouping reactions according to whether they are chemical or diffusive, and we will decide first whether the next reaction event is a chemical reaction or a diffusion. Conditional on this, we will then decide in which subvolume this reaction occurs, and conditional on this, we will decide exactly which reaction takes place. This is also a completely valid approach, also leading to an exact simulation algorithm. In order to keep the implementation as short and simple as possible, we will not consider any optimisations such as selective recomputation of hazards. An implementation of this algorithm is given in Figure 9.1. It is the analogue of the well-mixed version, `StepGillespie`, from Chapter 6. This function, `StepGillespie1D`, accepts as input an SPN object and a vector of diffusion coefficients (one for each species), and returns a function closure for advancing the state of the system. The function advances the state of the system using a Gillespie algorithm with reactions grouped as just described. The only variation on the theory described so far relates to diffusion at the boundary. This implementation assumes *periodic boundary conditions*, as this can help to reduce the prominence of boundary effects in simulated trajectories. Here, there are always two possible directions for molecules to diffuse, even in the boundary subvolumes. In this case, a diffusion 'left' from subvolume 1 will move the molecule into subvolume N. Similarly, a diffusion 'right' from subvolume N will move the molecule to subvolume 1. This wrap-around behaviour is typically not physically realistic, but is often convenient for synthetic simulations.

The function `StepGillespie1D` can be used to create a function closure for simulating from the transition kernel of a reaction–diffusion system. As in the well-mixed case, to simulate time course trajectories, we need a function analogous to `simTs` for repeatedly calling the function and storing the intermediate states in an appropriate data structure. In the well-mixed case, the state at each time point was a vector, and so we could store a sequence of states in a matrix (2-dimensional array). Here the state at each time point is itself a sequence of vectors, and hence can be regarded as a matrix. For time series simulation, we need to store a sequence of such matrices, and so we can store these in a 3-dimensional array. Fortunately this is very straightforward, since R has good support for multi-dimensional arrays. So our required function, `simTs1D`, shown in Figure 9.2, looks very like its well-mixed equivalent.

Some example code showing how to use `StepGillespie1D` and `simTs1D` together in order to simulate trajectories from reaction–diffusion systems is given in Figure 9.3, and the resulting plot is shown in Figure 9.4. Here we have simulated a trajectory from the LV predator–prey model previously considered in the well-mixed case. Visualising the results of reaction–diffusion simulations is more challenging here than in non-spatial settings. Here, since there are only two species, it is relatively convenient to slice the output into two matrices, and to then visualise the resulting matrices as 'images'. The shading of the image is used to indicate magnitude. In this case, the darker pixels of the image correspond to large numbers of molecules, and the light pixels correspond to small numbers of molecules. In these images, space is on the horizontal axis, and time increases on the vertical axis. A vertical slice through the image will give the time course behaviour at a single subvolume. In this case it is clear that when picking any subvolume, the time course behaviour will show the oscillatory behaviour that we are already familiar with from the well-mixed case. A peak in prey numbers is followed shortly by a peak in predator numbers, causing the prey numbers, and subsequently the predator numbers, to crash. However, looking only at vertical slices misses the interesting spatial behaviour.

The simulation was initialised with a few prey and predators in the central subvolume, with all other subvolumes being initially empty. As the number of prey builds, they start to diffuse out from the central volume in a 'wave'. Horizontal slices through the image give snapshots of the state of the system at any time. If we were to animate the state of the system over time, we would see waves of prey moving through space closely followed by waves of predators following the prey through space as well as time.

9.2.1 Grid and volume scaling issues

As previously explained, we should not regard the choice of subvolume size as arbitrary, or consider a limit as the volume size goes to zero. However, it is perfectly reasonable to want to be able to consider running simulations on somewhat different grid sizes, and to be able to obtain results which are broadly comparable, in that they are not especially sensitive to the choice of grid. For this we need to consider how rates of reactions and diffusion should scale as the grid and subvolume size changes. We have already discussed the scaling of stochastic reaction rate constants in Chapter 6. Second-order reactions should have a rate constant which scales inversely proportional to V_s. First-order reactions should have rates that are independent of V_s. Zeroth-order reactions should be considered carefully, but in principle should scale proportionally to V_s.

This scaling behaviour of second-order reactions illustrates why we can't consider taking the limit as V_s goes to zero, as in this case the rate of the second-order reactions within a subvolume would blow up to infinity. The problem is that for any finite second-order reaction rate, we will 'lose' the second-order reactions as the grid is refined. This is because in order for a second-order reaction to happen, the two reacting molecules must be present in the same subvolume. But as the subvolume size decreases, the probability of this happening becomes arbitrarily small. By making the rate scale inversely to V_s we compensate for this, but this does not work in the limit.

We also need to consider the scaling of the diffusion 'reactions'. Although these are essentially first-order reactions, they aren't 'real' chemical reactions, and so their scaling behaviour needs to be considered separately. We want the diffusion constants to scale so that the speed at which molecules diffuse through space is independent of the choice of grid. Note that the implication of this is that we want diffusion constants to scale with the *length* of a volume element, and not with the actual *volume* of the element. In the 1-d case these are the same, but in higher dimensions they are clearly not. So, on a regular cubical grid, we define l_s to be the length associated with the subvolume having volume V_s. In the 1-d case we have $l_s = V_s$, but in 2-d we will have $l_s^2 = V_s$, and in 3-d we will have $l_s^3 = V_s$. We now think about the position of a molecule in the grid as following a random walk over time. The number of diffusion events in each possible direction in time t will be Poisson with mean td. The displacement will therefore be a difference of independent Poisson random variables, having mean zero and variance $2td$, which is also the expected squared distance. Consequently, the expected distance from the origin (measured in subvolume lengths) will increase with the square root of t. This property of expected distance increasing with the square root of t is independent of the spatial dimension; in particular, it is also true in 2 and 3 dimensions. So, if we halve the grid size, l_s, it will take $2^2 = 4$ times as long to obtain the same expected physical distance, so we will need to increase the rate of diffusion by a factor of 4 in order to compensate. This tells us that the diffusion rate must scale as an inverse square of the subvolume length. That is, we should have

$$d = D/l_s^2,$$

where D is a macro-scopic diffusion constant.

We can illustrate these ideas by repeating our previous example on a grid that has been refined by a factor of 2. From the above analysis, we know that we should increase the diffusion constants by a factor of 4 and increase the rate of the second-order reaction by a factor of 2. Code to run this model is given in Figure 9.5, and the resulting plot is shown in Figure 9.6. We see that by adjusting the rate constants appropriately, the overall qualitative behaviour of the reaction–diffusion system is maintained.

9.2.2 *Approximations based on the CLE*

Discrete stochastic reaction–diffusion systems are relatively easy to build, and fairly straightforward to simulate. However, they become very computationally demanding to simulate, especially in 2 and 3 dimensions. Although there are computational savings which can be made by carefully optimising the implementation, and significant speedups that can be obtained by re-implementing in an efficient compiled programming language, simulation of interesting systems on large 2 and 3 dimensional grids is still problematic. Consequently, even more than is the case for well-mixed problems, there is considerable interest in developing fast approximate algorithms for stochastic reaction–diffusion models.

SDEs for diffusion

It is entirely possible to develop Poisson time-stepping algorithms for reaction–diffusion systems. However, the most significant computational savings come from relaxing discreteness and using CLE approximations. Here we can replace all of the reactions with CLE approximations to obtain a large set of coupled SDEs. Indeed, it turns out that this set of SDEs can be considered to be a discretisation of a *stochastic partial differential equation* (SPDE) model. This is interesting, because this discretisation *is* well-behaved as the grid size goes to zero, converging to the limiting SPDE model.

Let's begin by deriving the SDEs associated with diffusion of an arbitrary molecular species, $\mathcal{X}$. Let X_t^k be the number of molecules of $\mathcal{X}$ in volume k at time t, and let d_x be the (grid-dependent) diffusion rate. Consider first the flow of molecules from volume k to $k + 1$. Each molecule diffuses at rate d_x, so the combined rate is $X_t^k d_x$. In a small time interval δt, the number of molecules flowing from k to $k + 1$ will be approximately $Po(X_t^k d_x \, \delta t)$. However, the number of molecules flowing from $k + 1$ to k will be approximately $Po(X_t^{k+1} d_x \, \delta t)$, independently. So the *net* flow from k to $k + 1$ will be a random variable with expectation $(X_t^k - X_t^{k+1}) d_x \, \delta t$ and variance $(X_t^k + X_t^{k+1}) d_x \, \delta t$. Then, since the net flow from $k - 1$ to k is similar, it is clear that the expected change in the number of molecules in k is $(X_t^{k-1} - 2X_t^k + X_t^{k+1}) d_x \, \delta t$. If we now let W_t^k be independent Brownian motions (for each k) governing the net flow from $k + 1$ to k, we can write down a set of coupled SDEs governing the diffusion of $\mathcal{X}$ as

$$dX_t^k = (X_t^{k-1} - 2X_t^k + X_t^{k+1})d_x \, dt + \sqrt{(X_t^k + X_t^{k+1})d_x} \, dW_t^k$$
$$- \sqrt{(X_t^{k-1} + X_t^k)d_x} \, dW_t^{k-1}. \quad (9.1)$$

This clearly has the correct infinitesimal behaviour. Also note that due to the coupling of adjacent noise terms, the total number of molecules in the system is conserved, despite the random fluctuations of the diffusion process. We need to make some assumptions regarding the boundary conditions, $k = 1$ and $k = N$. We will again assume periodic boundary conditions in the implementation. It is illuminating to write this system of coupled SDEs as a multivariate SDE. For this, put $X_t = (X_t^1, \dots, X_t^N)^\mathsf{T}$ and $W_t = (W_t^1, \dots, W_t^N)^\mathsf{T}$, and introduce the *back-shift* matrix,

$$B = \begin{pmatrix} 0 & 0 & 0 & \cdots & 0 & 1 \\ 1 & 0 & 0 & \ddots & 0 & 0 \\ 0 & 1 & 0 & \ddots & 0 & 0 \\ \vdots & \ddots & \ddots & \ddots & \ddots & \vdots \\ 0 & 0 & 0 & \cdots & 1 & 0 \end{pmatrix},$$

noting that $B^{-1} = B^{\mathsf{T}}$ is the corresponding forward-shift. Also define $\nabla = I - B$, the backward difference matrix, and $\Delta = \nabla^2 B^{-1} = \nabla^2 B^{\mathsf{T}}$, the *discrete Laplacian*, so that

$$\Delta = \begin{pmatrix} -2 & 1 & 0 & 0 & \cdots & 0 & 1 \\ 1 & -2 & 1 & 0 & \ddots & 0 & 0 \\ 0 & 1 & -2 & 1 & \ddots & 0 & 0 \\ \vdots & \ddots & \ddots & \ddots & \ddots & \ddots & \vdots \\ 0 & 0 & 0 & 0 & \ddots & -2 & 1 \\ 1 & 0 & 0 & 0 & \cdots & 1 & -2 \end{pmatrix}.$$

Using this notation, we can write our system of coupled SDEs as the single multivariate SDE

$$dX_t = d_x \Delta X_t \, dt + \sqrt{d_x} \nabla \operatorname{diag} \left\{ \sqrt{(I + B^{\mathsf{T}})X_t} \right\} dW_t. \tag{9.2}$$

Now since ∇ (and hence Δ) conserve total mass*, it is again clear that this multivariate SDE will leave the total mass invariant, despite the dissipation and the fluctuations.

Before proceeding further, note that if we were to ignore the fluctuations and set the noise terms to zero, we would have the multivariate ODE

$$\frac{dx}{dt} = d_x \Delta x.$$

To anyone familiar with the numerical solution of *partial differential equations* (PDEs), it will be clear that this corresponds to a *method-of-lines* solution to the *heat equation*,

$$\frac{\partial u}{\partial t} - d_x \Delta u = 0,$$

where now $u(x, t)$ is a function of a continuous spatial coordinate x and Δ refers to the *Laplace operator*,

$$\Delta = \frac{\partial^2}{\partial x^2}.$$

As a multivariate SDE (complete with noise term), (9.2) represents a method-of-lines discretisation of a *stochastic partial differential equation* (SPDE) model of diffusion,

---

* The sum of the elements of ∇x is zero for any vector x. Consequently, the same is true for Δx.

the *stochastic heat equation*. SPDEs require significant technical machinery to explain clearly, so the details of this are beyond the scope of this text, but see Lord et al. (2014) for a comprehensive introduction to this topic.

Reaction–diffusion Langevin equation

We now know how to write an SDE representing stochastic diffusion on a 1-d grid, and we know how to write a CLE for each subvolume, so we can put these together to obtain a full Langevin equation for stochastic reaction–diffusion, sometimes known as the *spatial CLE* (Ghosh et al., 2015). If X_{jt}^k denotes the amount of $\mathcal{X}_j$ in volume k at time t, we can first write down the CLE corresponding to the reactions as components of (8.2):

$$dX_{jt}^k = S_j h(X_t^k)\, dt + S_j \operatorname{diag}\left\{\sqrt{h(X_t^k)}\right\} d\widetilde{W}_t^k,$$

where S_j is the jth row of S, and $\widetilde{W}_t^k$ is a v-dimensional Brownian motion for each k. We can now write the spatial CLE in component form as

$$dX_{jt}^k = \left[(X_{jt}^{k-1} - 2X_{jt}^k + X_{jt}^{k+1})d_j + S_j h(X_t^k)\right] dt +$$
$$\sqrt{(X_{jt}^k + X_{jt}^{k+1})d_j}\, dW_{jt}^k - \sqrt{(X_{jt}^{k-1} + X_{jt}^k)d_j}\, dW_{jt}^{k-1} +$$
$$S_j \operatorname{diag}\left\{\sqrt{h(X_t^k)}\right\} d\widetilde{W}_t^k. \quad (9.3)$$

This coupled system of uN SDEs driven by $(u + v)N$ Brownian motions is rarely analytically tractable, but it can be numerically integrated using any appropriate method, such as the Euler–Maruyama approximation previously considered. In practice, it is possible to perform the diffusion and reaction update steps separately, and apply them sequentially. Asymptotically, there is no difference in doing this, and anecdotally, for finite time steps it can lead to more stable behaviour. In some contexts, this technique is known as *operator splitting*, since formally it corresponds to splitting the forward operator of the Markov process into two separate terms. An R function to implement this, StepCLE1D, is shown in Figure 9.7. The only aspect that potentially requires comment is the rectify step. In principle, (9.3) defines a non-negative stochastic process, and hence no component should ever go negative. However, finite discretisations such as Euler-Maruyama will potentially drift to negative values, which will cause subsequent problems. A simple fix for this is to set any negative components to zero, though it is worth noting that there are other possibilities, such as reflection, which would instead replace each component with its absolute value. Example code, showing how to re-do Figure 9.5 using StepCLE1D in place of StepGillespie1D, is given in Figure 9.8, and the results are shown in Figure 9.9, which looks qualitatively similar to Figure 9.6. Since StepCLE1D runs much faster than StepGillespie1D, we can do a further grid refinement: code in Figure 9.10 (showing rate constants adjusted appropriately) and plots in Figure 9.11.

9.3 Two-dimensional reaction–diffusion systems

The study of reaction–diffusion systems with just one spatial dimension is useful and informative for understanding the essential concepts, but is of limited practical relevance. Most reaction–diffusion systems of interest in systems biology will involve two or three spatial dimensions. Here we will concentrate on the 2-d case as it is computationally cheaper and the results are simpler to visualise. The 3-d case is conceptually very similar, and the generalisation should be clear.

In 2-d we partition space using a square grid of subvolumes, which we can index with two numbers, k and l. So now in addition to having diffusion reactions which can move molecules 'left' and 'right', we now have additional reactions to move molecules 'up' and 'down'. Otherwise the situation is the same as for the 1-d case. We use $\mathcal{X}_j^{k,l}$ to represent species j in subvolume (k, l), and $X_{jt}^{k,l}$ to denote the number of molecules of $\mathcal{X}_j^{k,l}$ at time t. Since molecules can diffuse in 4 directions, we have 4 diffusion reactions associated with every species in every subvolume (assuming periodic boundary conditions). So, assuming an $N \times N$ grid, there will be a total of $4uN^2$ diffusion reactions in the system. There are also v chemical reactions per subvolume, so there will be a total of $N^2(v + 4u)$ reactions in the system. This will clearly be computationally challenging for large N.[†] If we assume *isotropic* diffusion, then we can use d_j as the reaction rate for diffusion in all 4 directions in every subvolume, but we still allow the rate to be specific to a molecular species.

The function StepGillespie1D is easily generalised to 2-d to obtain the function StepGillespie2D. To save space, the code is omitted from the text, but is included in the R package smfsb and can be inspected at the R command line. Similarly, the function simTs1D easily generalises to simTs2D, which now returns a 4-dimensional array of results rather than a 3-dimensional array. R code demonstrating how to use these functions to simulate a realisation from a 2-d reaction diffusion system is shown in Figure 9.12, and the resulting plots, showing a snapshot of the system state at the final time, is shown in Figure 9.13. Note that using the verb option of simTs2D, an animation of the state of the first species is shown on the console while the function runs. We see waves of prey followed by waves of predator, now propagating out from the centre of the 2-d region. Note that the code may take several hours to run on a typical personal computer.

9.3.1 Spatial CLE

The computationally intensive nature of discrete stochastic reaction–diffusion simulation in 2 and 3 dimensions makes fast approximations such as the spatial CLE even more attractive. Here we will concentrate on the 2-d case. The approximation proceeds similarly to the 1-d case. But now in addition to having a collection of Brownian motions governing diffusion in the horizontal direction, we now require an additional collection of Brownian motions governing diffusion in the vertical di-

---

[†] We can consider very briefly the 3-d case, assuming an $N \times N \times N$ square grid, there are now 6 directions in which one can diffuse (one for each face of the cubical volume), and so the total number of reactions will be $N^3(v + 6u)$. This is computationally demanding to work with for any non-trivial N.

rection. Again, the noise terms are carefully coupled to ensure that the diffusion fluctuations conserve mass. We can then generalise (9.3), and write down the components of our 2-d spatial CLE as

$$
\begin{aligned}
dX_{jt}^{k,l} = &\left[(X_{jt}^{k-1,l} + X_{jt}^{k,l-1} - 4X_{jt}^{k,l} + X_{jt}^{k+1,l} + X_{jt}^{k,l+1})d_j + S_j h(X_t^{k,l}) \right] dt + \\
&\sqrt{(X_{jt}^{k,l} + X_{jt}^{k+1,l})d_j}\, dW_{jt}^{k,l} - \sqrt{(X_{jt}^{k-1,l} + X_{jt}^{k,l})d_j}\, dW_{jt}^{k-1,l} + \\
&\sqrt{(X_{jt}^{k,l} + X_{jt}^{k,l+1})d_j}\, d\widehat{W}_{jt}^{k,l} - \sqrt{(X_{jt}^{k,l-1} + X_{jt}^{k,l})d_j}\, d\widehat{W}_{jt}^{k,l-1} + \\
&\qquad\qquad\qquad\qquad\qquad\qquad S_j \operatorname{diag}\left\{ \sqrt{h(X_t^{k,l})} \right\} d\widetilde{W}_t^{k,l}. \quad (9.4)
\end{aligned}
$$

We can simulate approximate realisations from this process using `StepCLE2D`, which is a straightforward generalisation of `StepCLE1D`. Again, it is omitted from the text to save space, but is included in the `smfsb` R package. We can re-do the previous example using `StepCLE2D` instead of `StepGillespie2D`. Code for this is shown in Figure 9.14 and the output in Figure 9.15. Note that this code is *much* quicker to run than the exact discrete version.

Since the CLE approximation runs much faster than an exact discrete event simulation approach, it is reasonable to consider simulation on much bigger and/or finer grids. Figure 9.16 shows the results of running the previous example on a much bigger grid, and Figure 9.17 shows the code required in order to produce it.

9.4 Exercises

1. Use the `smfsb` R package to reproduce the early figures from this Chapter (concentrating on the 1-d case, as these are quicker to generate).

2. Use the `spnModels`, `LV` (which is not volume-scaled) and `LVV` (which is) in order to investigate grid-size dependence and second-order reactions (again, start with the 1-d case). Use exact discrete stochastic simulation in order to be sure that results are not artefacts of additional approximations. See how things break if you don't re-scale diffusion constants and/or rate constants as the grid changes. Investigate problems that arise on very fine grids.

3. Investigate the sensitivity of the results of using `StepCLE1D` to the choice of algorithmic time step, `dt`.

4. Inspect the 2-d code omitted from the text but included in the `smfsb` R package: `StepGillespie2D`, `simTs2D`, `StepCLE2D`. Compare them to the corresponding 1-d versions.

5. Simulate a spatial SIR model, using the model `SIR` from the `spnModels` collection. Initialise your model with susceptibles distributed uniformly in space, and a few infectives in the middle (and no recovereds initially). Simulate in both one and two spatial dimensions, using both exact and approximate simulation algorithms. Investigate the effect of different grid sizes and parameter scalings. Experiment with visualisation of the results.

9.5 Further reading

For further analysis of spatial stochastic systems, Chapter 8 of Gardiner (2004) is worth consulting, though this text is aimed primarily at physicists. For more details on the next subvolume method for efficient exact stochastic simulation, see Elf and Ehrenberg (2004). Schnoerr et al. (2016) provide a nice overview of more recent developments in this field. Better theoretical understanding of these approaches to stochastic simulation of reaction–diffusion systems, and their relationship to single particle methods, is explored in a range of papers by Samuel Isaacson; see Isaacson (2008) and Isaacson and Isaacson (2009), for example. For more on SPDEs, see Lord et al. (2014).

```r
StepGillespie1D <- function (N, d)
{
    S = t(N$Post - N$Pre)
    v = ncol(S)
    u = nrow(S)
    return(function(x0, t0, deltat, ...) {
        t = t0
        x = x0
        n = dim(x)[2]
        termt = t + deltat
        repeat {
            hr = apply(x, 2, function(x) {
                N$h(x, t, ...)
            })
            hrs = apply(hr, 2, sum)
            hrss = sum(hrs)
            hd = x * (d * 2)
            hds = apply(hd, 2, sum)
            hdss = sum(hds)
            h0 = hrss + hdss
            if (h0 < 1e-10) t = 1e+99 else if (h0 > 1e+07) {
                t = 1e+99
                warning("Hazard too big - terminating!")
            } else t = t + rexp(1, h0)
            if (t > termt) return(x)
            if (runif(1, 0, h0) < hdss) {
                j = sample(n, 1, prob = hds)
                i = sample(u, 1, prob = hd[, j])
                x[i, j] = x[i, j] - 1
                if (runif(1) < 0.5) {
                    if (j > 1) x[i, j - 1] = x[i, j - 1] + 1
                        else x[i,
                        n] = x[i, n] + 1
                } else {
                    if (j < n) x[i, j + 1] = x[i, j + 1] + 1
                        else x[i,
                        1] = x[i, 1] + 1
                }
            } else {
                j = sample(n, 1, prob = hrs)
                i = sample(v, 1, prob = hr[, j])
                x[, j] = x[, j] + S[, i]
            }
        }
    })
}
```

Figure 9.1 *An R function to implement the Gillespie algorithm for a stochastic Petri net on a 1-d grid.*

```
simTs1D <- function (x0, t0 = 0, tt = 100, dt = 0.1,
    stepFun, verb = FALSE,
    ...)
{
    N = (tt - t0)%/%dt + 1
    u = nrow(x0)
    n = ncol(x0)
    arr = array(0, c(u, n, N))
    x = x0
    t = t0
    arr[, , 1] = x
    if (verb)
        cat("Steps", rownames(x), "\n")
    for (i in 2:N) {
        if (verb)
            cat(N - i, apply(x, 1, sum), "\n")
        t = t + dt
        x = stepFun(x, t, dt, ...)
        arr[, , i] = x
    }
    arr
}
```

Figure 9.2 *R code for repeatedly calling a 1-d transition kernel to obtain a time series of outputs from a 1-d reaction–diffusion model.*

```
data(spnModels)
N=20; T=30
x0=matrix(0,nrow=2,ncol=N)
rownames(x0)=c("x1","x2")
x0[,round(N/2)]=LV$M
stepLV1D = StepGillespie1D(LV,c(0.6,0.6))
xx = simTs1D(x0,0,T,0.2,stepLV1D,verb=TRUE)
op=par(mfrow=c(1,2))
image(xx[1,,],main="Prey",xlab="Space",ylab="Time",
    col=gray((32:0)/32))
image(xx[2,,],main="Predator",xlab="Space",ylab="Time",
    col=gray((32:0)/32))
par(op)
```

Figure 9.3 *R code showing how to use the functions* StepGillespie1D *and* simTs1D *together in order to simulate a realisation from an SPN on a 1-d grid. The resulting plot is shown in Figure 9.4.*

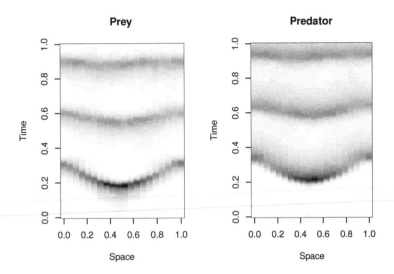

Figure 9.4 *A single simulated realisation from a 1-d discrete stochastic reaction–diffusion model corresponding to an LV predator–prey system, over a grid with N = 20 subvolumes. Code for generating this plot is given in Figure 9.3.*

```
data(spnModels)
N=40; T=30
x0=matrix(0,nrow=2,ncol=N)
rownames(x0)=c("x1","x2")
x0[,round(N/2)]=LV$M
stepLV1D = StepGillespie1D(LV,c(2.4,2.4))
xx = simTs1D(x0,0,T,0.2,stepLV1D,verb=TRUE,th=c(th1=1,
    th2=0.01,th3=0.6))
op=par(mfrow=c(1,2))
image(xx[1,,],main="Prey",xlab="Space",ylab="Time",
    col=gray((32:0)/32))
image(xx[2,,],main="Predator",xlab="Space",ylab="Time",
    col=gray((32:0)/32))
par(op)
```

Figure 9.5 *R code showing how to simulate the LV reaction–diffusion model on a fine grid. Diffusion and reaction rate constants are chosen to match the overall qualitative behaviour of the model considered in Figure 9.3. The resulting plot is shown in Figure 9.6.*

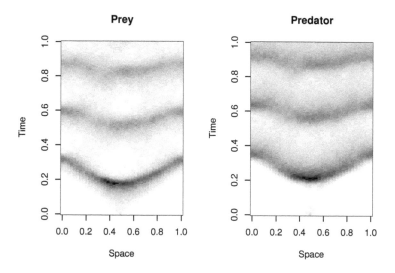

Figure 9.6 *A single simulated realisation from a 1-d discrete stochastic reaction–diffusion model corresponding to an LV predator–prey system, over a grid with $N = 40$ subvolumes. The plot shows qualitative behaviour similar to that in Figure 9.4. Code for generating this plot is given in Figure 9.5.*

```
StepCLE1D <- function (N, d, dt = 0.01)
{
    S = t(N$Post - N$Pre)
    v = ncol(S)
    u = nrow(S)
    sdt = sqrt(dt)
    forward <- function(m) cbind(m[, 2:ncol(m)], m[, 1])
    back <- function(m) {
        n = ncol(m)
        cbind(m[, n], m[, 1:(n - 1)])
    }
    laplacian <- function(m) forward(m) + back(m) - 2 * m
    rectify <- function(m) {
        m[m < 0] = 0
        m
    }
    diffuse <- function(m) {
        n = ncol(m)
        noise = matrix(rnorm(n * u, 0, sdt), nrow = u)
        m = m + d * laplacian(m) * dt + sqrt(d) * (sqrt(m +
            forward(m)) *
            noise - sqrt(m + back(m)) * back(noise))
        m = rectify(m)
        m
    }
    return(function(x0, t0, deltat, ...) {
        x = x0
        t = t0
        n = ncol(x0)
        termt = t0 + deltat
        repeat {
            x = diffuse(x)
            hr = apply(x, 2, function(x) {
                N$h(x, t, ...)
            })
            dwt = matrix(rnorm(n * v, 0, sdt), nrow = v)
            x = x + S %*% (hr * dt + sqrt(hr) * dwt)
            x = rectify(x)
            t = t + dt
            if (t > termt) return(x)
        }
    })
}
```

Figure 9.7 *R function for implementing the CLE approximation for a 1-d stochastic reaction diffusion system. Intended to be used as a fast approximation to* StepGillespie1D.

```
data(spnModels)
N=40; T=30
x0=matrix(0,nrow=2,ncol=N)
rownames(x0)=c("x1","x2")
x0[,round(N/2)]=LV$M
stepLV1D = StepCLE1D(LV,c(2.4,2.4),dt=0.05)
xx = simTs1D(x0,0,T,0.2,stepLV1D,verb=TRUE,th=c(th1=1,
    th2=0.01,th3=0.6))
op=par(mfrow=c(1,2))
image(xx[1,,],main="Prey",xlab="Space",ylab="Time",
    col=gray((32:0)/32))
image(xx[2,,],main="Predator",xlab="Space",ylab="Time",
    col=gray((32:0)/32))
par(op)
```

Figure 9.8 *R code demonstrating the use of* StepCLE1D *for the simulation of a 1-d reaction–diffusion model. The output of this code is shown in Figure 9.9.*

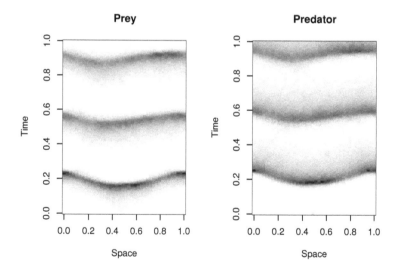

Figure 9.9 *A simulated realisation of a spatial CLE model, corresponding to the code in Figure 9.8.*

```
data(spnModels)
N=80; T=30
x0=matrix(0,nrow=2,ncol=N)
rownames(x0)=c("x1","x2")
x0[,round(N/2)]=LV$M
stepLV1D = StepCLE1D(LV,c(9.6,9.6),dt=0.01)
xx = simTs1D(x0,0,T,0.1,stepLV1D,verb=TRUE,th=c(th1=1,
    th2=0.02,th3=0.6))
op=par(mfrow=c(1,2))
image(xx[1,,],main="Prey",xlab="Space",ylab="Time",
    col=gray((32:0)/32))
image(xx[2,,],main="Predator",xlab="Space",ylab="Time",
    col=gray((32:0)/32))
par(op)
```

Figure 9.10 *R code to simulate a reaction diffusion system on a fine grid. Results are shown in Figure 9.11.*

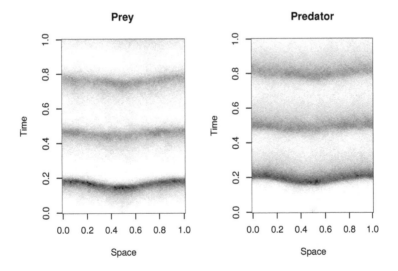

Figure 9.11 *Simulation of a spatial CLE on a fine grid, corresponding to the code in Figure 9.10.*

```r
data(spnModels)
m=20; n=30; T=10
x0=array(0,c(2,m,n))
dimnames(x0)[[1]]=c("x1","x2")
x0[,round(m/2),round(n/2)]=LV$M
stepLV2D = StepGillespie2D(LV,c(0.6,0.6))
xx = simTs2D(x0,0,T,0.2,stepLV2D,verb=TRUE)
N = dim(xx)[4]
op=par(mfrow=c(1,2))
image(xx[1,,,N],main="Prey",xlab="Space",ylab="Time",col
    =grey((32:0)/32))
image(xx[2,,,N],main="Predator",xlab="Space",ylab="Time"
    ,col=grey((32:0)/32))
par(op)
```

Figure 9.12 *R code showing how to use the functions* StepGillespie2D *and* simTs2D *together in order to simulate a realisation from an SPN on a 2-d grid. The resulting plot is shown in Figure 9.13.*

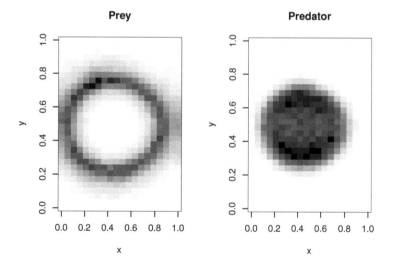

Figure 9.13 *A single simulated realisation from a 2-d discrete stochastic reaction–diffusion model corresponding to an LV predator–prey system, over a* 20×30 *grid. The plot shows a snapshot of the system state at time* $T = 10$. *Code for generating this plot is given in Figure 9.12.*

```
data(spnModels)
m=20; n=30; T=6
x0=array(0,c(2,m,n))
dimnames(x0)[[1]]=c("x1","x2")
x0[,round(m/2),round(n/2)]=LV$M
stepLV2D = StepCLE2D(LV,c(0.6,0.6),dt=0.1)
xx = simTs2D(x0,0,T,0.2,stepLV2D,verb=FALSE)
N = dim(xx)[4]
op=par(mfrow=c(1,2))
image(xx[1,,,N],main="Prey",xlab="Space",ylab="Time",col
    =grey((32:0)/32))
image(xx[2,,,N],main="Predator",xlab="Space",ylab="Time"
    ,col=grey((32:0)/32))
par(op)
```

Figure 9.14 *R code showing how to use the function* StepCLE2D *in order to simulate an approximate realisation from an SPN on a 2-d grid. The resulting plot is shown in Figure 9.15.*

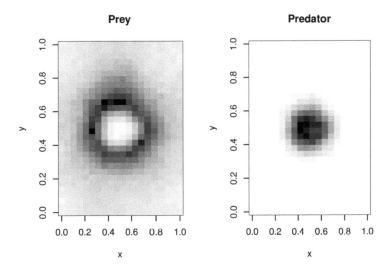

Figure 9.15 *A single simulated realisation from a 2-d spatial CLE reaction–diffusion model corresponding to an LV predator–prey system, over a 20 × 30 grid. The plot shows a snapshot of the system state at time $T = 6$. Code for generating this plot is given in Figure 9.14.*

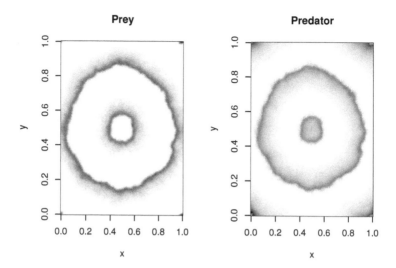

Figure 9.16 *A single simulated realisation from a 2-d spatial CLE reaction–diffusion model corresponding to an LV predator–prey system, over a 200 × 250 grid. Here the parameters have not been adjusted for the grid size — we simply consider the reaction taking place over a larger space. The plot shows a snapshot of the system state at time T = 30. Code for generating this plot is given in Figure 9.17.*

```
data(spnModels)
m=200; n=250; T=30
x0=array(0,c(2,m,n))
dimnames(x0)[[1]]=c("x1","x2")
x0[,round(m/2),round(n/2)]=LV$M
stepLV2D = StepCLE2D(LV,c(0.6,0.6),dt=0.1)
xx = simTs2D(x0,0,T,0.2,stepLV2D,verb=TRUE)
N = dim(xx)[4]
op=par(mfrow=c(1,2))
image(xx[1,,,N],main="Prey",xlab="Space",ylab="Time",col
    =grey((32:0)/32))
image(xx[2,,,N],main="Predator",xlab="Space",ylab="Time"
    ,col=grey((32:0)/32))
par(op)
```

Figure 9.17 *R code showing how to simulate an approximate realisation from an SPN on a fine 2-d grid. The resulting plot is shown in Figure 9.16.*

## PART IV

Bayesian inference

## CHAPTER 10

Bayesian inference and MCMC

10.1 Likelihood and Bayesian inference

The final part of this book turns attention away from probabilistic modelling and stochastic simulation tools appropriate for systems biology models and toward statistical inference for the parameters of such models on the basis of experimental data. Although most of the earlier parts of the book are intended to be largely self-contained, these last chapters are likely to be more accessible to readers with a basic familiarity with ideas of estimation and statistical inference, including the concepts of likelihood, Bayesian analysis, and multivariate statistics. A standard text on mathematical statistics such as Rice (2007) should provide most of the necessary background. However, any reader who has found the rest of the book straightforward should not have too much difficulty with the final chapters.

Chapter 11 examines the particular problem of inference for stochastic kinetic models using time-course experimental data. In this chapter we develop the methods of Bayesian inference and Markov chain Monte Carlo (MCMC) that will be required in Chapter 11. We will begin by examining the essential concepts of Bayesian inference and the reasons why analytic approaches to Bayesian inference in complex models are generally intractable, before moving on to MCMC and its application to Bayesian inference.

10.1.1 Bayesian inference

Often we are able to understand the probability of some *outcome*, X conditional on various possible *hypotheses*, H_i, $i = 1, 2, \ldots n$, where the H_i form a partition (Definition 3.6). We can then compute probabilities of the form $P(X = x'|H_i)$, $i = 1, \ldots, n$, $x' \in S_X$. However, when we actually *observe* some outcome $X = x$, we are interested in the probabilities of the hypotheses *conditional* on the outcome, $P(H_i|X = x)$. Bayes' theorem (Corollary 3.1) tells us how to compute these, but the answer also depends on the prior probabilities for the hypotheses, $P(H_i)$, and hence to use Bayes' theorem, these too must be specified. Thus Bayes' theorem provides us with a coherent way of updating our prior beliefs about the hypotheses $P(H_i)$ to $P(H_i|X = x)$, our posterior beliefs based on the occurrence of $X = x$, as

$$P(H_i|X = x) = \frac{P(X = x|H_i)\,P(H_i)}{\sum_{j=1}^{n} P(X = x|H_j)\,P(H_j)}, \quad i = 1, \ldots n.$$

Note that the probabilities $P(X = x|H_i)$ are known as *likelihoods*, and are often written $L(H_i; x)$, as they tend to be regarded as a function of the H_i for given fixed outcome x. Note, however, that the *likelihood function* does not represent a PMF for the H_i; in particular, there is no reason to suppose that it will sum to 1.

This is how it all works for purely discrete problems, but some adaptation is required before it can be used with continuous or mixed problems. Let us first stay with discrete outcome X and consider a continuum of hypotheses represented by a continuous parameter Θ. Our prior beliefs must now be represented by a density function, traditionally written as $\pi(\theta)$. Taking the continuous limit in the usual way, Bayes' theorem becomes

$$\pi(\theta|X = x) = \frac{\pi(\theta)\, P(X = x|\theta)}{\displaystyle\int_\Theta P(X = x|\theta')\, \pi(\theta')d\theta'}.$$

In this case, the likelihood function is $L(\theta; x) = P(X = x|\theta)$, regarded as a function of θ for given fixed x. Again, note that the likelihood function is not a density for θ, as it does not integrate to 1. Using this notation, we can rewrite Bayes' theorem as

$$\pi(\theta|X = x) = \frac{\pi(\theta)L(\theta; x)}{\displaystyle\int_\Theta \pi(\theta')L(\theta'; x)d\theta'},$$

and this is the way it is usually written in the context of Bayesian statistics, though the likelihood function $L(\theta; x)$ means slightly different things depending on the context. Note that the integral on the bottom line of Bayes' theorem is not a function of θ, and so simply represents a constant of proportionality. Thus, we can rewrite Bayes' theorem in the simpler form

$$\pi(\theta|X = x) \propto \pi(\theta)L(\theta; x), \tag{10.1}$$

giving rise to the Bayesian mantra *'the posterior is proportional to the prior times the likelihood'*.

Example

Suppose that for a particular gene in a particular cell, transcription events occur according to a Poisson process with rate θ per minute. Prior to carrying out an experiment, a biological expert specifies his opinion regarding θ in the form of a $Ga(a, b)$ distribution (Section 3.11). Suppose that for our expert, $a = 2$, $b = 1$. Counts of the number of transcript events are gathered from n separate one-minute intervals to get data $x = (x_1, x_2, \ldots, x_n)^\mathsf{T}$. In this case the likelihood for θ is

$$L(\theta; x) = \mathrm{P}(x|\theta)$$

$$= \prod_{i=1}^{n} \mathrm{P}(x_i|\theta)$$

$$= \prod_{i=1}^{n} \frac{\theta^{x_i} e^{-\theta}}{x_i!}$$

$$\propto \prod_{i=1}^{n} \theta^{x_i} e^{-\theta}$$

$$= \theta^{\sum_{i=1}^{n} x_i} e^{-n\theta}.$$

The second line follows from the first because the data are independent (given θ). The likelihood depends on the data only through n and $\bar{x}$, so n and $\bar{x}$ are said to be *sufficient statistics* for the likelihood function. Then since θ is gamma, we have

$$\pi(\theta) \propto \theta^{a-1} e^{-b\theta}$$

giving

$$\pi(\theta|x) \propto \pi(\theta) L(\theta; x)$$
$$\propto \theta^{a + \sum_{i=1}^{n} x_i - 1} e^{-(b+n)\theta}.$$

In other words,

$$\theta|x \sim Ga\left(a + \sum_{i=1}^{n} x_i, b + n\right).$$

So in this case, starting with a gamma prior results in a gamma posterior. Problems of this nature are said to be *conjugate*, and so in this case the gamma prior is said to be conjugate for the Poisson likelihood. In the context of our example, observing the data $x = (4, 2, 3)$ leads to a $Ga(11, 4)$ posterior distribution. This distribution (which has an expectation of $11/4$ and a variance of $11/16$) represents our belief about the value of θ having observed the data, and is shown in Figure 10.1.

The analysis is essentially the same when X is continuous rather than discrete, except that here the likelihood is evaluated using the PDF rather than the PMF.

10.1.2 Bayesian computation

In principle, the previous section covers everything we need to know about Bayesian inference — the posterior is nothing more (or less) than a conditional distribution for the parameters given the data. In practice, however, this may not be entirely trivial to work with.

The first problem one encounters is choosing the constant of proportionality so that the density integrates to 1. If the density is non-standard (as is usually the case for non-trivial problems), then the problem reduces to integrating the product of the likelihood and the prior (known as the *kernel* of the posterior) over the support of Θ.

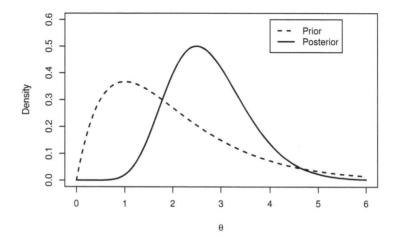

Figure 10.1 *Plot showing the prior and posterior for the Poisson rate example. Note how the prior is modified to give a posterior more consistent with the data (which has a sample mean of 3).*

If the support is infinite in extent, and/or multi-dimensional, then this is a highly non-trivial numerical problem.

Even if we have the constant of integration, if the parameter space is multi-dimensional, we will want to know what the marginal distribution of each component looks like. For each component, we have a very difficult numerical integration problem.

Example

Consider the case where we have a collection of observations, X_i, which we believe to be independent identically distributed (iid) normal with unknown mean and precision (the reciprocal of variance). We write

$$X_i|\mu, \tau \sim N(\mu, 1/\tau).$$

The likelihood for a single observation is

$$L(\mu, \tau; x_i) = f(x_i|\mu, \tau) = \sqrt{\frac{\tau}{2\pi}} \exp\left\{-\frac{\tau}{2}(x_i - \mu)^2\right\}$$

and so for n independent observations, $x = (x_1, \ldots, x_n)^{\mathsf{T}}$ is

$$L(\mu, \tau; x) = f(x|\mu, \tau) = \prod_{i=1}^{n} \sqrt{\frac{\tau}{2\pi}} \exp\left\{-\frac{\tau}{2}(x_i - \mu)^2\right\}$$

$$= \left(\frac{\tau}{2\pi}\right)^{\frac{n}{2}} \exp\left\{-\frac{\tau}{2}\left[(n-1)s^2 + n(\bar{x} - \mu)^2\right]\right\}$$

$$\propto \tau^{\frac{n}{2}} \exp\left\{-\frac{\tau}{2}\left[(n-1)s^2 + n(\bar{x} - \mu)^2\right]\right\}$$

where

$$\bar{x} = \frac{1}{n}\sum_{i=1}^{n} x_i \quad \text{and} \quad s^2 = \frac{1}{n-1}\sum_{i=1}^{n}(x_i - \bar{x})^2.$$

For a Bayesian analysis, we also need to specify prior distributions for the parameters, μ and τ. There is a conjugate analysis for this problem based on the specifications

$$\tau \sim Ga(a, b), \qquad \mu|\tau \sim N\left(c, \frac{1}{d\tau}\right).$$

However, this specification is rather unsatisfactory — μ and τ are not independent, and in many cases our prior beliefs for μ and τ will separate into independent specifications. For example, we may prefer to specify independent priors for the parameters:

$$\tau \sim Ga(a, b), \qquad \mu \sim N\left(c, \frac{1}{d}\right).$$

However, this specification is no longer conjugate, making analytic analysis intractable. Let us see why. We have

$$\pi(\mu) = \sqrt{\frac{d}{2\pi}} \exp\left\{-\frac{d}{2}(\mu - c)^2\right\} \propto \exp\left\{-\frac{d}{2}(\mu - c)^2\right\}$$

and

$$\pi(\tau) = \frac{b^a}{\Gamma(a)}\tau^{a-1} \exp\{-b\tau\} \propto \tau^{a-1} \exp\{-b\tau\},$$

so

$$\pi(\mu, \tau) \propto \tau^{a-1} \exp\left\{-\frac{d}{2}(\mu - c)^2 - b\tau\right\},$$

giving

$$\pi(\mu, \tau|x) \propto \tau^{a-1} \exp\left\{-\frac{d}{2}(\mu - c)^2 - b\tau\right\} \times \tau^{\frac{n}{2}} \exp\left\{-\frac{\tau}{2}\left[(n-1)s^2\right.\right.$$
$$\left.\left. + n(\bar{x} - \mu)^2\right]\right\}$$

$$= \tau^{a+\frac{n}{2}-1} \exp\left\{-\frac{\tau}{2}\left[(n-1)s^2 + n(\bar{x} - \mu)^2\right] - \frac{d}{2}(\mu - c)^2 - b\tau\right\}.$$

The posterior density for μ and τ certainly will not factorise (μ and τ are not independent *a posteriori*), and will not even separate into the form of the conditional normal-gamma conjugate form mentioned earlier.

So we have the kernel of the posterior for μ and τ, but it is not in a standard form. We can gain some idea of the likely values of (μ, τ) by plotting the bivariate surface (the integration constant is not necessary for that), but we cannot work out the posterior mean or variance, or the forms of the marginal posterior distributions for μ or τ, since we cannot integrate out the other variable. We need a way of understanding posterior densities which does not rely on being able to analytically integrate the posterior density.

In fact, there is nothing particularly special about the fact that the density represents a Bayesian posterior. Given any complex non-standard multivariate probability distribution, we need ways to understand it, to calculate its moments, and to compute its conditional and marginal distributions and their moments. Markov chain Monte Carlo (MCMC) algorithms such as the Gibbs sampler and the Metropolis–Hastings method provide a possible solution.

10.2 The Gibbs sampler

10.2.1 Introduction

The Gibbs sampler is a way of simulating from multivariate distributions based only on the ability to simulate from conditional distributions. In particular, it is appropriate when sampling from marginal distributions is not convenient or possible.

Example

Reconsider the problem of Bayesian inference for the mean and variance of a normally distributed random sample. In particular, consider the non-conjugate approach based on independent prior distributions for the mean and variance. The posterior took the form

$$\pi(\mu, \tau | x) \propto \tau^{a + \frac{n}{2} - 1} \exp\left\{ -\frac{\tau}{2} \left[(n-1)s^2 + n(\bar{x} - \mu)^2 \right] - \frac{d}{2}(\mu - c)^2 - b\tau \right\}.$$

As explained previously, this distribution is not in a standard form. However, while clearly not conjugate, this problem is often referred to as *semi-conjugate*, because the two *full conditional* distributions $\pi(\mu | \tau, x)$ and $\pi(\tau | \mu, x)$ *are* of standard form, and further, are of the same form as the independent prior specifications. That is, $\tau | \mu, x$ is gamma distributed and $\mu | \tau, x$ is normally distributed. In fact, by picking out terms in the variable of interest and regarding everything else as a constant of proportionality, we get

$$\tau | \mu, x \sim Ga\left(a + \frac{n}{2}, b + \frac{1}{2}\left[(n-1)s^2 + n(\bar{x} - \mu)^2 \right] \right),$$

$$\mu | \tau, x \sim N\left(\frac{cd + n\tau\bar{x}}{n\tau + d}, \frac{1}{n\tau + d} \right).$$

So providing that we can simulate normal and gamma quantities, we can simulate from the full conditionals. We therefore need a way to simulate from the joint density (and hence the marginals) based only on the ability to sample from the full-conditionals.

10.2.2 Sampling from bivariate densities

Consider a bivariate density $\pi(x, y)$. We have

$$\pi(x, y) = \pi(x)\pi(y|x),$$

and so we can simulate from $\pi(x, y)$ by first simulating $X = x$ from $\pi(x)$, and then simulating $Y = y$ from $\pi(y|x)$. On the other hand, if we can simulate from the marginal for y, we can write

$$\pi(x, y) = \pi(y)\pi(x|y)$$

and simulate $Y = y$ from $\pi(y)$ and then $X = x$ from $\pi(x|y)$. Either way we need to be able to simulate from one of the marginals. So let us just suppose that we can. That is, we have an $X = x$ from $\pi(x)$. Given this, we can now simulate a $Y = y$ from $\pi(y|x)$ to give a pair of points (x, y) from the bivariate density. However, in that case the y value must be from the marginal $\pi(y)$, and so we can simulate an $X' = x'$ from $\pi(x'|y)$ to give a new pair of points (x', y), also from the joint density. But now x' is from the marginal $\pi(x)$, and so we can keep going. This alternate sampling from conditional distributions defines a bivariate Markov chain, and we have just given an intuitive explanation for why $\pi(x, y)$ is its stationary distribution. The transition kernel for this bivariate Markov chain is

$$p((x, y), (x', y')) = \pi(x', y'|x, y) = \pi(x'|x, y)\pi(y'|x', x, y) = \pi(x'|y)\pi(y'|x').$$

10.2.3 The Gibbs sampling algorithm

Suppose the density of interest is $\pi(\theta)$, where $\theta = (\theta_1, \ldots, \theta_d)^\mathsf{T}$, and that the full conditionals

$$\pi(\theta_i|\theta_1, \ldots, \theta_{i-1}, \theta_{i+1}, \ldots, \theta_d) = \pi(\theta_i|\theta_{-i}) = \pi_i(\theta_i), \qquad i = 1, \ldots, d$$

are available for sampling. The Gibbs sampler can be summarised in the following algorithm:

1. Initialise the iteration counter to $j = 1$. Initialise the state of the chain to $\theta^{(0)} = (\theta_1^{(0)}, \ldots, \theta_d^{(0)})^\mathsf{T}$.

2. Obtain a new value $\theta^{(j)}$ from $\theta^{(j-1)}$ by successive generation of values

$$\theta_1^{(j)} \sim \pi(\theta_1|\theta_2^{(j-1)}, \ldots, \theta_d^{(j-1)})$$
$$\theta_2^{(j)} \sim \pi(\theta_2|\theta_1^{(j)}, \theta_3^{(j-1)}, \ldots, \theta_d^{(j-1)})$$
$$\vdots \qquad \vdots \qquad \vdots$$
$$\theta_d^{(j)} \sim \pi(\theta_d|\theta_1^{(j)}, \ldots, \theta_{d-1}^{(j)}).$$

3. Change counter j to $j + 1$, and return to step 2.

This clearly defines a homogeneous Markov chain, as each simulated value depends only on the previous simulated value, and not on any other previous values or the iteration counter j. However, we need to show that $\pi(\theta)$ is a stationary distribution of this chain. The transition kernel of the chain is

$$p(\theta, \phi) = \prod_{i=1}^{d} \pi(\phi_i | \phi_1, \dots, \phi_{i-1}, \theta_{i+1}, \dots, \theta_d).$$

Therefore, we just need to check that $\pi(\theta)$ is the stationary distribution of the chain with this transition kernel. Unfortunately, the traditional *fixed-sweep* Gibbs sampler just described is *not* reversible, and so we cannot check stationarity by checking for detailed balance (as detailed balance fails). We need to do a direct check of the stationarity of $\pi(\theta)$, that is, we need to check that

$$\pi(\phi) = \int_S p(\theta, \phi) \pi(\theta) \, d\theta.$$

See Section 5.3 for a recap of the relevant concepts. For the bivariate case, we have

$$
\begin{aligned}
\int_S p(\theta, \phi) \pi(\theta) \, d\theta &= \int_S \pi(\phi_1 | \theta_2) \pi(\phi_2 | \phi_1) \pi(\theta_1, \theta_2) \, d\theta_1 d\theta_2 \\
&= \pi(\phi_2 | \phi_1) \int_{S_1} \int_{S_2} \pi(\phi_1 | \theta_2) \pi(\theta_1, \theta_2) \, d\theta_1 d\theta_2 \\
&= \pi(\phi_2 | \phi_1) \int_{S_2} \pi(\phi_1 | \theta_2) \, d\theta_2 \int_{S_1} \pi(\theta_1, \theta_2) \, d\theta_1 \\
&= \pi(\phi_2 | \phi_1) \int_{S_2} \pi(\phi_1 | \theta_2) \pi(\theta_2) \, d\theta_2 \\
&= \pi(\phi_2 | \phi_1) \pi(\phi_1) \\
&= \pi(\phi_1, \phi_2) \\
&= \pi(\phi).
\end{aligned}
$$

The general case is similar. So, $\pi(\theta)$ is a stationary distribution of this chain. Discussions of uniqueness and convergence are beyond the scope of this book. In particular, these issues are complicated somewhat by the fact that the sampler described is not reversible.

10.2.4 Reversible Gibbs samplers

While the fixed-sweep Gibbs sampler itself is not reversible, each *component update* is, and hence there are many variations on the fixed-sweep Gibbs sampler which *are* reversible and do satisfy detailed balance. Let us start by looking at why each component update is reversible.

Suppose we wish to update component i, that is, we update θ by replacing θ_i with ϕ_i drawn from $\pi(\phi_i | \theta_{-i})$. All other components will remain unchanged. The transition kernel for this update is

$$p(\theta, \phi) = \pi(\phi_i|\theta_{-i})I(\theta_{-i} = \phi_{-i})$$

where

$$I(E) = \begin{cases} 1 & \text{if } E \text{ is true,} \\ 0 & \text{if } E \text{ is false.} \end{cases}$$

Note that the density is zero for any transition changing the other components. Now we may check for detailed balance:

$$\begin{aligned} \pi(\theta)p(\theta, \phi) &= \pi(\theta)\pi(\phi_i|\theta_{-i})I(\theta_{-i} = \phi_{-i}) \\ &= \pi(\theta_{-i})\pi(\theta_i|\theta_{-i})\pi(\phi_i|\theta_{-i})I(\theta_{-i} = \phi_{-i}) \\ &= \pi(\phi_{-i})\pi(\theta_i|\phi_{-i})\pi(\phi_i|\phi_{-i})I(\theta_{-i} = \phi_{-i}) \qquad (\text{as } \theta_{-i} = \phi_{-i}) \\ &= \pi(\phi)\pi(\theta_i|\phi_{-i})I(\theta_{-i} = \phi_{-i}) \\ &= \pi(\phi)p(\phi, \theta). \end{aligned}$$

Therefore, detailed balance is satisfied, and hence the update is reversible with stationary distribution $\pi(\cdot)$.

If this particular update is reversible and it preserves the equilibrium distribution of the chain, why bother updating any other component? The reason is that the chain defined by a single update is *reducible*, and hence will not converge to the stationary distribution from an arbitrary starting point. In order to ensure *irreducibility* of the chain, we need to make sure that we update each component sufficiently often. As we have seen, one way to do this is to update each component in a fixed order. The drawback of this method is that *reversibility* is lost.

An alternative to the *fixed-sweep* strategy is to pick a component at random at each stage and update that. This gives a reversible chain with the required stationary distribution, and is known as the *random scan* Gibbs sampler.

An even simpler way to restore the reversibility of the chain is to first scan through the components in fixed order, and then scan backward through the components. This does define a reversible Gibbs sampler. We can check that it works in the bivariate case as follows. The algorithm starts with (θ_1, θ_2) and then generates (ϕ_1, ϕ_2) as follows:

$$\begin{aligned} \theta_1' &\sim \pi(\theta_1'|\theta_2) \\ \phi_2 &\sim \pi(\phi_2|\theta_1') \\ \phi_1 &\sim \pi(\phi_1|\phi_2). \end{aligned}$$

Here θ_1' is an auxiliary variable that we are not interested in *per se* and which needs to be integrated out of the problem. The full transition kernel for a move from (θ_1, θ_2) to $(\theta_1', \phi_1, \phi_2)$ is

$$p(\theta, (\theta_1', \phi)) = \pi(\theta_1'|\theta_2)\pi(\phi_2|\theta_1')\pi(\phi_1|\phi_2),$$

and integrating out the auxiliary variable gives

$$p(\theta, \phi) = \int \pi(\theta_1'|\theta_2)\pi(\phi_2|\theta_1')\pi(\phi_1|\phi_2) \, d\theta_1'$$

$$= \pi(\phi_1|\phi_2) \int \pi(\theta_1'|\theta_2)\pi(\phi_2|\theta_1') \, d\theta_1'.$$

We can now check for detailed balance:

$$\pi(\theta)p(\theta, \phi) = \pi(\theta)\pi(\phi_1|\phi_2) \int \pi(\theta_1'|\theta_2)\pi(\phi_2|\theta_1') \, d\theta_1'$$

$$= \pi(\theta_2)\pi(\theta_1|\theta_2)\pi(\phi_1|\phi_2) \int \pi(\theta_1'|\theta_2)\pi(\phi_2|\theta_1') \, d\theta_1'$$

$$= \pi(\theta_1|\theta_2)\pi(\phi_1|\phi_2) \int \pi(\theta_2)\pi(\theta_1'|\theta_2)\pi(\phi_2|\theta_1') \, d\theta_1'$$

$$= \pi(\theta_1|\theta_2)\pi(\phi_1|\phi_2) \int \pi(\theta_1', \theta_2)\pi(\phi_2|\theta_1') \, d\theta_1'$$

$$= \pi(\theta_1|\theta_2)\pi(\phi_1|\phi_2) \int \pi(\theta_1')\pi(\theta_2|\theta_1')\pi(\phi_2|\theta_1') \, d\theta_1',$$

and, as this is symmetric in θ and ϕ, we must have

$$\pi(\theta)p(\theta, \phi) = \pi(\phi)p(\phi, \theta).$$

This chain is therefore reversible with stationary distribution $\pi(\cdot)$.

We have seen that there are ways of adapting the standard fixed-sweep Gibbs sampler in ways which ensure reversibility. However, reversibility is not a requirement of a useful algorithm — it simply makes it easier to determine the properties of the chain. In practice, the fixed-sweep Gibbs sampler often has as good as or better convergence properties than its reversible cousins. Given that it is slightly easier to implement and debug, it is often simpler to stick with the fixed-sweep scheme than to implement a more exotic version of the sampler.

10.2.5 Simulation and analysis

Suppose that we are interested in a multivariate distribution $\pi(\theta)$ (which may be a Bayesian posterior distribution), and that we are able to simulate from the full conditional distributions of $\pi(\theta)$. Simulation from $\pi(\theta)$ is possible by first initialising the sampler somewhere in the support of θ, and then running the Gibbs sampler. The resulting chain should be monitored for convergence, and the 'burn-in' period should be discarded for analysis. After convergence, the simulated values are all from $\pi(\theta)$. In particular, the values for a particular component will be simulated values from the marginal distribution of that component. A histogram of these values will give an idea of the 'shape' of the marginal distribution, and summary statistics such as the mean and variance will be approximations to the mean and variance of the marginal distribution. The accuracy of the estimates can be gauged using the techniques from the end of Chapter 3.

```
normgibbs <- function(N, n, a, b, cc, d, xbar, ssquared)
{
        mat = matrix(ncol = 2, nrow = N)
        mu = cc
        tau = a/b
        mat[1, ] = c(mu, tau)
        for (i in 2:N) {
                muprec = n*tau + d
                mumean = (d*cc + n*tau*xbar)/muprec
                mu = rnorm(1, mumean, sqrt(1/muprec))
                taub = b + 0.5*((n - 1)*ssquared + n*(
                    xbar - mu)^2)
                tau = rgamma(1, a + n/2, taub)
                mat[i, ] = c(mu, tau)
        }
        mat
}
```

Figure 10.2 *An R function to implement a Gibbs sampler for the simple normal random sample model. Example code for using this function is given in Figure 10.3.*

Example

Returning to the case of the posterior distribution for the normal model with unknown mean and precision, we have already established that the full conditional distributions are

$$\tau|\mu, x \sim Ga\left(a + \frac{n}{2}, b + \frac{1}{2}\left[(n-1)s^2 + n(\bar{x} - \mu)^2\right]\right),$$

$$\mu|\tau, x \sim N\left(\frac{cd + n\tau\bar{x}}{n\tau + d}, \frac{1}{n\tau + d}\right).$$

We can initialise the sampler anywhere in the half-plane where the posterior (and prior) has support, but convergence will be quicker if the chain is not started in the tails of the distribution. One possibility is to start the sampler near the posterior mode, though this can make convergence more difficult to diagnose. A simple strategy which is often easy to implement for problems in Bayesian inference is to start off the sampler at a point simulated from the prior distribution, or even at the mean of the prior distribution. Here, the prior mean for (τ, μ) is $(a/b, c)$. Once initialised, the sampler proceeds with alternate simulations from the full conditional distributions. The first few (hundred?) values should be discarded, and the rest can give information about the joint posterior distribution and marginals.

    An R function to implement a simple Gibbs sampler is given in Figure 10.2, some example code that uses it is shown in Figure 10.3, and the results of running the example code are shown in Figure 10.4; see the figure legends for further details.

    This example, though useful for illustrative purposes, is not a particularly com-

```
postmat=normgibbs(N=11000, n=15, a=3, b=11, cc=10, d
    =1/100, xbar=25, ssquared=20)
postmat=postmat[1001:11000,]
op=par(mfrow=c(3,3))
plot(postmat,col=1:10000)
plot(postmat,type="l")
plot.new()
plot(ts(postmat[,1]))
plot(ts(postmat[,2]))
plot(ts(sqrt(1/postmat[,2])))
hist(postmat[,1],40)
hist(postmat[,2],40)
hist(sqrt(1/postmat[,2]),40)
par(op)
```

Figure 10.3 *Example R code illustrating the use of the function* normgibbs *from Figure 10.2. The plots generated by running this code are shown in Figure 10.4. In this example, the prior took the form* $\mu \sim N(10, 100)$, $\tau \sim Ga(3, 11)$, *and the sufficient statistics for the data were* $n = 15$, $\bar{x} = 25$, $s^2 = 20$. *The sampler was run for 11,000 iterations with the first 1,000 discarded as burn-in, and the remaining 10,000 iterations used for the main monitoring run.*

pelling motivation for the deployment of MCMC methodology, due to the small number of dimensions and relatively simple structure. Before leaving the topic of Gibbs sampling, a slightly more substantial example will be examined.

Example

This example will be motivated by considering a biological experiment to estimate (the logarithm of) a biochemical reaction rate constant. The details of the experiment will not concern us here; we will assume simply that the experiment results in the generation of a single number, representing a (sensible) estimate of the rate constant. Several labs will conduct this experiment, and each lab will replicate the experiment several times. Thus, in totality, we will have a collection of estimates from a collection of labs. We will consider the problem of inference for the true rate constant on the basis of prior knowledge and all available data.

Consider the following simple *hierarchical* (or *one-way random effects*) model,

$$Y_{ij}|\theta_i, \tau \sim N(\theta_i, 1/\tau), \quad \text{independently}, \ i = 1, \dots, m, \ j = 1, \dots, n_i$$
$$\theta_i|\mu, \nu \sim N(\mu, 1/\nu), \quad i = 1, \dots, m.$$

Here there are m labs, and the ith lab replicates the experiment n_i times. Y_{ij} is the measurement made on the jth experiment by lab i. We assume that the measurements from lab i have mean θ_i and that the measurements are normally distributed. We also assume that the θ_i are themselves normally distributed around the true rate μ. Essentially, the model has the effect of inducing a correlation between replicates

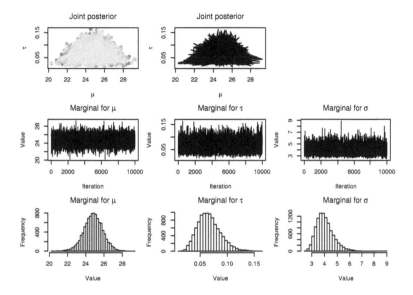

Figure 10.4 *Figure showing the Gibbs sampler output resulting from running the example code in Figure 10.3. The top two plots give an indication of the bivariate posterior distribution. The second row shows trace plots of the marginal distributions of interest, indicating a rapidly mixing MCMC algorithm. The final row shows empirical marginal posterior distributions for the parameters of interest.*

from a particular lab, due to the fact that we expect replicates from one lab to be more similar than replicates from different labs (due to hidden factors that are not being properly controlled and accounted for). Note that this generic scenario can be applied to a range of different problems and is particularly popular in the context of meta-analysis.

We will consider the most general (and quite typical) case where μ, τ, and ν are all unknown. We wish to make inferences about these parameters in addition to the unknown θ_i. Thus there are $m + 3$ parameters of interest in this model.

The specification of the model is completed with independent priors for μ, τ, and ν,

$$\mu \sim N(a, 1/b)$$
$$\tau \sim Ga(c, d)$$
$$\nu \sim Ga(e, f).$$

In principle we have now completely specified the model and can compute the posterior distribution. Of course, the posterior distribution is very high dimensional and, more importantly, not of a standard form. Here MCMC techniques can be used to study the posterior distribution.* The likelihood contribution for each observation y_{ij}

---

* It turns out that it is possible to use the linear Gaussian structure of the model to integrate out all variables except the two variance components, reducing the problem to a two-dimensional one; see

is

$$L(\theta_i, \tau; y_{ij}) = \sqrt{\frac{\tau}{2\pi}} \exp\left\{-\frac{\tau}{2}(y_{ij} - \theta_i)^2\right\},$$

and so the full likelihood is

$$L(\theta, \tau; y) = \prod_{i=1}^{m}\prod_{j=1}^{n_i} L(\theta_i, \tau; y_{ij})$$

$$= \left(\frac{\tau}{2\pi}\right)^{N/2} \exp\left\{-\frac{\tau}{2}\sum_{i=1}^{m}\left[(n_i - 1)s_i^2 + n_i(y_{i\cdot} - \theta_i)^2\right]\right\},$$

where

$$N = \sum_{i=1}^{m} n_i, \quad y_{i\cdot} = \frac{1}{n_i}\sum_{j=1}^{n_i} y_{ij}, \quad s_i^2 = \frac{1}{n_i - 1}\sum_{j=1}^{n_i}(y_{ij} - y_{i\cdot})^2.$$

The prior takes the form

$$\pi(\mu, \tau, \nu, \theta) = \pi(\mu)\pi(\tau)\pi(\nu)\pi(\theta|\mu, \nu),$$

where

$$\pi(\mu) \propto \exp\left\{-\frac{b}{2}(\mu - a)^2\right\}$$

$$\pi(\tau) \propto \tau^{c-1}\exp\{-d\tau\}$$

$$\pi(\nu) \propto \nu^{e-1}\exp\{-f\nu\}$$

$$\pi(\theta_i|\mu, \nu) = \sqrt{\frac{\nu}{2\pi}} \exp\left\{-\frac{\nu}{2}(\theta_i - \mu)^2\right\}$$

$$\Rightarrow \pi(\theta|\mu, \nu) \propto \nu^{m/2}\exp\left\{-\frac{\nu}{2}\sum_{i=1}^{m}(\theta_i - \mu)^2\right\},$$

and therefore,

$$\pi(\mu, \tau, \nu, \theta) \propto \nu^{e+m/2-1}\tau^{c-1}\exp\left\{-\frac{1}{2}\left[2d\tau + 2f\nu + b(\mu - a)^2 + \nu\sum_{i=1}^{m}(\theta_i - \mu)^2\right]\right\}.$$

Now we have the likelihood and the prior, and we can write down the posterior

---

Wilkinson and Yeung (2002) and Wilkinson and Yeung (2004) for details. However, for models without this linear Gaussian structure, such techniques are not always convenient or possible, and simple MCMC algorithms like the one about to be described are often the only straightforward way to work with the posterior distribution.

distribution,

$$\pi(\mu, \tau, \nu, \theta | y) \propto$$

$$\tau^{c+N/2-1} \nu^{e+m/2-1} \exp \left\{ -\frac{1}{2} \left[2d\tau + 2f\nu + b(\mu - a)^2 \right. \right.$$

$$\left. \left. + \sum_{i=1}^{m} \left(\nu(\theta_i - \mu)^2 + \tau(n_i - 1)s_i^2 + \tau n_i(y_{i\cdot} - \theta_i)^2 \right) \right] \right\}.$$

It is difficult to do anything analytic with this, so we will try to construct a Gibbs sampler in order to investigate it. This is fairly straightforward, and the full conditionals are as follows. Picking out terms in μ, the full conditional for μ is

$$\pi(\mu | \cdot) \propto \exp \left\{ -\frac{1}{2} \left[b(\mu - a)^2 + \sum_{i=1}^{m} \nu(\theta_i - \mu)^2 \right] \right\}$$

$$\propto \exp \left\{ -\frac{1}{2}(b + m\nu) \left(\mu - \frac{ba + m\nu\bar{\theta}}{b + m\nu} \right)^2 \right\}, \quad \text{where } \bar{\theta} = \frac{1}{m} \sum_{i=1}^{m} \theta_i,$$

that is,

$$\mu | \cdot \sim N \left(\frac{ba + m\nu\bar{\theta}}{b + m\nu}, \frac{1}{b + m\nu} \right).$$

The conditional for τ is

$$\pi(\tau | \cdot) \propto \tau^{c+N/2-1} \exp \left\{ -\tau \left[d + \frac{1}{2} \sum_{i=1}^{m} \left((n_i - 1)s_i^2 + n_i(y_{i\cdot} - \theta_i)^2 \right) \right] \right\},$$

that is,

$$\tau | \cdot \sim Ga \left(c + N/2, d + \frac{1}{2} \sum_{i=1}^{m} [(n_i - 1)s_i^2 + n_i(y_{i\cdot} - \theta_i)^2] \right).$$

The conditional for ν is

$$\pi(\nu | \cdot) \propto \nu^{e+m/2-1} \exp \left\{ -\nu \left[f + \frac{1}{2} \sum_{i=1}^{m} (\theta_i - \mu)^2 \right] \right\},$$

that is,

$$\nu | \cdot \sim Ga \left(e + m/2, f + \frac{1}{2} \sum_{i=1}^{m} (\theta_i - \mu)^2 \right).$$

The conditional for θ_i is

$$\pi(\theta_i|\cdot) \propto \exp\left\{-\frac{1}{2}\left[\nu(\theta_i - \mu)^2 + \tau n_i(y_{i.} - \theta_i)^2\right]\right\}$$

$$\propto \exp\left\{-\frac{1}{2}(\nu + n_i\tau)\left(\theta_i - \frac{\nu\mu + n_iy_{i.}\tau}{\nu + n_i\tau}\right)^2\right\},$$

that is,

$$\theta_i|\cdot \sim N\left(\frac{\nu\mu + n_iy_{i.}\tau}{\nu + n_i\tau}, \frac{1}{\nu + n_i\tau}\right), \quad i = 1, \ldots m.$$

Therefore, we have a Gibbs sampler with $m+3$ components. We just have to specify the prior parameters a, b, c, d, e, f and compute the data summaries $m, n_i, N, y_{i.}, s_i^2$, $i = 1, \ldots, m$. Then we initialise the sampler by simulating from the prior, or by starting off each component at its prior mean. The sampler is then run to convergence, and samples from the stationary distribution are used to understand the marginals of the posterior distribution. This model is of sufficient complexity that assessing convergence of the sampler to its stationary distribution is a non-trivial task. At the very least, multiple large simulation runs are required, with different starting points, and the first portion (say, a third) of any run should be discarded as 'burn-in'. As the complexity of the model increases, problems with assessment of the convergence of the sampler also increase. There are many software tools available for MCMC convergence diagnostics. R-CODA is an excellent package for R (package name coda on CRAN) which carries out a range of output analysis and diagnostic tasks, but its use is beyond the scope of this book.

It is clear that in principle at least, it ought to be possible to automate the construction of a Gibbs sampler from a specification containing the model, the prior, and the data. There are several freely available software packages that are able to do this for relatively simple models. Examples include, WinBUGS, OpenBugs, JAGS and Stan; see this book's website for links. Unfortunately, it turns out to be difficult to directly use these software packages for the stochastic kinetic models that will be considered in Chapter 11.

Of course, the Gibbs sampler tacitly assumes that we have some reasonably efficient mechanism for simulating from the full conditional distributions, and yet this is not always the case. Fortunately, the Gibbs sampler can be combined with Metropolis–Hastings algorithms when the full conditionals are difficult to simulate from.

10.3 The Metropolis–Hastings algorithm

Suppose that $\pi(\theta)$ is the density of interest. Suppose further that we have some (arbitrary) transition kernel $q(\theta, \phi)$ (known as the *proposal distribution*), which is easy to simulate from but does not (necessarily) have $\pi(\theta)$ as its stationary density. Consider the following algorithm:

1. Initialise the iteration counter to $j = 1$, and initialise the chain to $\theta^{(0)}$.

2. Generate a *proposed* value ϕ using the kernel $q(\theta^{(j-1)}, \phi)$.

3. Evaluate the *acceptance probability* $\alpha(\theta^{(j-1)}, \phi)$ of the proposed move, where

$$\alpha(\theta, \phi) = \min\left\{1, \frac{\pi(\phi)q(\phi, \theta)}{\pi(\theta)q(\theta, \phi)}\right\}.$$

4. Put $\theta^{(j)} = \phi$ with probability $\alpha(\theta^{(j-1)}, \phi)$, and put $\theta^{(j)} = \theta^{(j-1)}$ otherwise.

5. Change the counter from j to $j + 1$ and return to step 2.

In other words, at each stage, a new value is generated from the proposal distribution. This is either accepted, in which case the chain moves, or rejected, in which case the chain stays where it is. Whether or not the move is accepted or rejected depends on an acceptance probability which itself depends on the relationship between the density of interest and the proposal distribution. Note that the density of interest, $\pi(\cdot)$, only enters into the acceptance probability as a ratio, and so the method can be used when the density of interest is only known up to a scaling constant. This algorithm is essentially that of Hastings (1970), which is a generalisation of the algorithm introduced by Metropolis et al. (1953).

The Markov chain defined in this way is reversible and has stationary distribution $\pi(\cdot)$ irrespective of the choice of proposal distribution, $q(\cdot, \cdot)$. Let us see why. The transition kernel is clearly given by

$$p(\theta, \phi) = q(\theta, \phi)\alpha(\theta, \phi), \quad \text{if } \theta \neq \phi.$$

But there is also a finite probability that the chain will remain at θ. This is 1 minus the probability that the chain moves, and thus is given by

$$1 - \int q(\theta, \phi)\alpha(\theta, \phi)\, d\phi.$$

So, the transition kernel is part continuous and part discrete. We can easily write down the cumulative distribution form of the transition kernel as

$$P(\theta, \phi) = \int_{-\infty}^{\phi} q(\theta, \phi)\alpha(\theta, \phi)\, d\phi + I(\phi \geq \theta)\left[1 - \int q(\theta, \phi)\alpha(\theta, \phi)\, d\phi\right].$$

We then get the full density form of the kernel by differentiating with respect to ϕ as

$$p(\theta, \phi) = q(\theta, \phi)\alpha(\theta, \phi) + \delta(\theta - \phi)\left[1 - \int q(\theta, \phi)\alpha(\theta, \phi)\, d\phi\right],$$

where $\delta(\cdot)$ is the Dirac δ-function. Now that we have the transition kernel, we can check whether detailed balance is satisfied:

$$\pi(\theta)p(\theta,\phi) = \pi(\theta)q(\theta,\phi)\min\left\{1, \frac{\pi(\phi)q(\phi,\theta)}{\pi(\theta)q(\theta,\phi)}\right\}$$

$$+ \delta(\theta - \phi)\left[\pi(\theta) - \int \pi(\theta)q(\theta,\phi)\min\left\{1, \frac{\pi(\phi)q(\phi,\theta)}{\pi(\theta)q(\theta,\phi)}\right\} d\phi\right]$$

$$= \min\left\{\pi(\theta)q(\theta,\phi), \pi(\phi)q(\phi,\theta)\right\}$$

$$+ \delta(\theta - \phi)\left[\pi(\theta) - \int \min\left\{\pi(\theta)q(\theta,\phi), \pi(\phi)q(\phi,\theta)\right\} d\phi\right].$$

The first term is clearly symmetric in θ and ϕ. Also, the second term must be symmetric in θ and ϕ, because it is only non-zero precisely when $\theta = \phi$. Consequently, detailed balance is satisfied, and the Metropolis–Hastings algorithm defines a reversible Markov chain with stationary distribution $\pi(\cdot)$, irrespective of the form of $q(\cdot,\cdot)$.

Complete freedom in the choice of the proposal distribution $q(\cdot,\cdot)$ leaves us wondering what kinds of choices might be good, or generally quite useful. Some commonly used special cases are discussed below.

10.3.1 Symmetric chains (Metropolis method)

The simplest case is the Metropolis sampler, which is based on the use of a symmetric proposal with $q(\theta,\phi) = q(\phi,\theta)$, $\forall \theta, \phi$. We see then that the acceptance probability simplifies to

$$\alpha(\theta,\phi) = \min\left\{1, \frac{\pi(\phi)}{\pi(\theta)}\right\},$$

and hence does not involve the proposal density at all. Consequently proposed moves which will take the chain to a region of higher density are always accepted, while moves which take the chain to a region of lower density are accepted with probability proportional to the ratio of the two densities — moves which will take the chain to a region of very low density will be accepted with very low probability. Note that any proposal of the form $q(\theta,\phi) = f(|\theta - \phi|)$ is symmetric, where $f(\cdot)$ is an arbitrary density. In this case, the proposal will represent a symmetric displacement from the current value. This also motivates random walk chains.

10.3.2 Random walk chains

In this case, the proposed value ϕ at stage j is $\phi = \theta^{(j-1)} + w_j$ where the w_j are iid random variables (completely independent of the state of the chain). Suppose that the w_j have density $f(\cdot)$, which is easy to simulate from. We can then simulate an *innovation*, w_j, and set the *candidate* point to $\phi = \theta^{(j-1)} + w_j$. The transition kernel is then $q(\theta,\phi) = f(\phi - \theta)$, and this can be used to compute the acceptance probability. Of course, if $f(\cdot)$ is symmetric about zero, then we have a symmetric chain, and the acceptance probability does not depend on $f(\cdot)$ at all.

So suppose that it is decided to use a symmetric random walk chain with proposed mean zero innovations. There is still the question of how they should be distributed, and what variance they should have. A simple, easy-to-simulate-from distribution is always a good idea, such as uniform or normal (normal is generally better, but is a bit more expensive to simulate). The choice of variance will affect the acceptance probability, and hence the overall proportion of accepted moves. If the variance of the innovation is too low, then most proposed values will be accepted, but the chain will move very slowly around the space — the chain is said to be too 'cold'. On the other hand, if the variance of the innovation is too large, very few proposed values will be accepted, but when they are, they will often correspond to quite large moves — the chain is said to be too 'hot'. Experience suggests that an overall acceptance rate of around 30% is desirable, and so it is possible to 'tune' the variance of the innovation distribution to get an acceptance rate of around this level. This should be done using a few trial short runs, and then a single fixed value should be adopted for the main monitoring run.[†]

An R function implementing a simple Metropolis random walk sampler is given in Figure 10.5. In this example the target distribution is a $N(0, 1)$ random quantity, and the innovations are $U(-\alpha, \alpha)$. The results of running the algorithm for different values of α are shown in Figure 10.6. The auto-correlation function (ACF) plots are a useful diagnostic for assessing the rate of mixing of the chain.

10.3.3 Independence chains

In this case, reminiscent of the envelope rejection method (Section 4.6.1) and importance sampling (Section 4.2), the proposed transition is formed independently of the previous position of the chain, and so $q(\theta, \phi) = f(\phi)$ for some density $f(\cdot)$. Here the acceptance probability becomes

$$\alpha(\theta, \phi) = \min\left\{1, \frac{\pi(\phi)}{\pi(\theta)} \Big/ \frac{f(\phi)}{f(\theta)}\right\},$$

and we see that the acceptance probability can be increased by making $f(\cdot)$ as similar to $\pi(\cdot)$ as possible (in this case, the higher the acceptance probability, the better).

Bayes' theorem via independence chains

In the context of Bayesian inference, one possible choice for the proposal density is the prior density. The acceptance probability then becomes

$$\alpha(\theta, \phi) = \min\left\{1, \frac{L(\phi; x)}{L(\theta; x)}\right\},$$

and hence depends only on the likelihood ratio of the candidate point and the current value. This is attractive because it is usually very straightforward to simulate

---

[†] Although it sounds appealing to adaptively change the tuning parameter during the main monitoring run, this usually destroys the Markov property of the chain, and often affects the stationary distribution. This should therefore be avoided unless you *really* know what you are doing. The field of 'adaptive MCMC' is concerned with how to do this correctly (Andrieu and Moulines, 2006).

```
metrop <- function(n, alpha)
{
        vec = vector("numeric", n)
        x = 0
        vec[1] = x
        for (i in 2:n) {
                can = x+runif(1,-alpha,alpha)
                aprob = dnorm(can)/dnorm(x)
                u = runif(1)
                if (u < aprob)
                        x=can
                vec[i] = x
        }
        vec
}
```

Figure 10.5 *An R function to implement a Metropolis sampler for a standard normal random quantity based on $U(-\alpha, \alpha)$ innovations. So,* metrop(10000,1) *will execute a run of length 10,000 with an α of 1. See Figure 10.6 for results of running the algorithm with different values of α.*

from the prior distribution and very difficult to simulate from the posterior. Also, the acceptance probability depends only on the likelihood ratio, which is sometimes straightforward to calculate. Unfortunately this method tends to result in a badly mixing chain if the problem is high dimensional and the data are not in strong accordance with the prior. Also, for many hierarchical models, the (marginal) likelihood of the data is intractable.

10.4 Hybrid MCMC schemes

We have seen how we can use the Gibbs sampler to sample from multivariate distributions provided that we can simulate from the full conditionals. We have also seen how we can use Metropolis–Hastings methods to sample from awkward distributions (perhaps full conditionals). If we wish, we can combine these in order to form hybrid Markov chains whose stationary distribution is a distribution of interest.

Componentwise transition: Given a multivariate distribution with full conditionals that are awkward to sample from directly, we can define a Metropolis–Hastings scheme for each full conditional and apply them to each component in turn for each iteration. This is like the Gibbs sampler, but each component update is a Metropolis–Hastings update, rather than a direct simulation from the full conditional. This is in fact the original form of the Metropolis algorithm.

Metropolis within Gibbs: Given a multivariate distribution with full conditionals, some of which may be simulated from directly, and others which have Metropolis–Hastings updating schemes, the Metropolis within Gibbs algorithm goes through

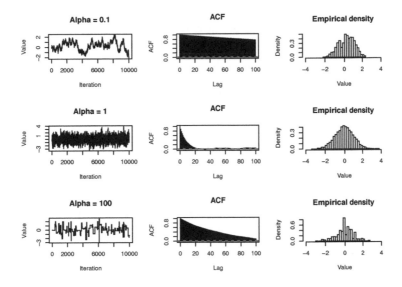

Figure 10.6 *Output from the Metropolis sampler given in Figure 10.5. The top row shows the result of running the chain with $\alpha = 0.1$, corresponding to a chain that is too cold. The middle row shows the results for $\alpha = 1$. This α is close to optimal, and the ACF plot shows auto-correlations in the sampled values decaying away rapidly to zero. The final row shows the results for $\alpha = 100$, representing a chain that is too hot, with many rejected proposed moves.*

each in turn, and simulates directly from the full conditional, or carries out a Metropolis–Hastings update as necessary.

Blocking: The components of a Gibbs sampler, and those of Metropolis–Hastings chains, can be vectors (or matrices) as well as scalars. For many high-dimensional problems, it can be helpful to group related parameters into blocks and use multi-variate simulation techniques to update those together if possible. This can greatly improve the mixing of the chain, at the expense of increasing the computational cost of each iteration.

Some of the methods discussed in this section will be illustrated in practice in Chapter 11.

10.5 Metropolis–Hastings algorithms for Bayesian inference

Let us now consider the generic problem of using a Metropolis–Hastings sampler in the context of inference for a parameter vector θ given some data x generated from a probability model of the form $\pi(x|\theta)$. We factorise the joint distribution as

$$\pi(\theta, x) = \pi(\theta)\pi(x|\theta)$$

and use this factorisation into 'prior' and 'likelihood' in order to compute the posterior distribution $\pi(\theta|x)$, which is just the joint density modulo a normalising constant. Consequently, we can construct a Metropolis–Hastings scheme which targets $\pi(\theta|x)$ using an essentially arbitrary proposal kernel $q(\theta, \theta^\star)$ for a proposed move from θ to $\theta^\star$ in conjunction with an acceptance probability of the form $\alpha(\theta, \theta^\star) = \min\{1, A\}$, where

$$A = \frac{\pi(\theta^\star)\pi(x|\theta^\star)q(\theta^\star, \theta)}{\pi(\theta)\pi(x|\theta)q(\theta, \theta^\star)}.$$

We have already seen that the independent transition kernel $q(\theta, \theta^\star) = \pi(\theta^\star)$ leads to a simple acceptance ratio, but that does not typically represent a good choice of proposal. Very often, proposals which randomly walk on the space of θ are more effective. Also note that if we were able to use the true posterior as our proposal, $q(\theta, \theta^\star) = \pi(\theta^\star|x)$, this would give an acceptance ratio of one, as one would intuitively expect.

Code for a generic implementation of Metropolis–Hastings MCMC for Bayesian inference is given in Figure 10.7. Note that the code works entirely with log-likelihoods and densities in order to prevent numerical underflow, and that if prior and/or proposal densities are omitted they will drop out of the acceptance ratio by default. Also note that likelihood evaluations are carried over to the next iteration. In the case of expensive likelihoods, this effectively doubles the speed of the algorithm. It also has other benefits, to be discussed later. A simple example of how to use it is given in Figure 10.8.

10.6 Bayesian inference for latent variable models

We have introduced Bayesian inference in the context of inference for a parameter vector θ given some data x generated from a probability model of the form $\pi(x|\theta)$. The joint distribution was factorised as

$$\pi(\theta, x) = \pi(\theta)\pi(x|\theta)$$

and this factorisation into 'prior' and 'likelihood' has been used in order to compute the posterior distribution $\pi(\theta|x)$.

At this point it is useful to consider a commonly encountered extension of the basic inferential framework described above. Suppose now that we cannot observe x, as x is not measured, and represents a 'latent', 'missing', or 'hidden' layer in our model. Instead we observe some aspect of x indirectly. We now denote the actual data by y, and it is modelled conditional on the missing data x, and possibly also depends on the parameter vector θ. An example of this was the random effects model considered at the end of Section 10.2.5, where (with a change of notation) the 3 top level parameters would form θ and the m random effects constitute the missing data, x. So now our joint distribution factorises as

$$\pi(\theta, x, y) = \pi(\theta)\pi(x|\theta)\pi(y|x, \theta).$$

We now consider using this new joint distribution as the basis of our inferential framework, which may be concerned with posterior distributions such as $\pi(\theta|y)$ or $\pi(\theta, x|y)$.

Before proceeding further it may be useful to have a relevant concrete example in

```r
metropolisHastings <- function (init, logLik, rprop, dprop =
    function (new, old, ...) { 1 }, dprior = function (x, ...) {
    1 }, iters = 10000, thin = 10, verb = TRUE)
{
    p = length (init)
    ll = -Inf
    mat = matrix (0, nrow = iters, ncol = p)
    colnames (mat) = names (init)
    x = init
    if (verb)
        message (paste (iters, "iterations"))
    for (i in 1:iters) {
        if (verb)
            message (paste (i, ""), appendLF = FALSE)
        for (j in 1:thin) {
            prop = rprop (x)
            if (dprior (prop, log = TRUE) > -Inf) {
                llprop = logLik (prop)
                a = (llprop - ll + dprior (prop, log = TRUE) -
                    dprior (x, log = TRUE) + dprop (x, prop, log =
                        TRUE) -
                    dprop (prop, x, log = TRUE))
                if (log (runif (1)) < a) {
                    x = prop
                    ll = llprop
                }
            }
        }
        mat [i, ] = x
    }
    if (verb)
        message ("Done.")
    mat
}
```

Figure 10.7 *An R function to implement a Metropolis–Hastings MCMC sampler for Bayesian inference.*

mind for motivation. In the next chapter we will consider how to apply the techniques developed in this chapter to the problem of making inference for the rate constants of a stochastic kinetic model given some discrete time measurements of the system state. In that case, the parameter vector, θ, will represent the vector of rate constants, x will represent the complete unobserved sample path of the entire process, and y will represent the actual measurements we obtain.

If we are really just interested in the marginal posterior $\pi(\theta|y)$ we know from the previous section that in principle we can construct a Metropolis–Hastings scheme

```
## First simulate some synthetic data
data = rnorm(250,5,2)
## Now use MH to recover the parameters
llik = function(x) { sum(dnorm(data,x[1],x[2],log=TRUE))
    }
prop = function(x) { rnorm(2,x,0.1) }
prior = function(x, log=TRUE) {
    l = dnorm(x[1],0,100,log=TRUE) + dgamma(x
        [2],1,0.0001,log=TRUE)
    if (log) l else exp(l)
}
out = metropolisHastings(c(mu=1,sig=1), llik, prop,
                         dprior=prior, verb=FALSE)
out = out[1000:10000,]
mcmcSummary(out, truth=c(5,2), rows=2, plot=FALSE)
```

Figure 10.8 *Code illustrating the use of the* metropolisHastings *function given in Figure 10.7.*

targeting this posterior using a proposal $q(\theta, \theta^\star)$ and using the acceptance ratio

$$A = \frac{\pi(\theta^\star)\pi(y|\theta^\star)q(\theta^\star, \theta)}{\pi(\theta)\pi(y|\theta)q(\theta, \theta^\star)},$$

where

$$\pi(y|\theta) = \int_X \pi(y|x, \theta)\pi(x|\theta)\, dx.$$

In practice, marginalising over x is often impossible, but for certain tractable distribution families it can be done, and is often easiest to compute using the basic marginal likelihood identity (BMI) (Chib, 1995)

$$\pi(y|\theta) = \frac{\pi(x|\theta)\pi(y|x, \theta)}{\pi(x|y, \theta)}.$$

Note that since the LHS of the above expression is independent of x, the RHS must also be, and hence can be evaluated at any convenient x. However, although there are special cases where this is possible, typically it is not the case, and so another strategy must be adopted. The usual approach is to focus instead on the full posterior distribution $\pi(\theta, x|y)$, since this is often of interest anyway, and if an MCMC scheme can be constructed which targets this, the marginal posterior $\pi(\theta|y)$ can in any case be obtained by considering only the sampled values of θ.

It is clear that the joint posterior $\pi(\theta, x|y)$ can be targeted by a chain with proposal $q((\theta, x), (\theta^\star, x^\star))$ by using the acceptance ratio

$$A = \frac{\pi(\theta^\star)\pi(x^\star|\theta^\star)\pi(y|x^\star, \theta^\star)q((\theta^\star, x^\star), (\theta, x))}{\pi(\theta)\pi(x|\theta)\pi(y|x, \theta)q((\theta, x), (\theta^\star, x^\star))}.$$

There are many possible ways to construct the proposal distribution $q(\cdot, \cdot)$. An interesting special case is where the proposed new $(\theta^\star, x^\star)$ is constructed in two stages: first, a new $\theta^\star$ is proposed from a kernel $f(\theta^\star|\theta)$, and then a new $x^\star$ is proposed from a kernel $g(x^\star|\theta^\star)$, conditional on the newly proposed $\theta^\star$. This allows the proposals to be constructed in such a way as to ensure that the $\theta^\star$ and $x^\star$ are consistent with one another. In this case the proposal is clearly

$$q((\theta, x), (\theta^\star, x^\star)) = f(\theta^\star|\theta)g(x^\star|\theta^\star),$$

and the acceptance ratio becomes

$$A = \frac{\pi(\theta^\star)\pi(x^\star|\theta^\star)\pi(y|x^\star, \theta^\star)f(\theta|\theta^\star)g(x|\theta)}{\pi(\theta)\pi(x|\theta)\pi(y|x, \theta)f(\theta^\star|\theta)g(x^\star|\theta^\star)}.$$

As usual, the proposal for $\theta^\star$, $f(\theta^\star|\theta)$ can be fairly arbitrary, and random walk distributions can often be effective. However, considerable care needs to be taken with the form of the proposal for $x^\star$, $g(x^\star|\theta^\star)$, since $x^\star$ is typically high-dimensional, and a poor choice will lead to a poorly mixing chain. There are two important special cases for $g(\cdot|\cdot)$ which lead to considerable simplification of the acceptance ratio. The first is the so-called *likelihood-free* MCMC (LF-MCMC) proposal, where the proposed new $X^\star$ is generated by simple forward simulation from the model using the newly proposed parameters $\theta^\star$. That is, $g(x^\star|\theta^\star) = \pi(x^\star|\theta^\star)$. This choice obviously leads to cancellation in the acceptance ratio, giving

$$A = \frac{\pi(\theta^\star)\pi(y|x^\star, \theta^\star)f(\theta|\theta^\star)}{\pi(\theta)\pi(y|x, \theta)f(\theta^\star|\theta)}.$$

Again, further simplification can be obtained by choosing $f(\theta^\star|\theta) = \pi(\theta^\star)$, leading to the even simpler acceptance ratio

$$A = \frac{\pi(y|x^\star, \theta^\star)}{\pi(y|x, \theta)},$$

but this will typically not represent an efficient choice. The likelihood-free approach is potentially attractive, as it is usually easier to simulate realisations from the model than to evaluate its likelihood, $\pi(x|\theta)$. However, in the case of high-dimensional y, or 'low noise' measurement scenarios, the proposal is very inefficient, leading to a poorly mixing chain. In the context of inference for systems biology models, it is possible to partly alleviate these difficulties by applying the technique sequentially — see Wilkinson (2011) for an example and application. However, this approach suffers from other issues, and so we will explore other likelihood free strategies here.

The other important special case is the 'optimal' choice $g(x^\star|\theta^\star) = \pi(x^\star|\theta^\star, y)$. This clearly gives the acceptance ratio

$$A = \frac{\pi(\theta^\star)\pi(x^\star|\theta^\star)\pi(y|x^\star, \theta^\star)f(\theta|\theta^\star)\pi(x|\theta, y)}{\pi(\theta)\pi(x|\theta)\pi(y|x, \theta)f(\theta^\star|\theta)\pi(x^\star|\theta^\star, y)},$$

but using the BMI this simplifies to

$$A = \frac{\pi(\theta^\star)\pi(y|\theta^\star)f(\theta|\theta^\star)}{\pi(\theta)\pi(y|\theta)f(\theta^\star|\theta)}.$$

It is clear therefore that if it is possible to use this proposal, the acceptance ratio does not depend on the sample path x, and we have a scheme exactly equivalent to that of the marginal updating scheme for $\pi(\theta|y)$. It is rarely possible to implement this scheme, but understanding it and its relationship with the marginal scheme is important for understanding some of the more advanced MCMC schemes that we will use later for inference in stochastic kinetic models.

10.6.1 The pseudo-marginal approach to 'exact approximate' MCMC

Let us now reconsider the latent variable problem

$$\pi(\theta, x, y) = \pi(\theta)\pi(x|\theta)\pi(y|x, \theta),$$

and the problem of designing an MCMC algorithm for the marginal posterior $\pi(\theta|y)$. We have already seen that using an arbitrary proposal $q(\theta, \theta^\star)$, we can target this posterior distribution using the acceptance ratio

$$A = \frac{\pi(\theta^\star)\pi(y|\theta^\star)q(\theta^\star, \theta)}{\pi(\theta)\pi(y|\theta)q(\theta, \theta^\star)}.$$

As already explained, the difficulty is computation of the marginal likelihood of the data

$$\pi(y|\theta) = \int_X \pi(y|x, \theta)\pi(x|\theta)\, dx.$$

Although this is often analytically intractable, it is often straightforward to construct a Monte Carlo estimate of it, and such estimates can be used as the basis of a (highly computationally intensive) MC within MCMC algorithm. For example, we know that if we simulate realisations $x_1, x_2, \ldots, x_n$ from the model $\pi(x|\theta)$ for a given θ, then the Monte Carlo estimate

$$\hat{\pi}(y|\theta) = \frac{1}{n}\sum_{i=1}^n \pi(y|x_i, \theta)$$

will converge to $\pi(y|\theta)$ as $n \to \infty$ by the law of large numbers. In that sense, $\hat{\pi}(y|\theta)$ is said to be a *consistent* estimator of $\pi(y|\theta)$. However, it is also clear that for any n,

$$\begin{aligned}
E(\hat{\pi}(y|\theta)) &= \frac{1}{n}\sum_{i=1}^n E(\pi(y|x_i, \theta)) \\
&= E(\pi(y|x_1, \theta)) \\
&= \int_X \pi(y|x_1, \theta)\pi(x_1|\theta)dx_1 \\
&= \pi(y|\theta)
\end{aligned}$$

and so in that sense, $\hat{\pi}(y|\theta)$ is said to be an *unbiased* estimator of $\pi(y|\theta)$. Let us now consider how such estimates can be used within an MCMC algorithm targeting $\pi(\theta|y)$. We know that the 'correct' acceptance ratio is

$$A = \frac{\pi(\theta^\star)\pi(y|\theta^\star)q(\theta^\star, \theta)}{\pi(\theta)\pi(y|\theta)q(\theta, \theta^\star)}.$$

If we now just plug in a Monte Carlo estimate of the marginal likelihood, we get the acceptance ratio

$$A = \frac{\pi(\theta^\star)\hat{\pi}(y|\theta^\star)q(\theta^\star,\theta)}{\pi(\theta)\hat{\pi}(y|\theta)q(\theta,\theta^\star)}.$$

It is clear that if our Monte Carlo estimate is good, then this ratio will be close to the correct ratio, and we could therefore hope that as a result, the target of the chain will be 'close' to the correct target in some appropriate sense. However, it turns out that if the Monte Carlo estimate that we plug in is *unbiased*, then the target is *exactly* the correct posterior, $\pi(\theta|y)$. To understand why, we must consider the marginal scheme as a joint update on a larger space which includes the Monte Carlo error in the marginal likelihood estimate. Specifically, we define the random variable W as

$$W = \frac{\hat{\pi}(y|\theta)}{\pi(y|\theta)}.$$

When our Monte Carlo estimate of likelihood is unbiased, we have $\mathrm{E}(W|\theta) = 1$, $\forall \theta$. We can write the distribution of W as $\pi(w|\theta)$, and our unbiasedness property implies that

$$\int_W w\,\pi(w|\theta)\,dw = 1$$

for all θ. This turns out to be key to understanding why the algorithm works. We can now regard our MCMC scheme as providing a joint update for θ and w, since we sample a new w at each MCMC iteration. In this case our proposal is $q((\theta,w),(\theta^\star,w^\star)) = q(\theta,\theta^\star)\pi(w^\star|\theta^\star)$. We can now re-write our acceptance ratio in the form

$$A = \frac{\pi(\theta^\star)\pi(y|\theta^\star)w^\star\pi(w^\star|\theta^\star)}{\pi(\theta)\pi(y|\theta)w\pi(w|\theta)} \times \frac{q(\theta^\star,\theta)\pi(w|\theta)}{q(\theta,\theta^\star)\pi(w^\star|\theta^\star)}.$$

From this it is clear that the target of the chain must be proportional to $\pi(\theta)\pi(y|\theta) \times w\pi(w|\theta)$. The corresponding marginal for θ can be obtained by integrating this target over the range of W. Using our unbiasedness property, this gives a density for θ proportional to $\pi(\theta)\pi(y|\theta)$, which is exactly the required posterior distribution $\pi(\theta|y)$. Our resulting Markov chain is therefore exact, despite the fact that an approximate Monte Carlo estimate of likelihood is used in the calculation of the acceptance ratio. See Beaumont (2003) and Andrieu and Roberts (2009) for further discussion of this technique. We will see how this technique can be used in practice for parameter inference in stochastic kinetic models in the next chapter. For now, note that the implication is that the `metropolisHastings` function presented in Figure 10.7 will be exact even in the case where the function `logLik` returns the log of a noisy unbiased estimate of likelihood, due to the function carrying forward the old log-likelihood evaluation.

10.7 Alternatives to MCMC

10.7.1 Approximate Bayesian computation (ABC)

There is a close connection between LF-MCMC methods and those of approximate Bayesian computation (ABC). To keep things simple, consider the case of a perfectly

```
abcRun <- function (n, rprior, rdist)
{
    v = vector ("list", n)
    p = mclapply(v, function (x) {
        rprior()
    })
    d = mclapply(p, rdist)
    pm = t(sapply (p, identity))
    if (dim (pm) [1] == 1)
        pm = as.vector (pm)
    dm = t(sapply (d, identity))
    if (dim (dm) [1] == 1)
        dm = as.vector (dm)
    list (param = pm, dist = dm)
}
```

Figure 10.9 *An R function to run forwards simulations and compute statistics for ABC-based inference.*

observed system, so that there is no latent variable layer. Then there are model parameters θ described by a prior $\pi(\theta)$, and a forwards-simulation model for the data x, defined by $\pi(x|\theta)$. It is clear that a simple algorithm for simulating from the desired posterior $\pi(\theta|x)$ can be obtained as follows. First simulate from the joint distribution $\pi(\theta, x)$ by simulating $\theta^\star \sim \pi(\theta)$ and then $x^\star \sim \pi(x|\theta^\star)$. This gives a sample $(\theta^\star, x^\star)$ from the joint distribution. A simple rejection algorithm which rejects the proposed pair unless $x^\star$ matches the true data x clearly gives a sample from the required posterior distribution. However, in many problems this will lead to an intolerably high rejection rate. The 'approximation' is to accept values provided that $x^\star$ is 'sufficiently close' to x. In the simplest case, this is done by forming a (vector of) summary statistic(s), $s(x^\star)$ (ideally a *sufficient statistic*), and accepting provided that $\|s(x^\star) - s(x)\| < \varepsilon$ for some suitable choice of norm and ε (Beaumont et al., 2002). The samples generated by this procedure are no longer samples from the exact posterior distribution, $\pi(\theta|x)$, but the procedure is 'honest' in the sense that it clearly generates samples from $\pi(\theta \mid \|s(x^\star) - s(x)\| < \varepsilon)$. In certain circumstances the 'tolerance', ε can be interpreted as a measurement error model (Wilkinson, 2013). A basic function for running ABC algorithms is shown in Figure 10.9. Essentially, it samples from a prior, and then applies a function which calculates a distance from the observed summary stats. However, it uses the function `mclapply` in order to run in parallel. An attractive feature of ABC algorithms is that they parallelise well on multi-core machines. Examples of using this code in practice will be deferred to the next chapter.

Rather than attempting to jump from samples of the prior to samples from the posterior in one go, ABC may be applied sequentially (Sisson et al., 2007). Sequential ABC approaches have been applied to systems biology problems by Toni et al. (2009). Further, it is well known that ABC approaches can be combined with MCMC to get approximate LF-MCMC schemes, as described in Marjoram et al. (2003). We will examine this approach in detail in the next chapter.

10.7.2 Importance resampling for Bayesian inference

Consider again the problem of generating an (approximate) sample from the posterior $\pi(\theta|x)$. We can use importance resampling (Section 4.7, p. 112) using the prior distribution $\pi(\theta)$ as an auxiliary. Explicitly, we sample $\theta_1, \theta_2, \ldots, \theta_n \sim \pi(\theta)$, then compute unnormalised weights

$$w_i = \pi(x|\theta_i), \qquad i = 1, 2, \ldots, n.$$

These weights can be normalised via

$$\tilde{w}_i = \frac{w_i}{\sum_{j=1}^{n} w_j},$$

and the normalised weights can be used to resample a new sample of size n which will then be approximately distributed according to $\pi(\theta|x)$. This procedure will not be very effective if the posterior is very different from the prior, but when used sequentially, it forms the basis of the bootstrap particle filter we will consider in the next chapter. It should also be noted that the average of the unnormalised weights

$$\bar{w} = \frac{1}{n} \sum_{j=1}^{n} w_j$$

is an unbiased estimator of the marginal likelihood $\pi(y)$. We therefore get estimates of marginal likelihood as a by-product of importance resampling. This too has implications that we will consider in further detail later. The sequential application of ABC methods to inference also relies on importance sampling, and we will also examine that in more detail in the next chapter.

10.8 Exercises

1. Modify the simple Metropolis code given in Figure 10.5 in order to compute the overall acceptance rate of the chain. Write another R function which uses this modified function in order to automatically find a tuning parameter giving an overall acceptance rate of around 30%.

2. Re-write the Gibbs sampling code from Figure 10.2 in a faster language such as C, Java, or Scala. Use an appropriate free scientific library for the language for random number generation. Real MCMC algorithms run too slowly in R, so it is necessary to build up an MCMC code base in a more efficient language.

3. Install the `coda` R package for MCMC output analysis and diagnostics and learn how it works. Try it out on the examples you have been studying.

4. Download some automatic MCMC software such as JAGS or Stan (linked from this book's website). Learn how these packages work and try one on some simple models.

10.9 Further reading

For a more leisurely and more comprehensive introduction to Bayesian inference and MCMC, it is worth starting with O'Hagan and Forster (2004) for the background on Bayesian inference, and then moving on to Gamerman and Lopes (2006) for further information regarding MCMC.

CHAPTER 11

Inference for stochastic kinetic models

11.1 Introduction

This chapter provides an introduction to the computational statistical techniques that can be used in order to identify stochastic kinetic models from experimental data. At this point the experimental framework envisioned is worth mentioning. Essentially, the techniques to be developed require high-resolution time-course measurements of levels of a subset of model species at the single-cell level, and will ideally be calibrated. At the time of writing, such data are relatively difficult to obtain, so it is reasonable to question the practical utility of such methods to the average systems biologist. The reality is that such data are going to be *required* to identify stochastic kinetic models, and so experimental biology needs to change to make such data more readily available. In the meantime, however, the development of these inference techniques sheds light on the issues involved in identifying stochastic kinetic models, and can in any case be adapted to other (less perfect) data sources, including population-averaged data.* The methods are particularly useful for highlighting model identification issues. For example, it might be intended to identify a small network containing 15 unknown rate constants using data on three key protein species. The inference techniques here described might tell you that it is impossible to identify some of the rate constants even using *perfect* information on those three proteins. The methodology should also be able to generate a minimal set of species that need to be measured in order to identify all 15 rate constants satisfactorily and give some indication of the amount of data on these species that will need to be collected in order to have reliable estimates. Thus, even in the absence of high-quality single-cell data, the methods allow investigation of the vital but difficult experimental design issues that need to be addressed in order to get maximum benefit from costly wet-biology programmes. In any case, work on data acquisition at the single-cell level is progressing rapidly, and so good single-cell data is much more readily available than when the first edition of this book was written, and is now a realistically attainable target for any systems biology centre.

This chapter can provide only a brief introduction to the ideas and techniques required. The aim of the chapter is not to make the reader an expert in Bayesian inference for stochastic kinetic models, but simply to make the (fairly technical)

---

* The problem with population-averaged data is that it essentially provides information about the mean of the stochastic process, and effectively masks all other information about the process behaviour. We saw in Chapter 1 an example of a stochastic process (the linear birth–death process) that cannot be identified by its mean. Because it involves only first-order reactions, the deterministic solution is the mean of the stochastic process (Section 6.8), and it was explained in Chapter 1 why it is not possible to use the deterministic model to identify both rate constants.

literature in this area more accessible. Appropriate further reading will be highlighted where relevant.

11.2 Inference given complete data

It turns out to be helpful to first consider the problem of inference given perfect data on the state of the model over a finite time interval $[0, T]$. That is, we will assume that the entire sample path of each species in the model is known over the time period $[0, T]$. This is equivalent to assuming that we have been given discrete-event output from a Gillespie simulator, and are then required to figure out the rate constants that were used on the basis of the output. Although it is completely unrealistic to assume that experimental data of this quality will be available in practice, understanding this scenario is central to understanding the more general inference problem. In any case, it is clear that if we cannot solve even this problem, then inference from data sources of lower quality will be beyond our reach.

It will be helpful to assume the model notation from Chapter 6, with species $\mathcal{X}_1, \ldots, \mathcal{X}_u$, reactions $\mathcal{R}_1, \ldots, \mathcal{R}_v$, rate constants $c = (c_1, \ldots, c_v)^\mathsf{T}$, reaction hazards $h_1(x, c_1), \ldots, h_v(x, c_v)$, and combined hazard

$$h_0(x, c) = \sum_{i=1}^{v} h_i(x, c_i).$$

It is necessary to explicitly consider the state of the system at a given time, and this will be denoted $x(t) = (x_1(t), \ldots, x_u(t))^\mathsf{T}$. Our observed sample path will be written

$$x = \{x(t) : t \in [0, T]\}.$$

As we have complete information on the sample path, we also know the time and type of each reaction event (in fact, this is what we really mean by complete data). It is helpful to use the notation r_j for the number of reaction events of type $\mathcal{R}_j$ that occurred in the sample path x, $j = 1, \ldots, v$, and to define $n = \sum_{j=1}^{v} r_j$ to be the total number of reaction events occurring in the interval $[0, T]$. We will now consider the time and type of each reaction event, (t_i, ν_i), $i = 1, \ldots, n$, where the t_i are assumed to be in increasing order and $\nu_i \in \{1, \ldots, v\}$. It is notationally convenient to make the additional definitions $t_0 = 0$ and $t_{n+1} = T$.

In order to carry out model-based inference for the process, we need the likelihood function. A formal approach to the development of a rigorous theory of likelihood for continuous sample paths is beyond the scope of a text such as this, but it is straightforward to compute the likelihood in an informal way by considering the terms in the likelihood that arise from constructing the sample path according to Gillespie's direct method. Here, the term in the likelihood corresponding to the ith event is just the joint density of the time and type of that event. That is,

$$h_0(x(t_{i-1}), c) \exp\{-h_0(x(t_{i-1}), c)[t_i - t_{i-1}]\} \times \frac{h_{\nu_i}(x(t_{i-1}), c_{\nu_i})}{h_0(x(t_{i-1}), c)}$$

$$= \exp\{-h_0(x(t_{i-1}), c)[t_i - t_{i-1}]\} h_{\nu_i}(x(t_{i-1}), c_{\nu_i}).$$

The full likelihood is the product of these terms, together with a final term representing the information in the fact that there is no event in the final interval $(t_n, T]$. This is just given by the probability of that event, which is

$$\exp\{-h_0(x(t_n), c)[T - t_n]\},$$

giving the combined likelihood

$$L(c; \boldsymbol{x}) = \pi(\boldsymbol{x}|c)$$

$$= \left\{\prod_{i=1}^{n} h_{\nu_i}(x(t_{i-1}), c_{\nu_i}) \exp\{-h_0(x(t_{i-1}), c)[t_i - t_{i-1}]\}\right\}$$

$$\times \exp\{-h_0(x(t_n), c)[T - t_n]\}$$

$$= \left\{\prod_{i=1}^{n} h_{\nu_i}(x(t_{i-1}), c_{\nu_i})\right\} \left\{\prod_{i=1}^{n+1} \exp\{-h_0(x(t_{i-1}), c)[t_i - t_{i-1}]\}\right\}$$

$$= \left\{\prod_{i=1}^{n} h_{\nu_i}(x(t_{i-1}), c_{\nu_i})\right\} \exp\left\{\sum_{i=1}^{n+1} -h_0(x(t_{i-1}), c)[t_i - t_{i-1}]\right\}$$

$$= \left\{\prod_{i=1}^{n} h_{\nu_i}(x(t_{i-1}), c_{\nu_i})\right\} \exp\left\{-\sum_{i=0}^{n} h_0(x(t_i), c)[t_{i+1} - t_i]\right\}. \quad (11.1)$$

This expression for the likelihood (11.1) can be used directly for computing the likelihood from complete data and is therefore known as the *complete-data likelihood*. It is interesting to note that due to the piece-wise constant nature of the combined hazard function, the sum in the exponential term is actually an integral, and so the complete-data likelihood can be written more neatly in the following way.

Theorem 11.1 *The complete-data likelihood for a stochastic kinetic model on the time interval $[0, T]$ takes the form*

$$L(c; \boldsymbol{x}) = \pi(\boldsymbol{x}|c) = \left\{\prod_{i=1}^{n} h_{\nu_i}(x(t_{i-1}), c_{\nu_i})\right\} \exp\left\{-\int_0^T h_0(x(t), c) \, dt\right\}.$$
$$(11.2)$$

From this informal perspective, the occurrence of the integral of the combined hazard function in (11.2) seems slightly mysterious. In fact, when viewed from a more advanced perspective, developing the notion of likelihood through stochastic calculus and the Radon-Nikodym derivative, the form of (11.2) is seen to be completely intuitive and typical of all Markov jump processes; the first (product) term represents all of the information in the jumps, and the second (integral) term represents all of the information in the period of full measure where no jumps occur.

For completely general hazard functions, Equations (11.1) and (11.2) represent the complete-data likelihood in its simplest form. However, in the case of the simple mass-action kinetic rate laws typically used in this context, it turns out that the likelihood factorises in a particularly convenient way. The key requirement is that the hazard functions can be written in the form $h_j(x, c_j) = c_j g_j(x)$, $j = 1, \ldots, v$.

Substituting into (11.2) and simplifying then gives

$$
L(c; x) = \left\{ \prod_{i=1}^{n} c_{\nu_i} g_{\nu_i}(x(t_{i-1})) \right\} \exp \left\{ -\int_0^T \sum_{j=1}^{v} c_j g_j(x(t)) \, dt \right\}
$$

$$
\propto \left\{ \prod_{j=1}^{v} c_j^{r_j} \right\} \exp \left\{ -\sum_{j=1}^{v} \int_0^T c_j g_j(x(t)) \, dt \right\}
$$

$$
= \prod_{j=1}^{v} c_j^{r_j} \exp \left\{ -c_j \int_0^T g_j(x(t)) \, dt \right\}
$$

$$
= \prod_{j=1}^{v} L_j(c_j; x),
$$

where the component likelihoods are defined by

$$
L_j(c_j; x) = c_j^{r_j} \exp \left\{ -c_j \int_0^T g_j(x(t)) \, dt \right\}, \quad j = 1, \ldots, v. \tag{11.3}
$$

This factorisation of the complete-data likelihood has numerous important consequences for inference. It means that in the complete data scenario, information regarding each rate constant is independent of the information regarding the other rate constants. That is, inference may be carried out for each rate constant separately. For example, in a maximum likelihood framework (where parameters are chosen to make the likelihood as large as possible), the likelihood can be maximised for each parameter separately. So, by partially differentiating (11.3) with respect to c_j and equating to zero, we obtain the maximum likelihood estimate of c_j as

$$
\widehat{c_j} = \frac{r_j}{\int_0^T g_j(x(t)) \, dt}, \quad j = 1, \ldots, v. \tag{11.4}
$$

In the context of Bayesian inference, the factorisation means that if independent prior distributions are adopted for the rate constants, then this independence will be retained *a posteriori*. It is also clear from the form of (11.3) that the complete-data likelihood is conjugate to an independent gamma prior for the rate constants. Thus, adopting priors for the rate constants of the form

$$
\pi(c) = \prod_{j=1}^{v} \pi(c_j), \quad c_j \sim Ga(a_j, b_j), \, j = 1, \ldots v,
$$

we can use Bayes' theorem (10.1) to obtain

$$\pi_j(c_j|\boldsymbol{x}) \propto \pi(c_j)L_j(c_j;\boldsymbol{x})$$

$$\propto c_j^{a_j-1}\exp\{-b_jc_j\}c_j^{r_j}\exp\left\{-c_j\int_0^T g_j(x(t))dt\right\}$$

$$= c_j^{a_j+r_j-1}\exp\left\{-c_j\left[b_j+\int_0^T g_j(x(t))dt\right]\right\},$$

that is

$$c_j|\boldsymbol{x} \sim Ga\left(a_j+r_j, b_j+\int_0^T g_j(x(t))dt\right), \quad j=1,\ldots v. \tag{11.5}$$

So, in the context of complete data, the inference problem is straightforward. However, even in the context of less than complete data, the distributions in (11.5) represent full-conditionals for the rate constants, and thus potentially form an important part of an MCMC algorithm for inferring the rate constants given discrete-time observations.

11.3 Discrete-time observations of the system state

It is, of course, unrealistic to suppose that it is possible to perfectly observe the entire sample path of the process, and so this assumption can be gradually weakened, and its effects on the inference process considered. It seems sensible to begin by considering the case of perfect observation of the system state at a finite collection of equally spaced times. However, as we will see later, there are a number of computational simplifications which arise in the case of noisy measurements, and therefore if inference from noisy observations is of primary interest, this rather technical section may be skipped without significant loss. In order to simplify notation, we will assume without loss of generality that we observe the process at integer times $t = 0, 1, \ldots, T$. That is, we have a total of $T+1$ observations on $x(t)$. We have seen in the previous section how to sample from the full-conditionals for the rate constants given a complete sample path on $[0, T]$, so an obvious strategy is to develop an MCMC algorithm which includes all of the missing parts of the sample path, and then simulate and average over the set of all plausible sample paths in the correct way.

Conditional on the $T+1$ observations (and the rate constants), the stochastic process breaks up into T conditionally independent processes on unit intervals with given fixed end points. The idea is to construct an MCMC algorithm which cycles through each interval in turn, sampling an appropriate path from its relevant full-conditional distribution, and then completing the iteration by sampling the rates conditional on the full sample path. We therefore need to be able to sample paths on an interval of unit length given its end points. For notational convenience we will consider without loss of generality the interval $[0, 1]$. Of course we know several exact methods for sampling the path conditional only on the left-hand end-point, $x(0)$ (including Gillespie's direct method). However, here we need to sample from the interval given both $x(0)$ and $x(1)$. This turns out to be *much* more difficult to do directly, and so a simpler strategy is adopted. The idea is that because we are working in the context of an MCMC algorithm, we can use a Metropolis–Hastings move to update the sample path, resulting in a Metropolis-within-Gibbs style overall algorithm. We therefore need only a method that will sample from a distribution over the space of bridging sample paths that has the same support as the true bridging process. Then we can use an appropriate Metropolis–Hastings acceptance probability in order to correct for the approximate step. An outline of the proposed MCMC algorithm can be stated as follows.

1. Initialise the algorithm with a valid sample path consistent with the observed data.
2. Sample rate constants from their full conditionals given the current sample path.
3. For each of the T intervals, propose a new sample path consistent with its endpoints and accept/reject it with a Metropolis–Hastings step.
4. Output the current rate constants.
5. Return to step 2.

In order to make progress with this problem, some notation is required. To keep the notation as simple as possible, we will now redefine some notation for the unit interval $[0, 1]$ which previously referred to the entire interval $[0, T]$. So now

$$\boldsymbol{x} = \{x(t) : t \in [0, 1]\}$$

denotes the 'true' sample path that is only observed at times $t = 0$ and $t = 1$, and

$$\boldsymbol{X} = \{X(t) : t \in [0, 1]\}$$

represents the stochastic process that gives rise to $\boldsymbol{x}$ as a single observation. Our problem is that we would like to sample directly from the distribution $(\boldsymbol{X}|x(0), x(1), c)$, but this is difficult, so instead we will content ourselves with constructing a Metropolis–Hastings update that has $\pi(\boldsymbol{x}|x(0), x(1), c)$ as its target distribution. Let us also re-define $r = (r_1, \ldots, r_v)^\mathsf{T}$ to be the numbers of reaction events in the interval $[0, 1]$, and $n = \sum_{j=1}^{v} r_j$. It is clear that knowing both $x(0)$ and $x(1)$ places some constraints on r, but it will not typically determine it completely. It turns out to be easiest to sample a new interval in two stages: first pick an r consistent with the end constraints and then sample a new interval conditional on $x(0)$ and r. So, ignoring the problem of sampling r for the time being, we would ideally like to be able to sample from $\pi(\boldsymbol{x}|x(0), r, c)$, but this is still quite difficult to do directly. At this point it is helpful to think of the u-component sample path $\boldsymbol{X}$ as being a function of the v-component point process of reaction events. This point process is hard to simulate directly as its hazard function is random, but the hazards are known at the end-points $x(0)$ and $x(1)$, and so they can probably be reasonably well approximated by v independent inhomogeneous Poisson processes whose rates vary linearly between the rates at the end points. In order to make this work, we need to be able to sample from an inhomogeneous Poisson process conditional on the number of events. This requires some Poisson process theory not covered in Chapter 5.

Lemma 11.1 *For given fixed $\lambda, p > 0$, consider $N \sim Po(\lambda)$ and $X|N \sim Bin(N, p)$. Then marginally we have*

$$X \sim Po(\lambda p).$$

Proof.

$$P(X = k) = \sum_{i=0}^{\infty} P(X = k|N = i) \, P(N = i)$$

$$= \sum_{i=k}^{\infty} P(X = k|N = i) \, P(N = i)$$

$$= \sum_{i=k}^{\infty} \frac{i!}{k!(i-k)!} p^k (1-p)^{i-k} \times \frac{\lambda^i e^{-\lambda}}{i!}$$

$$= \sum_{i=k}^{\infty} \frac{\lambda^i e^{-\lambda} p^k (1-p)^{i-k}}{k!(i-k)!}$$

$$= \sum_{i=0}^{\infty} \frac{\lambda^{i+k} e^{-\lambda} p^k (1-p)^i}{k!i!}$$

$$= \frac{\lambda^k e^{-\lambda} p^k}{k!} \sum_{i=0}^{\infty} \frac{\lambda^i (1-p)^i}{i!}$$

$$= \frac{\lambda^k e^{-\lambda} p^k}{k!} e^{\lambda(1-p)}$$

$$= \frac{(\lambda p)^k e^{-\lambda p}}{k!}, \qquad k = 0, 1, \ldots$$

□

Proposition 11.1 *Consider a homogeneous Poisson process with rate λ on the finite interval $[0, T]$. Let $R \sim Po(\lambda T)$ be the number of events. Then conditional on R, the (unsorted) event times are $U(0, T)$ random variables. In other words, the ordered event times correspond to R uniform order statistics.*

Proof. We will just give a sketch proof of this fact that is anyway intuitively clear. The key property of the Poisson process is that in a finite interval $[a, b] \subseteq [0, T]$, the number of events is $Po(\lambda[b - a])$. So we will show that if a point process is constructed by first sampling R and then scattering R points uniformly, then the correct number of events will be assigned to the interval $[a, b]$. It is clear that for any particular event, the probability that it falls in the interval $[a, b]$ is $(b - a)/T$. Now, letting X denote the number of events in the interval $[a, b]$, it is clear that

$$(X|R = r) \sim Bin(r, (b - a)/T)$$

and so by the previous Lemma,

$$X \sim Po(\lambda T \times (b - a)/T)$$
$$= Po(\lambda[b - a]).$$

□

So we now have a way of simulating a homogeneous Poisson process conditional on the number of events. However, as we saw in Chapter 5, one way of thinking about an inhomogeneous Poisson process is as a homogeneous Poisson process with time rescaled in a non-uniform way.

Proposition 11.2 *Let X be a homogeneous Poisson process on the interval $[0,1]$ with rate $\mu = (h_0 + h_1)/2$, and let Y be an inhomogeneous Poisson process on the same interval with rate $\lambda(t) = (1-t)h_0 + th_1$, for given fixed $h_0 \neq h_1$, $h_0, h_1 > 0$. A realisation of the process Y can be obtained from a realisation of the process X by applying the time transformation*

$$t := \frac{\sqrt{h_0^2 + \{h_1^2 - h_0^2\}t} - h_0}{h_1 - h_0}$$

to the event times of the X process.

Proof. Process X has cumulative hazard $M(t) = t(h_0 + h_1)/2$, while process Y has cumulative hazard

$$\Lambda(t) = \int_0^t [(1-t)h_0 + th_1]dt = h_0 t + \frac{t^2}{2}(h_1 - h_0).$$

Note that the cumulative hazards for the two processes match at both $t = 0$ and $t = 1$, and so one process can be mapped to the other by distorting time to make the cumulative hazards match also at intermediate times. Let the local time for the X process be s and the local time for the Y process be t. Then setting $M(s) = \Lambda(t)$ gives

$$\frac{s}{2}(h_0 + h_1) = h_0 t + \frac{t^2}{2}(h_1 - h_0)$$

$$\Rightarrow 0 = \frac{t^2}{2}(h_1 - h_0) + h_0 t - \frac{s}{2}(h_0 + h_1)$$

$$\Rightarrow t = \frac{-h_0 + \sqrt{h_0^2 + (h_1 - h_0)(h_0 + h_1)s}}{h_1 - h_0}.$$

□

So, we can sample an inhomogeneous Poisson process conditional on the number of events by first sampling a homogeneous Poisson process with the average rate conditional on the number of events and then transforming time to get the correct inhomogeneity.

In order to correct for the fact that we are not sampling from the correct bridging process, we will need a Metropolis–Hastings acceptance probability that will depend both on the likelihood of the sample path under the true model and the likelihood of the sample path under the approximate model. We have already calculated the likelihood under the true model (the complete-data likelihood). We now need the likelihood under the inhomogeneous Poisson process model.

Proposition 11.3 *The complete data likelihood for a sample path x on the interval $[0, 1]$ under the approximate inhomogeneous Poisson process model is given by*

$$L_A(c; x) = \left\{ \prod_{i=1}^n \lambda_{\nu_i}(t_i) \right\} \exp \left\{ -\frac{1}{2}[h_0(x(0), c) + h_0(x(1), c)] \right\},$$

where $\lambda_j(t) = (1-t)h_j(x(0), c) + t\, h_j(x(1), c)$, $j = 1, \ldots, v$.

Proof. Again, we will compute this in an informal way. Define the cumulative rates

$$\Lambda_j(t) = \int_0^t \lambda_j(t) dt,$$

the combined rate $\lambda_0(t) = \sum_{j=1}^v \lambda_j(t)$, and the cumulative combined rate $\Lambda_0(t) = \sum_{j=1}^v \Lambda_j(t)$. Considering the ith event, the density of the time and type is given by

$$\lambda_0(t_i) \exp\{-[\Lambda_0(t_i) - \Lambda_0(t_{i-1})]\} \times \frac{\lambda_{\nu_i}(t_i)}{\lambda_0(t_i)}$$

$$= \lambda_{\nu_i}(t_i) \exp\{-[\Lambda_0(t_i) - \Lambda_0(t_{i-1})]\},$$

and the probability of no event after time t_n is given by

$$\exp\{-[\Lambda_0(T) - \Lambda_0(t_n)]\}.$$

This leads to a combined likelihood of the form

$$L_A(c; \boldsymbol{x}) = \left\{\prod_{i=1}^n \lambda_{\nu_i}(t_i) \exp\{-[\Lambda_0(t_i) - \Lambda_0(t_{i-1})\}\right\} \times \exp\{-[\Lambda_0(T) - \Lambda_0(t_n)]\}$$

$$= \left\{\prod_{i=1}^n \lambda_{\nu_i}(t_i)\right\} \left\{\prod_{i=1}^{n+1} \exp\{-[\Lambda_0(t_i) - \Lambda_0(t_{i-1})]\}\right\}$$

$$= \left\{\prod_{i=1}^n \lambda_{\nu_i}(t_i)\right\} \exp\left\{-\sum_{i=1}^{n+1} [\Lambda_0(t_i) - \Lambda_0(t_{i-1})]\right\}$$

$$= \left\{\prod_{i=1}^n \lambda_{\nu_i}(t_i)\right\} \exp\left\{-\Lambda_0(T)\right\}$$

$$= \left\{\prod_{i=1}^n \lambda_{\nu_i}(t_i)\right\} \exp\left\{-\frac{1}{2}[h_0(x(0), c) + h_0(x(1), c)]\right\}.$$

$\square$

As we will see, the complete-data likelihoods occur in the Metropolis–Hastings acceptance probability in the form of the ratio $L(c; \boldsymbol{x})/L_A(c; \boldsymbol{x})$. This ratio is clearly just

$$\frac{L(c; \boldsymbol{x})}{L_A(c; \boldsymbol{x})} = \left\{\prod_{i=1}^n \frac{h_{\nu_i}(x(t_{i-1}), c_{\nu_i})}{\lambda_{\nu_i}(t_i)}\right\}$$

$$\times \exp\left\{\frac{1}{2}[h_0(x(0), c) + h_0(x(1), c)] - \int_0^1 h_0(x(t), c)dt\right\}.$$

From a more advanced standpoint, this likelihood ratio is seen to be the Radon–Nikodym derivative $\frac{d\mathbb{P}}{d\mathbb{Q}}(\boldsymbol{x})$ of the true Markov jump process ($\mathbb{P}$) with respect to the linear inhomogeneous Poisson process approximation ($\mathbb{Q}$), and may be derived primitively and directly from the properties of the jump processes in an entirely rigorous way. The Radon–Nikodym derivative measures the 'closeness' of the approximating process to the true process, in the sense that the more closely the processes match, the closer the derivative will be to 1.

We are now in a position to state the basic form of the Metropolis–Hastings update of the interval $[0, 1]$. First a proposed new r vector will be sampled from an appropriate proposal distribution with PMF $f(r^*|r)$ (we will discuss appropriate ways of constructing this later). Then, conditional on r^*, we will sample a proposed sample path $\boldsymbol{x}^*$ from the approximate process and accept the pair $(r^*, \boldsymbol{x}^*)$ with probability $\min\{1, A\}$ where

$$A = \frac{\pi(\boldsymbol{x}^*|x(0), x(1), c)}{\pi(\boldsymbol{x}|x(0), x(1), c)} \bigg/ \frac{f(r^*|r)\pi_A(\boldsymbol{x}^*|x(0), r^*, c)}{f(r|r^*)\pi_A(\boldsymbol{x}|x(0), r, c)}$$

$$= \frac{\dfrac{\pi(\boldsymbol{x}^*|x(0), c)}{\pi_A(\boldsymbol{x}^*|x(0), c)}}{\dfrac{\pi(\boldsymbol{x}|x(0), c)}{\pi_A(\boldsymbol{x}|x(0), c)}} \times \frac{\dfrac{q(r^*)}{f(r^*|r)}}{\dfrac{q(r)}{f(r|r^*)}}$$

$$= \frac{\dfrac{L(c; \boldsymbol{x}^*)}{L_A(c; \boldsymbol{x}^*)}}{\dfrac{L(c; \boldsymbol{x})}{L_A(c; \boldsymbol{x})}} \times \frac{\dfrac{q(r^*)}{f(r^*|r)}}{\dfrac{q(r)}{f(r|r^*)}},$$

where $q(r)$ is the PMF of r under the approximate model. That is,

$$q(r) = \prod_{j=1}^{v} q_j(r_j),$$

where $q_j(r_j)$ is the PMF of a Poisson with mean $[h_j(x(0), c) + h_j(x(1), c)]/2$. Again, we could write this more formally as

$$A = \frac{\dfrac{d\mathbb{P}}{d\mathbb{Q}}(\boldsymbol{x}^*)}{\dfrac{d\mathbb{P}}{d\mathbb{Q}}(\boldsymbol{x})} \times \frac{\dfrac{q(r^*)}{f(r^*|r)}}{\dfrac{q(r)}{f(r|r^*)}}.$$

So now the only key aspect of the MCMC algorithm that has not yet been discussed is the choice of the proposal distribution $f(r^*|r)$. Again, ideally we would like to sample directly from the true distribution of r given $x(0)$ and $x(1)$, but this is not straightforward. Instead we simply want to pick a proposal that effectively explores the space of rs consistent with the end points. Recalling the discussion of Petri nets from Section 2.3, to a first approximation, the set of r that we are interested in is the set of all non-negative integer solutions in r to

$$x(1) = x(0) + Sr$$
$$\Rightarrow Sr = x(1) - x(0). \tag{11.6}$$

There will be some solutions to this equation that do not correspond to possible sample paths, but there will not be many of these. Note that given a valid solution r, then $r + r'$ is another valid solution, where r' is any T-invariant of the Petri net. Thus, the set of all solutions is closely related to the set of all T-invariants of the associated Petri net. Assuming that S is of full rank (if S is not of full rank, the dimension-reducing techniques from Section 7.3 can be used to reduce the model until it is), then the space of solutions will have dimension $v - u$. One way to explore this space is to permute the columns of S so that the first u columns represent an invertible $u \times u$ matrix, $\tilde{S}$. Denoting the remaining columns of S by $\vec{S}$, we partition S as $S = (\tilde{S}, \vec{S})$. We similarly partition $r = \binom{\tilde{r}}{\vec{r}}$, with $\tilde{r}$ representing the first u elements of r. We then have

$$Sr = (\tilde{S}, \vec{S})\binom{\tilde{r}}{\vec{r}} = \tilde{S}\tilde{r} + \vec{S}\vec{r} = x(1) - x(0)$$

$$\Rightarrow \tilde{r} = \tilde{S}^{-1}\left[x(1) - x(0) - \vec{S}\vec{r}\right]. \tag{11.7}$$

Equation (11.7) suggests a possible strategy for exploring the solution space. Starting from a valid solution r, one can perturb the elements corresponding to $\vec{r}$ in an essentially arbitrary way, and then set the elements of $\tilde{r}$ accordingly. Of course, there is a chance that some element will go negative, but such moves will be immediately rejected. One possible choice of symmetric proposal for updating the elements of $\vec{r}$ is to add to each element the difference between two independent Poisson random quantities with the same mean, ω. If the tuning parameter ω is chosen to be independent of the current state, then the proposal distribution is truly symmetric and the relevant PMF terms cancel out of the Metropolis–Hastings acceptance probability. However, it is sometimes useful to allow ω to be a function of the current state of the chain (for example, allowing ω to be larger when the state is larger), and in this case the PMF for this proposal is required. The PMF is well known and is given by $p(y) = e^{-2\omega}I_y(2\omega)$, where $I_y(\cdot)$ is a regular modified Bessel function of order y; see Johnson and Kotz (1969) and Abramowitz and Stegun (1984) for further details.

Now, given a correctly initialised Markov chain, we have everything we need to implement an MCMC algorithm that will have as its equilibrium distribution the exact posterior distribution of the rate constants given the (perfect) discrete time observations. However, it turns out that even the problem of initialising the chain is not entirely straightforward. Here we need a valid solution to (11.6) for each of the T unit intervals. The constraints encoded by (11.6) correspond to the constraints of an integer linear programming problem. By adopting an (essentially arbitrary) objective function $\min\{\sum_j r_j\}$, we can use standard algorithms and software libraries (such as glpk) for the solution of integer linear programming problems in order to initialise the chain.

This MCMC strategy was examined in detail for the special case of the stochastic kinetic Lotka–Volterra model in Boys et al. (2008). In addition to the 'exact' MCMC algorithm presented here, an exact reversible jump MCMC (RJ-MCMC) strategy is also explored, though this turns out to be rather inefficient. The paper also explores a fast approximate algorithm which drops the Radon–Nikodym derivative correction factor from the acceptance ratio. For further details of the approximation and its effectiveness, see the discussion in Boys et al. (2008). The paper also discusses the extension of these ideas to deal with partial observation of the system state. It should also be possible to develop better strategies for sampling from the conditioned process; see Golightly and Wilkinson (2015) for some initial ideas in this direction.

11.4 Diffusion approximations for inference

The discussion in the previous section demonstrates that it is possible to construct exact MCMC algorithms for inference in discrete stochastic kinetic models based on discrete time observations (and it is possible to extend the techniques to more realistic data scenarios than those directly considered). The discussion gives great insight into the nature of the inferential problem and its conceptual solution. However, there is a slight problem with the techniques discussed there in the context of the relatively large and complex models of genuine interest to systems biologists. It should be clear that each iteration of the MCMC algorithm described in the previous section is more computationally demanding than simulating the process exactly using Gillespie's direct method (for the sake of argument, let us say that it is one order of magnitude more demanding). For satisfactory inference, a large number of MCMC iterations will be required. For models of the complexity discussed in the previous section, it is not uncommon for 10^7–10^8 iterations to be required for satisfactory convergence to the true posterior distribution. Using such methods for inference therefore has a computational complexity of 10^8–10^9 times that required to simulate the process. As if this were not bad enough, it turns out that MCMC algorithms are particularly difficult to parallelise effectively (Wilkinson, 2005). One possible approach to improving the situation is to approximate the algorithm with a much faster one that is less accurate, as discussed in Boys et al. (2008). Unfortunately, even that approach does not scale up well to genuinely interesting problems, so a different approach is required.

A similar problem was considered in Chapter 8, from the viewpoint of simulation rather than inference. We saw there how it was possible to approximate the true Markov jump process by the chemical Langevin equation (CLE), which is the diffusion process that behaves most like the true jump process. It was explained that simulation of the CLE can be orders of magnitude faster than an exact algorithm. This suggests the possibility of using the CLE as an approximate model for inferential purposes. It turns out that the CLE provides an excellent model for inference, even in situations where it does not perform particularly well as a simulation model. This observation at first seems a little counter-intuitive, but the reason is that in the context of inference, one is conditioning on data from the true model, and this helps to calibrate the approximate model and stop MCMC algorithms from wandering off into parts of the space that are plausible in the context of the approximate model, but not in the context of the true model.

What is required is a method for inference for general non-linear multivariate diffusion processes observed partially, discretely, and possibly also with measurement error. This too turns out to be a highly non-trivial problem, and is still the subject of ongoing research. Such inference problems arise often in financial mathematics and econometrics, and so much of the literature relating to this problem can be found in that area; see Durham and Gallant (2002) for an overview.

The problem with diffusion processes is that any finite sample path contains an infinite amount of information, and so the concept of a complete-data likelihood does not exist in general. We will illustrate the problem in the context of high-resolution time-course data on the CLE. Starting with the CLE in the form of (8.3), define $\mu(x, c) = Sh(x, c)$ and $\beta(x, c) = S \operatorname{diag}\{h(x, c)\} S^\mathsf{T}$ to get the u-dimensional diffusion process

$$dX_t = \mu(X_t, c)dt + \sqrt{\beta(X_t, c)}dW_t.$$

We will assume that for some small time-step Δt we have data $x = (x_0, x_{\Delta t}, x_{2\Delta t}, \ldots, x_{n\Delta t})$ (that is, $n+1$ observations), and that the time-step is sufficiently small for the Euler–Maruyama approximation to be valid, leading to the difference equation

$$\Delta X_t \equiv X_{t+\Delta t} - X_t \simeq \mu(X_t, c)\Delta t + \sqrt{\beta(X_t, c)}\Delta W_t. \tag{11.8}$$

Equation (11.8) corresponds to the distributional statement

$$X_{t+\Delta t}|X_t, c \sim N(X_t + \mu(X_t, c)\Delta t, \beta(X_t, c)\Delta t),$$

where $N(\cdot, \cdot)$ here refers to the multivariate normal distribution, parametrised by its mean vector and covariance matrix. The probability density associated with this increment is given by

$$
\begin{aligned}
\pi(x_{t+\Delta t}|x_t, c) &= N(x_{t+\Delta t}; x_t + \mu(x_t, c)\Delta t, \beta(x_t, c)\Delta t) \\
&= (2\pi)^{-u/2}|\beta(x_t, c)\Delta t|^{-1/2} \\
&\quad \times \exp\Big\{ -\frac{1}{2}(\Delta x_t - \mu(x_t, c)\Delta t)^\mathsf{T}[\beta(x_t, c)\Delta t]^{-1}(\Delta x_t \\
&\qquad\qquad\qquad - \mu(x_t, c)\Delta t)\Big\} \\
&= (2\pi\Delta t)^{-u/2}|\beta(x_t, c)|^{-1/2} \\
&\quad \times \exp\Big\{ -\frac{\Delta t}{2}\Big(\frac{\Delta x_t}{\Delta t} - \mu(x_t, c)\Big)^\mathsf{T} \beta(x_t, c)^{-1}\Big(\frac{\Delta x_t}{\Delta t} \\
&\qquad\qquad\qquad - \mu(x_t, c)\Big)\Big\}.
\end{aligned}
$$

If S is not of full rank, then $\beta(x_t, c)$ will not be invertible. This can be tackled either by reducing the dimension of the model so that S is of full rank, or by using the Moore–Penrose generalised inverse of $\beta(x_t, c)$ wherever the inverse occurs in a multivariate normal density. It is important to understand that diffusion sample paths are not differentiable, so the quantity $\Delta x_t/\Delta t$ does not have a limit as Δt tends to zero. It is now possible to derive the likelihood associated with this set of observations as

$$L(c; x) = \pi(x|c)$$

$$= \pi(x_0|c) \prod_{i=0}^{n-1} \pi(x_{(i+1)\Delta t}|x_{i\Delta t}, c)$$

$$= \pi(x_0|c) \prod_{i=0}^{n-1} (2\pi\Delta t)^{-u/2} |\beta(x_{i\Delta t}, c)|^{-1/2}$$

$$\times \exp\left\{ -\frac{\Delta t}{2} \left(\frac{\Delta x_{i\Delta t}}{\Delta t} - \mu(x_{i\Delta t}, c) \right)^{\mathsf{T}} \beta(x_{i\Delta t}, c)^{-1} \left(\frac{\Delta x_{i\Delta t}}{\Delta t} \right. \right.$$

$$\left. \left. - \mu(x_{i\Delta t}, c) \right) \right\}.$$

Now assuming that $\pi(x_0|c)$ is in fact independent of c, we can simplify the likelihood as

$$L(c; x) \propto \left\{ \prod_{i=0}^{n-1} |\beta(x_{i\Delta t}, c)|^{-1/2} \right\} \times$$

$$\exp\left\{ -\frac{1}{2} \sum_{i=0}^{n-1} \left(\frac{\Delta x_{i\Delta t}}{\Delta t} - \mu(x_{i\Delta t}, c) \right)^{\mathsf{T}} \beta(x_{i\Delta t}, c)^{-1} \left(\frac{\Delta x_{i\Delta t}}{\Delta t} - \mu(x_{i\Delta t}, c) \right) \right.$$

$$\left. \Delta t \right\}. \quad (11.9)$$

Equation (11.9) is the closest we can get to a complete-data likelihood for the CLE, as it does not have a limit as Δt tends to zero.[†] In the case of perfect high-resolution observations on the system state, (11.9) represents the likelihood for the problem, which could be maximised (numerically) in the context of maximum likelihood estimation or combined with a prior to form the kernel of a posterior distribution for c. In this case there is no convenient conjugate analysis, but it is entirely straightforward to implement a Metropolis random walk MCMC sampler to explore the posterior distribution of c.

Of course it is unrealistic to assume perfect observation of all states of a model, and in the biological context, it is usually unrealistic to assume that the sampling frequency will be sufficiently high to make the Euler approximation sufficiently accurate. So, just as for the discrete case, MCMC algorithms can be used in order to 'fill-in' all of the missing information in the model.

MCMC algorithms for multivariate diffusions can be considered conceptually in a similar way to those discussed in the previous section. Given (perfect) discrete-time

---

[†] If it were the case that $\beta(x, c)$ was independent of c, then we could drop terms no longer involving c to get an expression that is well behaved as Δt tends to zero. In this case, the complete data likelihood is the exponential of the sum of two integrals, one of which is a regular Riemann integral, and the other is an Itô stochastic integral. Unfortunately this rather elegant result is of no use to us here, as the diffusion matrix of the CLE depends on c in a fundamental way.

observations, the diffusion process breaks into a collection of multivariate diffusion bridges that need to be sampled and averaged over in the correct way. However, there is a complication in the context of diffusion processes that does not occur in the context of Markov jump processes due to the infinite amount of information in a finite continuous diffusion sample path. This means both that it is impossible to fully impute a sample path, and also that even if it were possible, the full conditionals for the diffusion parameters would be degenerate, leading to a reducible MCMC algorithm. An informative discussion of this problem, together with an elegant solution in the context of univariate diffusions, can be found in Roberts and Stramer (2001). Fortunately, it turns out that by working with a time discretisation of the CLE (such as the Euler discretisation), both of these problems can be side-stepped (but not actually solved). Here the idea is to introduce a finite number of time points between each pair of observations, and implement an MCMC algorithm to explore the space of 'skeleton' bridges between the observations. This is relatively straightforward, and there are a variety of different approaches that can be used for implementation purposes. Then because the sample path is represented by just a finite number of points, the full conditionals for the diffusion parameters are not degenerate, and the MCMC algorithm will have as its equilibrium distribution the exact posterior distribution of the rate parameters given the data, conditional on the Euler approximation to the CLE being the true model. Of course neither the CLE nor the Euler approximation to it are actually the true model, but this will hopefully have little effect on inferences. Such an approach to inferring the rate constants of stochastic kinetic models is discussed in Golightly and Wilkinson (2005), to which the reader is referred for further details. This paper also discusses the case of partial observation of the system state and includes an application to a genetic regulatory network.

As mentioned in the previous paragraph, the technique of discretising time has the effect of side-stepping the technical problems of inference, but does not actually solve them. In particular, it is desirable to impute many latent points between each pair of observations, so that the discretisation bias is minimised and the skeleton bridges represent a good approximation to the true process on the same skeleton of points. Unfortunately, using a fine discretisation has the effect of making the full-conditionals for the diffusion parameters close to singular, which in turn makes the MCMC algorithm mix extremely poorly. This puts the modeller in the unhappy position of having to choose between a poor approximation to the CLE model or a poorly mixing MCMC algorithm. Clearly a more satisfactory solution is required.

At least two different methods are possible for solving this problem. First, it turns out that by adopting a sequential MCMC algorithm (where the posterior distribution is updated one time point at a time, rather than with a global algorithm that incorporates the information from all time points simultaneously), it is possible to solve the mixing problems by jointly updating the diffusion parameters and diffusion bridges, thereby breaking the dependence between them. The sampling mechanism relies heavily on the *modified diffusion bridge* construct of Durham and Gallant (2002). A detailed discussion of this algorithm is given in Golightly and Wilkinson (2006a), and its application to inference for stochastic kinetic models is explored in Golightly and Wilkinson (2006b). This last paper discusses the most realistic setting of multiple data sets that are partial, discrete, and include measurement error. One of the advantages of using a sequential algorithm is that it is very convenient to use in the context of multiple data sets from different cells or experiments, possibly measuring different species in each experiment. A discussion of this issue in the context of an illustrative example is given in Golightly and Wilkinson (2006b).

However, there are other problems with sequential Monte Carlo approaches to inference for static model parameters, which make the above technique impractical to use with very large numbers of time points. It turns out to be possible to instead construct a global (non-sequential) MCMC algorithm that is irreducible by using a careful re-parametrisation of the process, again exploiting the modified diffusion bridge construction. This approach is introduced in Golightly and Wilkinson (2008) and generalised in Wilkinson and Golightly (2010). In the absence of measurement error, this approach is arguably the most effective approach to Bayesian inference for non-linear multivariate diffusion process models currently available, but see Fuchs (2013) for additional technical details and Stramer and Bognar (2011) for a review.

11.5 Likelihood-free methods

11.5.1 Partially observed Markov process models

Although the introduction of measurement error into the problem in many ways represents a modest extension to the model complexity, it turns out that new simple and powerful approaches to the problem become possible in this case. Reconsider the latent variable modelling approach from Section 10.6. We have the factorisation

$$\pi(\theta, x, y) = \pi(\theta)\pi(x|\theta)\pi(y|x, \theta),$$

where θ represents a vector of model parameters (typically θ will consist of a vector of rate constants, c, together with any additional parameters of the model, including parameters of the measurement model), x represents the full true state trajectory of the stochastic kinetic model, and y represents the available data (typically discrete time observations of the system state subject to measurement error). The presence of measurement error allows likelihood-free approaches based on forward simulation from the model to be used without the MCMC scheme degenerating.

Suppose that we wish to construct an MCMC scheme targeting the full posterior distribution $\pi(\theta, x|y)$. An LF-MCMC scheme can be implemented by constructing a proposal in two stages: first sample $\theta^\star$ from an essentially arbitrary proposal distribution $f(\theta^\star|\theta)$, and then generate a complete sample path $x^\star$ from the simulation model $\pi(x^\star|\theta^\star)$. We know from our discussion in the previous chapter that such a proposal should be have acceptance ratio

$$A = \frac{\pi(\theta^\star)\pi(y|x^\star, \theta^\star)f(\theta|\theta^\star)}{\pi(\theta)\pi(y|x, \theta)f(\theta^\star|\theta)}. \tag{11.10}$$

Measurement error is necessary to ensure that the term $\pi(y|x^\star, \theta^\star)$ in the numerator of the acceptance ratio is not typically degenerate. However, this simple forward simulation approach will typically perform poorly if there is a non-trivial amount of data, y, since the forward simulation step ignores the data. There are several possible approaches to deal with this problem, which we now examine in detail.

In order to keep the notation simple, we will start by considering observations at integer times, $1, \ldots, T$, though extension to arbitrary observation times is trivial and will be considered later. We assume that we have observed data $y_1, \ldots, y_T$, determined by some measurement process $\pi(y_t|x(t), \theta)$, where $x(t)$ is the true unobserved state of the process at time t. So, in the case of observing just the ith species subject to Gaussian measurement error, we would have

$$\pi(y_t|x(t), \theta) = N(y_t; x_i(t), \sigma^2),$$

where σ might be 'known', or could be an element of the parameter vector, θ. Assuming conditional independence of the observations and using the Markov property of the kinetic model, we have the following factorisation of the joint density

$$\pi(\theta, \boldsymbol{x}, y) = \pi(\theta)\pi(x(0)|\theta) \prod_{t=1}^{T} [\pi(\boldsymbol{x}_t|x(t-1), \theta)\pi(y_t|x(t), \theta)],$$

where $\boldsymbol{x}_t = \{x(s)|t-1 < s \leq t\}$. We can use this factorisation to expand (11.10) as

$$A = \frac{\pi(\theta^\star)f(\theta|\theta^\star) \displaystyle\prod_{t=1}^{T} \pi(y_t|x(t)^\star, \theta^\star)}{\pi(\theta)f(\theta^\star|\theta) \displaystyle\prod_{t=1}^{T} \pi(y_t|x(t), \theta)}$$

for the simple LF-MCMC scheme. It is clear that if T is large, the product term in the numerator will be very small, leading to a poorly mixing MCMC chain. One way to deal with this problem is to adopt a sequential approach, running an LF-MCMC algorithm at each time point, thereby ensuring that only a single term from the likelihood occurs in the acceptance ratio at any one time. The approach is explored in detail in Wilkinson (2011). However, as previously mentioned, sequential Monte Carlo approaches to inference for static model parameters are problematic, due to issues of particle degeneracy, so it is worth exploring the possibility of developing efficient global MCMC algorithms. It turns out that a pseudo-marginal approach to this problem is possible, by exploiting sequential Monte Carlo methods within an MCMC sampler, and so we investigate this approach in detail now.

11.5.2 Pseudo-marginal approach

Let us suppose that we are not interested in the unobserved state of the process, and wish to construct an MCMC algorithm targeting $\pi(\theta|y)$. In that case, we know from Section 10.6.1 that this can be accomplished using a proposal $f(\theta^\star|\theta)$ together with an acceptance ratio of the form

$$A = \frac{\pi(\theta^\star)f(\theta|\theta^\star)\hat{\pi}(y|\theta^\star)}{\pi(\theta)f(\theta^\star|\theta)\hat{\pi}(y|\theta)}$$

where $\hat{\pi}(y|\theta)$ is an unbiased Monte Carlo estimate of the marginal likelihood $\pi(y|\theta)$. Obviously, in order to implement this scheme, we need an effective method for generating unbiased estimates of the likelihood of the data. It turns out that it is very easy to generate such estimates using a simple likelihood-free sequential Monte Carlo (SMC) method known as the *bootstrap particle filter* (Doucet et al., 2001).

11.5.3 Bootstrap particle filter

The bootstrap particle filter is a simple sequential importance resampling technique for estimating the unobserved state of the Markov process conditional on the observations (and the model parameters). This is interesting and useful in its own right, and we will look later at how to fully exploit the method in the context of particle MCMC methods (Andrieu et al., 2010). For now we will just exploit the fact that it gives an unbiased estimate of the likelihood of the data as a by-product. The particle filter assumes fixed model parameters, so we will drop θ from the notation for this section, accepting that all densities are implicitly conditioned on θ.

1. Put $t := 0$. The procedure is initialised with a sample $x_0^k \sim \pi(x_0)$, $k = 1, \ldots, M$ with uniform normalised weights $w'^k_0 = 1/M$.

2. Suppose by induction that we are currently at time t, and that we have a weighted sample $\{x_t^k, w'^k_t | k = 1, \ldots, M\}$ from $\pi(x_t|y_{1:t})$.

3. Using this weighted sample, next generate an equally weighted sample by resampling with replacement M times to obtain $\{\tilde{x}_t^k | k = 1, \ldots, M\}$ (giving an approximate random sample from $\pi(x_t|y_{1:t})$). Note that each sample is independently drawn from the discrete distribution $\sum_{i=1}^{M} w'^i_t \delta(x - x_t^i)$ (ideally using an efficient lookup algorithm).

4. Next, propagate each particle forward according to the Markov process model by sampling $x_{t+1}^k \sim \pi(x_{t+1}|\tilde{x}_t^k)$, $k = 1, \ldots, M$ (giving an approximate random sample from $\pi(x_{t+1}|y_{1:t})$).

5. Then for each of the new particles, compute a weight $w_{t+1}^k = \pi(y_{t+1}|x_{t+1}^k)$, and then a normalised weight $w'^k_{t+1} = w_{t+1}^k / \sum_i w_{t+1}^i$.

   It is clear from our understanding of importance resampling that these weights are appropriate for representing a sample from $\pi(x_{t+1}|y_{1:t+1})$, and so the particles and weights can be propagated forward to the next time point.

6. Put $t := t + 1$ and if $t < T$, return to Step 2.

It is clear that the average weight at each time gives an estimate of the marginal likelihood of the current data point given the data so far. So we define

$$\hat{\pi}(y_t|y_{1:t-1}) = \frac{1}{M} \sum_{k=1}^{M} w_t^k$$

and

$$\hat{\pi}(y_{1:T}) = \hat{\pi}(y_1) \prod_{t=2}^{T} \hat{\pi}(y_t|y_{1:t-1}).$$

Again, from our understanding of importance resampling, it should be reasonably clear that $\hat{\pi}(y_{1:T})$ is a *consistent* estimator of $\pi(y_{1:T})$, in that it will converge to $\pi(y_{1:T})$ as $M \longrightarrow \infty$. It is much less clear, but nevertheless true that this estimator is also *unbiased* for $\pi(y_{1:T})$. The standard reference for this fact is Del Moral (2004), but this is a rather technical monograph. A much more accessible proof (for a very general particle filter) is given in Pitt et al. (2012). The proof is straightforward, but somewhat involved, so we will not discuss it further here. The important thing to appreciate is that we have a simple sequential likelihood-free algorithm which generates an unbiased estimate of $\pi(y)$. An R function for constructing a function closure for marginal likelihood estimation based on a bootstrap particle filter in given in Figure 11.1.

This implementation works fine in many cases, but it has an important flaw, in that it works with 'raw' weights rather than log–weights. We know that it is better to work with log-likelihoods rather than actual likelihoods, since actual likelihoods have a tendency to numerically underflow. But here it isn't immediately obvious how to avoid raw likelihoods, since we need to sum them, and use them for resampling. It turns out that we can nevertheless work mainly with log-likelihoods and avoid the risk of underflow by using a technique often referred to as the 'log–sum–exp trick' (Murphy, 2012). Suppose that we have log (unnormalised) weights, $l_i = \log w_i$, but we want to compute the (log of the) sum of the raw weights. We can obviously exponentiate the log weights and sum them, but this carries the risk that all of the log weights will underflow (or, possibly, overflow). So, we want to compute

$$L = \log \sum_i \exp(l_i),$$

without the risk of underflow (hence 'log–sum–exp'). We can do this by first computing $m = \max_i l_i$ and then noting that

$$L = \log \sum_i \exp(l_i) = \log \sum_i \exp(m) \exp(l_i - m)$$

$$= \log \left[\exp(m) \sum_i \exp(l_i - m) \right] = m + \log \sum_i \exp(l_i - m).$$

We are now safe, since $l_i - m$ cannot be bigger than zero, and hence no term can overflow. Also, since we know that $l_i - m = 0$ for some i, we know that at least one term will not underflow. So we are guaranteed to get a sensible finite answer in a numerically stable way. Note that exactly the same technique works for computing a sample mean as opposed to just a sum. Also, since the $\exp(l_i - m)$ terms are just rescaled raw weights, we can safely feed them into a sampling function. We can use this to slightly re-write our bootstrap particle filter to work with log weights and avoid the risk of underflow, as shown in Figure 11.2. An example of use is given in Figure 11.3 — see the figure captions for further details.

Once we have a mechanism for generating unbiased Monte Carlo estimates of marginal likelihood, this can be embedded in a pseudo-marginal MCMC algorithm for making inference for model parameters from time course data.

```
pfMLLik1 <- function (n, simx0, t0, stepFun, dataLik,
    data)
{
    times = c(t0, as.numeric(rownames(data)))
    deltas = diff(times)
    return(function(...) {
        xmat = simx0(n, t0, ...)
        ll = 0
        for (i in 1:length(deltas)) {
            xmat = t(apply(xmat, 1, stepFun, t0 = times[
                i], deltat = deltas[i], ...))
            w = apply(xmat, 1, dataLik, t = times[i +
                1], y = data[i,], log = FALSE, ...)
            if (max(w) < 1e-20) {
                warning("Particle filter bombed")
                return(-1e+99)
            }
            ll = ll + log(mean(w))
            rows = sample(1:n, n, replace = TRUE, prob =
                w)
            xmat = xmat[rows, ]
        }
        ll
    })
}
```

Figure 11.1 *An R function to create a function closure for marginal likelihood estimation using a bootstrap particle filter. Note that it has the possibility of failing due to numerical underflow. This flaw is corrected in Figure 11.2.*

11.5.4 Case study: inference for the Lotka–Volterra model

In this section we will examine the problem of inferring stochastic kinetic model parameters from data using the Lotka–Volterra model as an example. Although relatively simple, it is analytically intractable and involves multiple species and reactions, so it illustrates many of the issues one encounters in practice in the context of more complex models. We will assume the presence of measurement error in the observation process, allowing us to use the likelihood-free MCMC techniques previously discussed. We will concentrate on a pseudo-marginal MCMC algorithm, using an unbiased estimate of marginal likelihood computed using a bootstrap particle filter. We will first examine the simplest case, where both prey and predator species are observed at discrete times, and then go on to examine more challenging missing-data scenarios.

The smfsb R package contains some simulated data sets based on the Lotka–Volterra model, subject to differing amounts of noise and partial observation. It is

```
pfMLLik <- function (n, simx0, t0, stepFun, dataLik,
    data)
{
    times = c(t0, as.numeric(rownames(data)))
    deltas = diff(times)
    return(function(...) {
        xmat = simx0(n, t0, ...)
        ll = 0
        for (i in 1:length(deltas)) {
            xmat[] = t(apply(xmat, 1, stepFun, t0 =
                times[i],
                deltat = deltas[i], ...))
            lw = apply(xmat, 1, dataLik, t = times[i +
                1], y = data[i,
                ], log = TRUE, ...)
            m = max(lw)
            rw = lw - m
            sw = exp(rw)
            ll = ll + m + log(mean(sw))
            rows = sample(1:n, n, replace = TRUE, prob =
                sw)
            xmat[] = xmat[rows, ]
        }
        ll
    })
}
```

Figure 11.2 *An R function to create a function closure for marginal likelihood estimation using a bootstrap particle filter. Note that this filter does not require observations on a regular time grid. An example of use is shown in Figure 11.3.*

hoped that these data sets will prove useful for the development, testing, and comparison of new inference algorithms in the future. The data sets can be loaded by first loading the smfsb R package and then typing **data**(LVdata) at the R command prompt. Here we will use the data sets LVnoise10 and LVpreyNoise10, but there are many other examples which can be listed by typing **data**(**package="**smfsb").

The dataset LVnoise10 is shown and described in Figure 11.4. R code implementing an MCMC algorithm to infer the three model parameters using this data is given in Figure 11.5. Similar code is included in the smfsb package demo, "PMCMC", which can easily be customised for different observation scenarios. No prior or proposal density terms are included in the Metropolis–Hastings sampler, so since the proposal kernel is symmetric on $\log(\theta)$, this corresponds to the assumption of a flat prior on $\log(\theta)$. Different priors can be accommodated by providing the ap-

```
noiseSD=5
# first simulate some data
truth=simTs(c(x1=50,x2=100),0,20,2,stepLVc)
data=truth+rnorm(prod(dim(truth)),0,noiseSD)
data=as.timedData(data)
# measurement error model
dataLik <- function(x,t,y,log=TRUE,...)
{
        ll=sum(dnorm(y,x,noiseSD,log=TRUE))
        if (log)
                return(ll)
        else
                return(exp(ll))
}
# now define a sampler for the prior on the initial
    state
simx0 <- function(N,t0,...)
{
        mat=cbind(rpois(N,50),rpois(N,100))
        colnames(mat)=c("x1","x2")
        mat
}
mLLik=pfMLLik(1000,simx0,0,stepLVc,dataLik,data)
print(mLLik())
print(mLLik(th=c(th1 = 1, th2 = 0.005, th3 = 0.6)))
print(mLLik(th=c(th1 = 1, th2 = 0.005, th3 = 0.5)))
```

Figure 11.3 *An R session showing how to use the function pfMLLik from Figure 11.2. Note that the data matrix is expected as a timed data matrix such as produced by simTimes. The function as.timedData converts an R time series object (such as produced by simTs) into this format. Functions are also required for sampling from the prior on the initial state, for evaluating the likelihood of the data, and for simulating from the Markov process model. The function also requires the number of particles to be used, and the time corresponding to the initial state. Note that stepLVc is a function provided as part of the smfsb package which is a wrapper over a native C implementation of a Gillespie simulator for the Lotka–Volterra model. This is much faster than a pure R implementation, and this is useful for testing inference algorithms. Note that the function stepLVc therefore provides an example of how to link native simulation codes into R in a manner that will allow them to be used with the R-based simulation and inference codes provided in the smfsb package. This is useful as for many complex models, forward simulation from the model is the main computational bottleneck.*

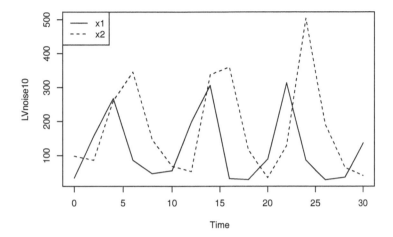

Figure 11.4 *Simulated time series data set,* `LVnoise10`, *consisting of 16 equally spaced observations of a realisation of a stochastic kinetic Lotka–Volterra model subject to Gaussian measurement error with a standard deviation of 10.*

propriate functions. Note that here we are assuming that we know that the standard deviation of the measurement noise is 10. We will examine the effect of relaxing this assumption later.

The results of running this code for 10,000 iterations with a thinning of 100 in conjunction with a particle filter based on 100 particles results in marginal posterior distributions for the components of θ as shown in Figure 11.6. The trace plots and auto-correlation plots indicate that the MCMC chain is mixing well, and it is also clear that the true parameters used to simulate the model, $\theta = (1, 0.005, 0.6)$ are well within the marginal posterior distributions, which are narrowly concentrated, and hence the parameters are well identified by the data. Plots such as these, together with quantitative summaries:

```
N = 10000 iterations
        th1                 th2                    th3
  Min.    :0.8386    Min.      :0.004289    Min.      :0.5431
  1st Qu.:0.9324     1st Qu.:0.004760       1st Qu.:0.6016
  Median :0.9545     Median :0.004863       Median :0.6160
  Mean    :0.9548    Mean      :0.004862    Mean      :0.6162
  3rd Qu.:0.9768     3rd Qu.:0.004964       3rd Qu.:0.6303
  Max.    :1.0982    Max.      :0.005457    Max.      :0.6954
Standard deviations:
        th1                 th2                    th3
0.0331779738 0.0001485412 0.0210022573
```

```
## load the reference data
data (LVdata)
## assume known measurement SD of 10
noiseSD=10
## now define the data likelihood function
dataLik <- function (x,t,y,log=TRUE,...)
{
        ll=sum (dnorm (y,x,noiseSD,log=TRUE))
        if (log) ll else exp (ll)
}
## now define a sampler for the prior on the initial state
simx0 <- function (N,t0,...)
{
        mat=cbind (rpois (N,50),rpois (N,100))
        colnames (mat)=c ("x1","x2")
        mat
}
LVdata=as.timedData (LVnoise10)
LVpreyData=as.timedData (LVpreyNoise10)
colnames (LVpreyData)=c ("x1")
## create marginal log-likelihood functions, based on a
    particle filter
mLLik=pfMLLik (100,simx0,0,stepLVc,dataLik,LVdata)
## Now create an MCMC algorithm...
th=c (th1 = 1, th2 = 0.005, th3 = 0.6)
p = length (th)
rprop = function (th, tune=0.01) { th*exp (rnorm (p,0,tune)) }
thmat = metropolisHastings (th,mLLik,rprop,iters=1000,thin=10)
## Compute and plot some basic summaries
mcmcSummary (thmat,truth=th)
```

Figure 11.5 *R code implementing an 'exact–approximate' MCMC sampler for fully Bayesian inference for the stochastic Lotka–Volterra model using time course data.*

can be generated using the function mcmcSummary, included as part of the smfsb R package. More sophisticated MCMC diagnostics may be computed using the coda R package, available from CRAN.

For most real systems biology models, observation of all species described by the model will be a quite unrealistic prospect, so it is interesting to investigate the effect on inference of observing just the prey species, again subject to a known measurement error model. The previously shown MCMC code can be modified by supplying a different data set and measurement error model to the pfMLLik function, and the rest of the MCMC code remains unchanged. The data set LVpreyNoise10 is a univariate R time series object consisting of just the first component of the LVnoise10 data set. Since univariate time series objects in R do not have labels, this data should be converted to a timed data object and labelled using the commands

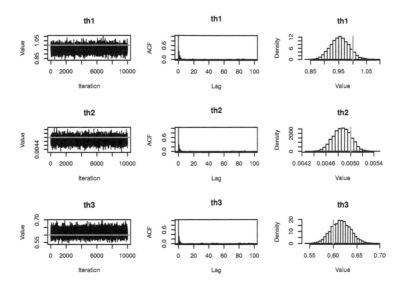

Figure 11.6 *Marginal posterior distributions for the parameters of the Lotka–Volterra model, based on the data given in Figure 11.4. Note that the mixing of the MCMC sampler is good, and that the true parameters, θ = (1, 0.005, 0.6) are well identified by the data.*

```
LVpreyData=as.timedData(LVpreyNoise10)
colnames(LVpreyData)=c("x1")
```

before being passed into the `pfMLLik` function. Similarly, the measurement error function can be defined by:

```
dataLik <- function(x,t,y,log=TRUE,...)
{
        with(as.list(x),{
                return(dnorm(y,x1,noiseSD,log))
        })
}
```

Running the MCMC scheme with just the prey data gives the plots shown in Figure 11.7, and quantitative summaries:

```
N = 10000 iterations
      th1                 th2                 th3
 Min.   :0.6707    Min.   :0.003499    Min.   :0.4030
 1st Qu.:0.8633    1st Qu.:0.004624    1st Qu.:0.5704
 Median :0.9141    Median :0.004936    Median :0.6133
 Mean   :0.9164    Mean   :0.004984    Mean   :0.6201
 3rd Qu.:0.9679    3rd Qu.:0.005297    3rd Qu.:0.6621
 Max.   :1.2492    Max.   :0.007557    Max.   :0.9858
Standard deviations:
          th1                 th2                 th3
```

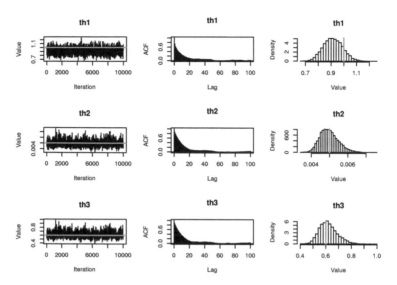

Figure 11.7 *Marginal posterior distributions for the parameters of the Lotka–Volterra model, based only on observations of prey species levels. Note that the mixing of the MCMC sampler is reasonable, and that the true parameters, $\theta = (1, 0.005, 0.6)$ are quite well identified by the data.*

```
0.0750796412  0.0005119223  0.0707775169
```

There are several interesting things worth noting about the MCMC output in relation to the fully observed case. The first is that the auto-correlation plots show that the mixing of the chain is not as good as in the case of full observation (but still reasonable). This is something that is typical of most MCMC algorithms — the greater the proportion of 'missing data', the worse the mixing of the chain. If necessary, one can just increase the thinning of the chain until satisfactory mixing is obtained. The next thing to note is that all three parameters have been well identified by the data, despite the fact that only prey observations were available. It is perhaps surprising that it is possible to make inferences for the predator death rate without making any observations on predators, but it is nevertheless true. However, looking at the spread of the marginal posterior distributions (reflected in the posterior standard deviations), it is unsurprising that we have learned somewhat less about the parameters than we did using observations on both species. Again, this is quite typical of most Bayesian analyses — the more data you have, the more you are able to learn.

Thus far we have been assuming that the measurement error model is completely known, but in practice this will not always be the case. For example, it will often be the case that there will be uncertainty regarding the standard deviation of the measurement error. In this case, we may treat any unknown parameters of the measurement error model just like unknown model parameters, and include them in the MCMC sampler. To investigate identification of the measurement error parameter, we will revert to using observations on both species, as parameter identifiability issues become more prevalent in this case. This is because we are trying to separate the noise in the stochastic process from the measurement error noise. This is a difficult problem, but can be done, in principle, due to the different auto-correlation structure of the two noise processes. Again, this new inference problem can be tackled with only modest changes to the original MCMC algorithm. Essentially, the parameter vector needs to be extended to include an sd component, and the data likelihood needs to be modified to use it, *viz*

```
dataLik <- function(x,t,y,log=TRUE,...)
{
        ll=sum(dnorm(y,x,th["sd"],log=TRUE))
        if (log) ll else exp(ll)
}
```

Re-running the algorithm with the same number of iterations results in a very poorly mixing chain, due to the lack of information in the data regarding the measurement error standard deviation. In this case it is helpful to use an informative prior for the measurement standard deviation. Assuming the prior,

$$\log(\sigma) \sim U(\log 5, \log 50),$$

simply imposes parameter bounds, and this is often a convenient mechanism for describing fairly vague, yet proper informative prior distributions. With this new prior distribution on the standard deviation, re-running the algorithm leads to an MCMC sample graphically summarised in the plots shown in Figure 11.8, and quantitative summaries:

```
N = 10000 iterations
      th1                th2                th3                sd
 Min.    :0.7990   Min.    :0.004126   Min.    :0.4808   Min.    : 5.000
 1st Qu.:0.9324    1st Qu.:0.004764    1st Qu.:0.6022    1st Qu.: 5.827
 Median :0.9549    Median :0.004863    Median :0.6185    Median : 7.864
 Mean   :0.9551    Mean   :0.004869    Mean   :0.6172    Mean   :12.280
 3rd Qu.:0.9771    3rd Qu.:0.004970    3rd Qu.:0.6337    3rd Qu.:14.873
 Max.   :1.1075    Max.   :0.005817    Max.   :0.7377    Max.   :49.970
Standard deviations:
       th1            th2           th3            sd
0.0357864038 0.0001633788 0.0243075447 9.6573167820
```

It is clear from the plots that the mixing of the sampler is still very poor, and that really a larger amount of thinning is required. As ever, simply increasing the amount of thinning used will allow for the generation of less correlated samples at the expense of increased computation time. To further investigate mixing and convergence issues, here it is helpful to look at the log-parameters. Since the parameters are non-negative, our proposal is symmetric on the log scale, and our prior is flat on the log scale, it can sometimes be the case that looking at the sampled MCMC values on the log scale shows features that are difficult to see on the original scale. In many ways, the log scale is a more natural scale for studying non-negative quantities such as rate constants and standard deviations. The log output is shown in Figure 11.9. Although the auto-correlation plots look similar, it is slightly easier to see the slow mixing behaviour in the trace plots on the log scale. This is especially true for the standard deviation parameter. It is also worth noting that although the true model

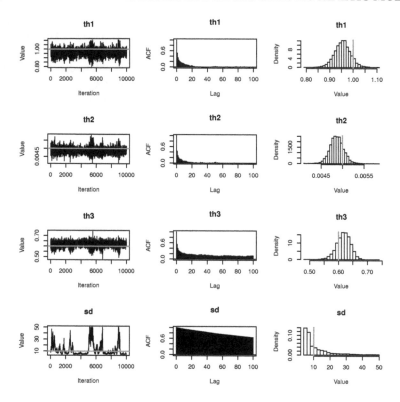

Figure 11.8 *Marginal posterior distributions for the parameters of the Lotka–Volterra model with unknown measurement error standard deviation. Note that the mixing of the MCMC sampler is poor, and that the true parameters, $\theta = (1, 0.005, 0.6, 10)$ are less well identified by the data than in the case of a fully specified measurement error model.*

standard deviation is within the support of the posterior distribution and the posterior mean is not far from the true value, this is partly due to the prior (bounds) adopted for this quantity. Certainly the shape of the posterior distribution suggests that there is information in the data consistent with a smaller value of the measurement error than the true value of 10. We can see from the log plots that the marginal posterior for the standard deviation is not simply an artefact of the prior adopted, as the prior is uniform on the log scale. Separation of different sources of noise from coarsely observed time course data is a notoriously difficult problem. However, here a much longer MCMC run is required in order to be confident that our sample is a good summary of the true posterior distribution.

Here we have just begun to explore the issues of parameter inference for stochastic kinetic models, but the implications are broad. First, we have seen that a pseudo-marginal MCMC algorithm can be used in order to carry out fully Bayesian inference for the exact discrete stochastic kinetic model (in the presence of measurement error), and that the techniques can be readily applied to a range of data-poor partial

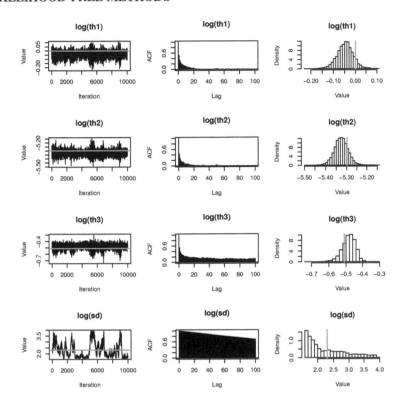

Figure 11.9 *Marginal posterior distributions for the log-parameters of the Lotka–Volterra model with unknown measurement error standard deviation. Note that the mixing of the MCMC sampler is poor, and that the true parameters are* $\log(\theta) = (0, -5.23, -0.51, 2.30)$.

observation scenarios. One issue we have not examined here is that of 'uncalibrated data', where measurements are made on a scale which does not directly translate to a number of molecules. Again, this can be handled with the introduction of a scaling parameter, though there can often be identifiability issues associated with this — see the end-of-chapter exercises.

We have not yet considered computation time and the computational expense of SMC-within-MCMC algorithms. Some of the MCMC runs included in this chapter took more than two full days of computation time on a powerful laptop computer. This is despite the fact that the principal bottle-neck, forward simulation from the model, was implemented in compiled C code. It will generally be impractical to run MCMC algorithms like those considered here for non-trivial models implemented in the R language. For real problems, the algorithms presented here will need to be implemented in an efficient statically typed programming language, as dynamic languages such as R are too slow for performance-critical code. The functional approach adopted here for the structuring of the simulation and inference codes will port most easily to a functional language such as Scala (Odersky et al., 2008); see the Scala library, `scala-smfsb`, associated with this book, described in Appendix B.4. However, it will also port easily to any reasonable object-oriented language, such as Java (Flanagan, 2005) or C++ (Satir and Brown, 1995). For example, in Java, the function closures used here become *instances* of *classes* implementing a single *method* conforming to an *interface*. It is also important to remember that forward simulation from the model is often the computational bottle-neck for likelihood-free inference methods. If one is prepared to sacrifice exactness, huge computational gains can be made by using a fast approximate simulation strategy, such as using a crude and simple Euler–Maruyama integration method in conjunction with a CLE approximation to the true model (Golightly and Wilkinson, 2011). Such strategies will be necessary for large and complex models.

11.5.5 The particle marginal Metropolis–Hastings (PMMH) particle MCMC algorithm

We now reconsider the SMC-within-MCMC algorithm that we have been using for parameter inference in stochastic kinetic models in light of the recently developed particle marginal Metropolis–Hastings (PMMH) algorithm (Andrieu et al., 2010), which belongs to a class of algorithms known as 'particle MCMC' (pMCMC). We will see shortly that the algorithm we have been using is a special case of the PMMH algorithm, and further, that the full PMMH algorithm is more general, in that it targets the (exact) full joint posterior distribution $\pi(\theta, x|y)$, and not just the marginal posterior $\pi(\theta|y)$. This distinction is very important, as only by studying the full joint posterior can Bayesian inferences be made for the full unobserved sample path, x, including unobserved components, and the model's initial conditions, $x(0)$.

The PMMH algorithm is an MCMC algorithm for partially observed Markov process models jointly updating θ and $x_{0:T}$. First, a proposed new $\theta^\star$ is generated from a proposal $f(\theta^\star|\theta)$, and then a corresponding $x_{0:T}^\star$ is generated by running a bootstrap particle filter using the proposed new model parameters, $\theta^\star$, and selecting a single trajectory by sampling once from the final set of particles using the final set of weights. This proposed pair $(\theta^\star, x_{0:T}^\star)$ is accepted using the Metropolis–Hastings ratio

$$A = \frac{\hat{\pi}_{\theta^\star}(y_{1:T})\pi(\theta^\star)f(\theta|\theta^\star)}{\hat{\pi}_\theta(y_{1:T})\pi(\theta)f(\theta^\star|\theta)},$$

where $\hat{\pi}_{\theta^\star}(y_{1:T})$ is the particle filter's (unbiased) estimate of marginal likelihood. Note that if the particle filter were to be run using just a single particle ($M = 1$), the 'particle filter' just blindly forward simulates from $\pi_\theta(x_{0:T}^\star)$. In this case the filter's estimate of marginal likelihood is just the observed data likelihood $\pi_\theta(y_{1:T}|x_{0:T}^\star)$, leading precisely to the simple LF-MCMC scheme considered earlier. To understand for an arbitrary finite number of particles, $M > 1$, one needs to think carefully about the structure of the particle filter.

To understand why PMMH works, it is necessary to think about the joint distribution of all random variables used in the bootstrap particle filter. To this end, it is helpful to re-visit the particle filter, thinking carefully about the resampling and propagation steps. First introduce notation for the 'particle cloud', consisting of states

$\mathbf{x}_t = \{x_t^k | k = 1, \ldots, M\}$, normalised weights, $\boldsymbol{\pi}_t = \{\pi_t^k | k = 1, \ldots, M\}$, and weighted particles $\tilde{\mathbf{x}}_t = \{(x_t^k, \pi_t^k) | k = 1, \ldots, M\}$.

1. Initialise the particle filter with $\tilde{\mathbf{x}}_0$, where $x_0^k \sim \pi(x_0)$ and $\pi_0^k = 1/M$ (note that the unnormalised weights w_0^k are not defined).

2. Now suppose at time t we have a weighted sample from $\pi(x_t | y_{1:t})$: $\tilde{\mathbf{x}}_t$. First resample by sampling $a_t^k \sim \mathcal{F}(a_t^k | \boldsymbol{\pi}_t)$, $k = 1, \ldots, M$. Here we use $\mathcal{F}(\cdot | \boldsymbol{\pi})$ for the discrete distribution on $1 : M$ with probability mass function $\boldsymbol{\pi}$. The a_t^k represent the indices of the selected particles.

3. Next sample $x_{t+1}^k \sim \pi(x_{t+1}^k | x_t^{a_t^k})$.

4. Set $w_{t+1}^k = \pi(y_{t+1} | x_{t+1}^k)$ and $\pi_{t+1}^k = w_{t+1}^k / \sum_{i=1}^M w_{t+1}^i$.

5. Finally, propagate $\tilde{\mathbf{x}}_{t+1}$ to the next step.

We define the filter's estimate of likelihood as

$$\hat{\pi}(y_t | y_{1:t-1}) = \frac{1}{M} \sum_{i=1}^M w_t^i$$

and

$$\hat{\pi}(y_{1:T}) = \prod_{i=1}^T \hat{\pi}(y_t | y_{1:t-1}).$$

See Doucet et al. (2001) for further theoretical background on particle filters and SMC more generally. Describing the filter carefully as above allows us to write down the joint density of all random variables in the filter as

$$\tilde{q}(\mathbf{x}_0, \ldots, \mathbf{x}_T, \mathbf{a}_0, \ldots, \mathbf{a}_{T-1}) = \left[\prod_{k=1}^M \pi(x_0^k)\right] \left[\prod_{t=0}^{T-1} \prod_{k=1}^M \pi_t^{a_t^k} \pi(x_{t+1}^k | x_t^{a_t^k})\right].$$

For PMMH we also sample a final index k' from $\mathcal{F}(k' | \boldsymbol{\pi}_T)$ giving the joint density

$$\tilde{q}(\mathbf{x}_0, \ldots, \mathbf{x}_T, \mathbf{a}_0, \ldots, \mathbf{a}_{T-1}) \pi_T^{k'},$$

as this index is used to select the sampled trajectory. We write the final selected trajectory as

$$x_{0:T}^{k'} = (x_0^{b_0^{k'}}, \ldots, x_T^{b_T^{k'}}),$$

where $b_t^{k'} = a_t^{b_{t+1}^{k'}}$, and $b_T^{k'} = k'$. If we now think about the structure of the PMMH algorithm, our proposal on the space of all random variables in the problem is in fact

$$f(\theta^\star | \theta) \tilde{q}_{\theta^\star}(\mathbf{x}_0^\star, \ldots, \mathbf{x}_T^\star, \mathbf{a}_0^\star, \ldots, \mathbf{a}_{T-1}^\star) \pi_T^{k'^\star}$$

and by considering the proposal and the acceptance ratio, it is clear that detailed balance for the chain is satisfied by the target with density proportional to

$$\pi(\theta) \hat{\pi}_\theta(y_{1:T}) \tilde{q}_\theta(\mathbf{x}_0, \ldots, \mathbf{x}_T, \mathbf{a}_0, \ldots, \mathbf{a}_{T-1}) \pi_T^{k'}.$$

We want to show that this target marginalises down to the correct posterior $\pi(\theta, x_{0:T} |$

$y_{1:T}$) when we consider just the parameters and the selected trajectory. But if we consider the terms in the joint distribution of the proposal corresponding to the trajectory selected by k', this is given by

$$\pi_\theta(x_0^{b_0^{k'}}) \left[\prod_{t=0}^{T-1} \pi_t^{b_t^{k'}} \pi_\theta(x_{t+1}^{b_{t+1}^{k'}} | x_t^{b_t^{k'}}) \right] \pi_T^{k'} = \pi_\theta(x_{0:T}^{k'}) \prod_{t=0}^{T} \pi_t^{b_t^{k'}}$$

which, by expanding the $\pi_t^{b_t^{k'}}$ in terms of the unnormalised weights, simplifies to

$$\frac{\pi_\theta(x_{0:T}^{k'}) \pi_\theta(y_{1:T} | x_{0:T}^{k'})}{M^{T+1} \hat{\pi}_\theta(y_{1:T})}.$$

It is worth dwelling on this result, as this is the key insight required to understand why the PMMH algorithm works. The whole point is that the terms in the joint density of the proposal corresponding to the selected trajectory exactly represent the required joint distribution modulo a couple of normalising constants, one of which is the particle filter's estimate of marginal likelihood. Thus, by including $\hat{\pi}_\theta(y_{1:T})$ in the acceptance ratio, we knock out the normalising constant, allowing all of the other terms in the proposal to be marginalised away. In other words, the target of the chain can be written as proportional to

$$\frac{\pi(\theta) \pi_\theta(x_{0:T}^{k'}, y_{1:T})}{M^{T+1}} \times \text{(Other terms)}.$$

The other terms are all probabilities of random variables which do not occur elsewhere in the target, and hence can all be marginalised away to leave the correct posterior,

$$\pi(\theta, x_{0:T} | y_{1:T}).$$

Thus the PMMH algorithm targets the correct posterior for any number of particles, M. Also note the implied uniform distribution on the selected indices in the target. See Andrieu et al. (2010) for further details and generalisations.

It is therefore clear that the MCMC algorithm we have been using is a special case of the PMMH algorithm, where we have not bothered to sample or store the particle filter's simulated trajectory. This also gives further insight into why the pseudo-marginal algorithm works, and why relatively few particles are needed for the particle filter. The latter is because there only needs to be sufficiently many particles in order to be able to generate *one* plausible sample path of the process, as that is all that is utilised by the procedure. In more conventional particle filtering scenarios, one is typically attempting to estimate the full filtering distribution, which clearly requires many more particles. It is also clear that it would be very straightforward to modify the R code we have been using for particle filtering and MCMC in order to implement a full PMMH scheme.

11.5.6 Non-Bayesian approaches for partially observed Markov process models

Before concluding this section, it is worth pointing out that not all approaches to inference for the parameters of partially observed Markov process (POMP) models are Bayesian. Iterated filtering (Ionides et al., 2006) is a likelihood-free (plug-and-play) particle filtering technique designed to lead to maximum likelihood estimates for model parameters using time course data. It has much in common with the computational Bayesian approaches in that it uses forwards simulation from the model in conjunction with particle filtering in order to avoid the need to explicitly calculate discrete time transition densities for complex Markov processes. Interested readers are referred to Ionides et al. (2006) for further details. It is worth noting that there is an R package on CRAN called `pomp` (King et al., 2008) which implements iterated filtering for POMP models, in addition to several other likelihood-free methods, including the simple version of the PMMH algorithm discussed previously.

11.6 Approximate Bayesian computation (ABC) for parameter inference

11.6.1 Naive ABC

The likelihood free methods discussed in the previous section break down in the case where there is no measurement error. In this case we can consider approximate likelihood free methods such as ABC. We will start with a very simple rejection-based ABC sampler before moving on to slightly more sophisticated ABC approaches. We will use the data set `LVperfect` in order to estimate the parameters of the generating LV model. We will use the function `abcRun` from Figure 10.9, and in the first instance we will just use a naive Euclidean distance between a simulated time course and the data as our ABC distance metric. This is not optimal, but it will give us a baseline for comparative purposes.

The choice of prior distribution is particularly important for ABC methods. Using very vague priors will typically lead to a poorly performing ABC method. In general one should use relevant background information in order to specify a prior capturing genuine belief about the likely parameter values. Here we are testing our algorithm using synthetic data, so we will use a prior that is not too informative, but nevertheless, not completely vague, and certainly giving reasonable mass in the vicinity of the true generating parameter values. We will adopt the prior distributions

$$\log \theta_1 \sim U(-3, 3)$$
$$\log \theta_2 \sim U(-8, -2)$$
$$\log \theta_1 \sim U(-4, 2).$$

These are relatively vague, in the sense that they each have a width of 6 on the log scale, but are certainly informative. Code using `abcRun` to implement a simple rejection-based ABC sampler is given in Figure 11.10. Note that `abcRun` uses the `parallel` package to run computations over multiple cores. The number of cores to use can be chosen with, e.g., `options(mc.cores=4)`.

Marginal summary statistics for the raw and log marginal distributions are as follows:

```r
data(LVdata)
rprior <- function() { exp(c(runif(1, -3, 3), runif
    (1,-8,-2), runif(1,-4,2))) }
rmodel <- function(th) { simTs(c(50,100), 0, 30, 2,
    stepLVc, th) }
sumStats <- identity
ssd = sumStats(LVperfect)
distance <- function(s) {
    diff = s - ssd
    sqrt(sum(diff*diff))
}
rdist <- function(th) { distance(sumStats(rmodel(th))) }
out = abcRun(1000000, rprior, rdist)
q=quantile(out$dist, c(0.01, 0.05, 0.1))
print(q)
accepted = out$param[out$dist < q[1],]
print(summary(accepted))
print(summary(log(accepted)))
op=par(mfrow=c(3,1))
hist(log(accepted[,1]),xlim=c(-3,3))
abline(v=log(1),col=2,lwd=2)
hist(log(accepted[,2]),xlim=c(-8,-2))
abline(v=log(0.005),col=2,lwd=2)
hist(log(accepted[,3]),xlim=c(-4,2))
abline(v=log(0.6),col=2,lwd=2)
par(op)
```

Figure 11.10 *R code implementing a simple ABC rejection sampler for parameter inference using time course data. One million samples are generated from the forward model initialised using random samples from the prior. The 10,000 samples that most closely match the data are kept as an approximate sample from the posterior. Marginal posterior summaries are shown in Figure 11.11.*

th1		th2		th3	
Min. :	0.04987	Min. :	0.0003365	Min. :	0.01832
1st Qu.:	0.15710	1st Qu.:	0.0011451	1st Qu.:	0.03741
Median :	0.29438	Median :	0.0020464	Median :	0.11260
Mean :	0.65061	Mean :	0.0044851	Mean :	0.33076
3rd Qu.:	0.82161	3rd Qu.:	0.0056622	3rd Qu.:	0.39161
Max. :	14.51874	Max. :	0.0972073	Max. :	6.95414

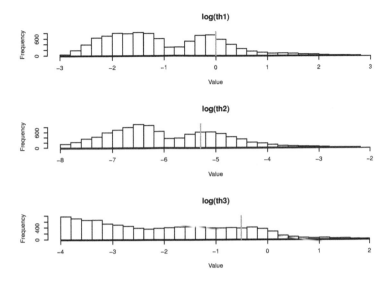

Figure 11.11 *Marginal summaries of a simple ABC posterior corresponding to the code in Figure 11.10. Very little has been learned about the true parameter values that generated the data.*

```
       log(th1)                  log(th2)                  log(th3)
Min.    :-2.9983    Min.    :-7.997    Min.    :-3.9999
1st Qu.:-1.8509    1st Qu.:-6.772    1st Qu.:-3.2859
Median :-1.2229    Median :-6.192    Median :-2.1839
Mean    :-1.0187    Mean    :-5.986    Mean    :-2.0498
3rd Qu.:-0.1965    3rd Qu.:-5.174    3rd Qu.:-0.9375
Max.    : 2.6754    Max.    :-2.331    Max.    : 1.9393
```

This very simple approach has provided some information about the parameter values that are more consistent with the data, but very little, and in general the above approach is unlikely to work well. The output of a stochastic kinetic model is, by definition, noisy, and so even if we forward sample from a model with the correct parameters there is no reason to suppose that the output will be close to the target data. Also, if the target data is high dimensional, we are unlikely to get close to it even if the output is not particularly noisy. For ABC to work well we need to come up with a set of summary statistics which are discriminative in the sense that they change as the parameters generating the model change, but are not as noisy as the raw time series output, and hence a more reasonable target for distance-based ABC matching.

11.6.2 ABC with calibrated summary statistics

Simple Euclidean distance between the observed time course trajectory and the true time course trajectory is unsatisfactory partly because it is high-dimensional, and also because it is a very noisy statistic. We can improve the situation by using some statistics typically used to characterise time series data. The sample mean and standard deviation are statistics used to characterise any noisy data. Auto-correlations at various lags are often used to understand the dependency structure in time series. For multivariate data, including time series, the cross-correlations between the series are also of interest. We can use these to create a collection of summary statistics which should adequately describe the simulation output, but not be too noisy so as to make matching unrealistic. So here we will take 5 statistics per component (mean, log variance, and 3 auto-correlations), together with the cross-correlation, giving a total of 11 summary statistics to be used for matching purposes. Having decided on a collection of summary statistics, there is still an issue as to how they should be scaled in order to build a distance function. It is important to scale summary statistics, as some vary on very different scales to others. Without any scaling, the statistics with the largest variance would dominate the distance measure, preventing other statistics from helping to distinguish good parameter combinations. The problem of how to select and scale summary statistics is an active research area beyond the scope of this text. For guidance on this topic, see Blum et al. (2013) and Fearnhead and Prangle (2012). Here we will adopt the simplest reasonable approach, which is to re-scale each statistic by its standard deviation, to give all components vaguely comparable weight in the distance metric. It should be emphasised that this approach is not optimal, but it is sufficient to get us started. We will use a short 'pilot run' in order to estimate the variance of the summary statistics. R code illustrating the use of this approach with rejection ABC is shown in Figure 11.12. Marginal summary statistics are given below.

```
        th1                    th2                    th3
Min.    : 0.2839    Min.    :0.0005469    Min.    :0.1288
1st Qu.: 0.6206    1st Qu.:0.0028255    1st Qu.:0.4886
Median : 0.8018    Median :0.0050543    Median :0.6532
Mean    : 0.9580    Mean    :0.0067307    Mean    :0.7786
3rd Qu.: 1.0479    3rd Qu.:0.0089381    3rd Qu.:0.8691
Max.    :10.3805    Max.    :0.0668414    Max.    :7.2226


     log(th1)               log(th2)               log(th3)
Min.    :-1.25916    Min.    :-7.511    Min.    :-2.0495
1st Qu.:-0.47711    1st Qu.:-5.869    1st Qu.:-0.7163
Median :-0.22092    Median :-5.288    Median :-0.4259
Mean    :-0.17763    Mean    :-5.289    Mean    :-0.3989
3rd Qu.: 0.04683    3rd Qu.:-4.717    3rd Qu.:-0.1403
Max.    : 2.33993    Max.    :-2.705    Max.    : 1.9772
```

These statistics clearly show a large improvement over the naive rejection ABC approach, with posterior means and medians quite close to the true generating values. Plots of the marginal distributions are shown in Figure 11.13, and these clearly look much improved over Figure 11.11. Pairs plots giving some insight into the joint posterior distribution are shown in Figure 11.14. These confirm that the posterior is concentrating around the generating values, but multiple modes can be identified in

```
distance <- function(s) {
    diff = s - ssd
    sqrt(sum(diff*diff))
}
ss1d <- function(vec) {
    acs=as.vector(acf(vec, lag.max=3, plot=FALSE)$acf)
        [2:4]
    c(mean(vec), log(var(vec)+1), acs)
}
ssi <- function(ts) {
    c(ss1d(ts[,1]), ss1d(ts[,2]), cor(ts[,1],ts[,2]))
}
cat("Pilot run\n")
out = abcRun(100000, rprior, function(th) { ssi(rmodel(
    th)) })
sds = apply(out$dist, 2, sd)
print(sds)
cat("Main run with calibrated summary stats\n")
sumStats <- function(ts) { ssi(ts)/sds }
ssd = sumStats(LVperfect)
rdist <- function(th) { distance(sumStats(rmodel(th))) }
out = abcRun(1000000, rprior, rdist)
q=quantile(out$dist, c(0.01, 0.05, 0.1))
print(q)
accepted = out$param[out$dist < q[1],]
colnames(accepted)=c("th1","th2","th3")
print(summary(accepted))
print(summary(log(accepted)))
```

Figure 11.12 *R code implementing an ABC rejection sampler for parameter inference using time course data using calibrated summary statistics. Following a pilot run to estimate the variances of the uncalibrated summary statistics, one million samples are generated from the forward model initialised using random samples from the prior. The 10,000 samples that most closely match the data (according to the calibrated summary statistics) are kept as an approximate sample from the posterior. Marginal posterior summaries are shown in Figure 11.13.*

these plots, which may just be artefacts of the rejection ABC procedure. As previously discussed, attempting to jump straight from the prior distribution to the posterior distribution in one go may be too much to expect for non-trivial problems, and so a sequential approach to ABC may work better in practice.

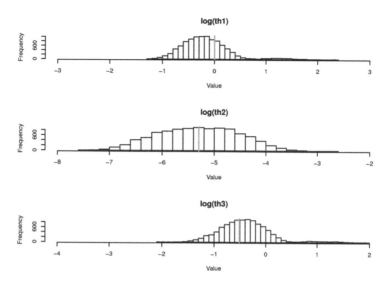

Figure 11.13 *Marginal summaries of an ABC posterior corresponding to the code in Figure 11.12. A significant amount has been learned about the true parameter values that generated the data.*

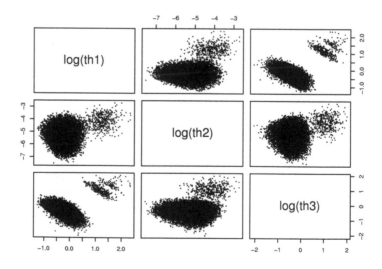

Figure 11.14 *Pairs plots for the ABC posterior corresponding to the code in Figure 11.12. Although a significant amount has been learned about the true parameter values that generated the data, multiple modes are present in the ABC posterior which may be artefacts of the procedure.*

11.6.3 ABC–SMC: a sequential ABC approach

Here we will examine an approach to sequential ABC often referred to as ABC–SMC. Sequential Monte Carlo (SMC) samplers for static parameter problems were formally introduced in Del Moral et al. (2006). An early attempt to use SMC for ABC problems was described in Sisson et al. (2007). The presentation here largely follows the approach of Toni et al. (2009), who used the acronym ABC–SMC to describe their algorithm.

The basic idea behind ABC–SMC is fairly straightforward. We know that for a rejection-based ABC sampler, the approximate Bayesian posterior is actually

$$\pi\left(\theta \mid \|s(x^\star) - s(x)\| < \varepsilon\right),$$

and that we better approximate the true posterior by making ε small. However, we also know that setting a very small value for ε will lead to a very high rejection rate. The idea behind ABC–SMC is to have a decreasing sequence of ε: $\infty = \varepsilon_0 > \varepsilon_1 > \varepsilon_2 > \cdots > \varepsilon_n > 0$, corresponding to a collection of ABC posterior distributions

$$\pi_t(\theta) = \pi\left(\theta \mid \|s(x^\star) - s(x)\| < \varepsilon_t\right), \quad t = 0, 1, \ldots, n,$$

with $\pi_0(\theta)$ corresponding to the prior and $\pi_n(\theta)$ corresponding to the ABC posterior of primary interest. Starting with a sample from the prior π_0, we sequentially resample from $\pi_{t-1}(\theta)$ to obtain $\pi_t(\theta)$, where in each case the distributions $\pi_{t-1}(\theta)$ and $\pi_t(\theta)$ should be sufficiently similar to make importance resampling a reasonably effective approach. All that remains is to understand how to resample a sample from $\pi_{t-1}(\theta)$ in order to obtain a sample from $\pi_t(\theta)$.

In principle we could directly resample from $\pi_{t-1}(\theta)$ to obtain samples from $\pi_t(\theta)$, but such an approach will lead to highly degenerate samples very quickly as t increases, since there is no mechanism for particle rejuvenation. So for SMC samplers we typically introduce some kind of *innovation* or *perturbation kernel* in order to 'noise up' samples from the source distribution before resampling to create a sample from the target distribution. This leads to a much more robust resampling strategy, less prone to degeneracy issues, at the expense of complicating the resampling procedure.

At the start of sweep t we assume that we have a weighted sample $(\theta^i_{t-1}, \tilde{w}^i_{t-1})$, $i = 1, \ldots, N$ from $\pi_{t-1}(\theta)$, where the weights are normalised so that they sum to one. At the end of the sweep, we want to have a weighted sample from $\pi_t(\theta)$. Start with $i = 1$.

1. We begin by drawing a candidate $\theta^\star_i$ from $\pi_{t-1}(\cdot)$ (by drawing from our empirical sample using the appropriate weights).

2. In order to combat degeneracy, we propagate through a perturbation kernel to 'noise up' our proposal: $\theta^{\star\star}_i \sim K(\cdot | \theta^\star_i)$.

3. We use this noised up parameter to initialise our forward model, and generate a synthetic data set: $x^\star_i \sim f(\cdot | \theta^{\star\star}_i)$.

4. If $\|s(x^\star_i) - s(x)\| < \varepsilon_t$, keep $\theta^{\star\star}_i$, otherwise reject this $\theta^\star_i$ and return to step 1.

5. Keep $\theta^{\star\star}_i$ as θ^i_t and compute its unnormalised weight w^i_t (discussed below).

6. Increment i. If $i \leq N$ return to step 1.

7. Compute the normalised weights $\tilde{w}_t^i$ and return the sample $(\theta_t^i, \tilde{w}_t^i)$, $i = 1, \ldots, N$ from $\pi_t(\theta)$.

This algorithm is entirely reasonable, but everything hinges on our ability to calculate an appropriate importance sampling weight w_t^i, which we now consider. At the end of step 3. we have a joint density on $(\theta_i^\star, \theta_i^{\star\star}, x_i^\star)$,

$$\pi_{t-1}(\theta_i^\star)K(\theta_i^{\star\star}|\theta_i^\star)f(x_i^\star|\theta_i^{\star\star}).$$

As we are not interested in $\theta_i^\star$, we can marginalise it out to obtain a density on $(\theta_i^{\star\star}, x_i^\star)$,

$$\int_\Theta \pi_{t-1}(\theta)K(\theta_i^{\star\star}|\theta)f(x_i^\star|\theta_i^{\star\star})\, d\theta.$$

Step 4. truncates this distribution, giving a density proportional to

$$I(\|s(x_i^\star) - s(x)\| < \varepsilon_t)f(x_i^\star|\theta_i^{\star\star})\int_\Theta \pi_{t-1}(\theta)K(\theta_i^{\star\star}|\theta)\, d\theta.$$

On the other hand, our target distribution on $(\theta_i^{\star\star}, x_i^\star)$ is proportional to

$$\pi(\theta_i^{\star\star})f(x_i^\star|\theta_i^{\star\star})I(\|s(x_i^\star) - s(x)\| < \varepsilon_t),$$

leading to the importance sampling ratio,

$$w_t^i = \frac{\pi(\theta_i^{\star\star})}{\int_\Theta \pi_{t-1}(\theta)K(\theta_i^{\star\star}|\theta)\, d\theta}.$$

For our empirical distribution $\pi_{t-1}(\theta)$, this is just

$$w_t^i = \frac{\pi(\theta_i^{\star\star})}{\sum_{j=1}^N \tilde{w}_{t-1}^j K(\theta_i^{\star\star}|\theta_{t-1}^j)}. \tag{11.11}$$

We use (11.11) in order to compute the weight required in step 5. of the above algorithm. R code to implement a sweep of ABC–SMC is presented in Figure 11.15. Rather than explicitly specifying ε_t, this algorithm uses a parameter factor, where $1/\text{factor}$ is the probability of a proposed $\theta_i^{\star\star}$ being accepted at step 4. This is easier to tune for complex models. Note that use is made of the previously discussed 'log–sum–exp' trick to prevent numerical underflow and overflow. We can embed this algorithm within a full ABC–SMC algorithm, and code for this is given in Figure 11.16. Again, the parallel package is used to do the sampling in parallel on multiple cores.

Figure 11.17 shows how to define an appropriate perturbation kernel and use the abcSmc function for carrying out parameter inference for our LV system. Marginal summary statistics are as follows.

```
         log(th1)              log(th2)              log(th3)
Min.    :-1.16045     Min.    :-7.120      Min.    :-1.7464
1st Qu.:-0.45754      1st Qu.:-5.828       1st Qu.:-0.7204
Median :-0.20140      Median :-5.217       Median :-0.4334
Mean    :-0.15227     Mean    :-5.240      Mean    :-0.3869
```

```
abcSmcStep <- function (dprior, priorSample, priorLW,
    rdist, rperturb, dperturb,
    factor = 10)
{
    n = length(priorSample)
    mx = max(priorLW)
    rw = exp(priorLW - mx)
    prior = sample(priorSample, n * factor, replace =
        TRUE, prob = rw)
    prop = mcMap(rperturb, prior)
    dist = mcMap(rdist, prop)
    qCut = quantile(unlist(dist), 1/factor)
    new = prop[dist < qCut]
    lw = mcMap(function(th) {
        terms = priorLW + sapply(priorSample, function(x
            ) {
            dperturb(th, x, log = TRUE)
        })
        mt = max(terms)
        denom = mt + log(sum(exp(terms - mt)))
        dprior(th, log = TRUE) - denom
    }, new)
    lw = unlist(lw)
    mx = max(lw)
    lw = lw - mx
    nlw = log(exp(lw)/sum(exp(lw)))
    list(sample = new, lw = nlw)
}
```

Figure 11.15 *R function to execute one sweep of an ABC–SMC algorithm.*

```
3rd Qu.: 0.06514    3rd Qu.:-4.671    3rd Qu.:-0.1404
Max.   : 2.00980    Max.   :-3.129    Max.   : 1.6249
```

Plots showing the marginal distributions of the posterior components are shown in Figure 11.18, and pairs plots giving some insight into the joint ABC–SMC posterior are shown in Figure 11.19. The overall pattern is similar to that obtained using rejection ABC, but the posterior is very slightly more concentrated around the true generating values and there appear to be fewer sampling artefacts. The choice of perturbation kernel can dramatically affect the performance of ABC–SMC approaches. Discussion of this issue is beyond the scope of this text, but see Filippi et al. (2013) for guidance. For further discussion of the relative merits of various likelihood free methods for inference for Markov processes, see Owen et al. (2015a). It is also possible to combine ABC–SMC methods with PMCMC algorithms to better exploit parallel hardware; see Owen et al. (2015b) for details.

```
abcSmc <- function (N, rprior, dprior, rdist, rperturb,
    dperturb, factor = 10,
    steps = 15, verb = FALSE)
{
    priorLW = log(rep(1/N, N))
    priorSample = mclapply(as.list(priorLW), function(x)
        {
        rprior()
    })
    for (i in steps:1) {
        if (verb)
            message(paste(i, ""), appendLF = FALSE)
        out = abcSmcStep(dprior, priorSample, priorLW,
            rdist,
            rperturb, dperturb, factor)
        priorSample = out[[1]]
        priorLW = out[[2]]
    }
    if (verb)
        message("")
    t(sapply(sample(priorSample, N, replace = TRUE, prob
        = exp(priorLW)),
        identity))
}
```

Figure 11.16 *R function for running an ABC–SMC algorithm, which calls on* abcSmcStep *to execute each sweep.*

Here we have only briefly explored some of the simplest ABC algorithms. It should be noted that there are a huge range of software tools and libraries available for conducting ABC analyses. Interesting R packages on CRAN include abc, abctools, and EasyABC. It is worth exploring the capabilities of these software libraries if you are interested in ABC approaches.

11.7 Network inference and model comparison

At this point it is worth saying a few words regarding network inference. It is clearly desirable to be able to deduce the structure of biochemical networks *ab initio* from routinely available experimental data. While it is possible to develop computational algorithms which attempt to do this, the utility of doing so is not at all clear due to the fact that there will typically be a very large number of distinct network structures, all of which are consistent with the available experimental data (large numbers of these will be biologically implausible, but a large number will also be quite plausible). In this case the 'best fitting' network is almost certainly incorrect. See De Smet

```
rprior <- function() { c(runif(1, -3, 3), runif(1, -8,
   -2), runif(1, -4, 2)) }
dprior <- function(x, ...) { dunif(x[1], -3, 3, ...) +
   dunif(x[2], -8, -2, ...) + dunif(x[3], -4, 2, ...) }
rmodel <- function(th) { simTs(c(50,100), 0, 30, 2,
   stepLVc, exp(th)) }
## pilot run
out = abcRun(100000, rprior, function(th) { ssi(rmodel(
   th)) })
sds = apply(out$dist, 2, sd)
print(sds)
sumStats <- function(ts) { ssi(ts)/sds }
ssd = sumStats(LVperfect)
rdist <- function(th) { distance(sumStats(rmodel(th))) }
## now ABC-SMC
rperturb <- function(th){th + rnorm(3, 0, 0.5)}
dperturb <- function(thNew, thOld, ...){sum(dnorm(thNew,
   thOld, 0.5, ...))}
out = abcSmc(10000, rprior, dprior, rdist, rperturb,
            dperturb, verb=TRUE, steps=8, factor=5)
colnames(out)=c("log(th1)","log(th2)","log(th3)")
print(summary(out))
```

Figure 11.17 *R code showing how to use the function* abcSmc *for ABC–SMC sampling for parameter inference (using the calibrated summary statistics already obtained from the previous section). Note the switch to working directly on the log scale, which is often more convenient for algorithms incorporating a perturbation step. The perturbation kernel has a standard deviation of 0.5 on the log scale. Marginal posterior summaries are shown in Figure 11.18.*

and Marchal (2010) for a review of work in this area. In many cases it will be more prudent to restrict attention to the less ambitious goal of comparing the experimental support for a small number of competing network structures. Typically this will concern a relatively well-characterised biochemical network where there is some uncertainty as to whether one or two of the potential reaction steps actually take place. In this case, deciding whether or not a given reaction is present is a problem of discrimination between two competing network structures. There are a number of ways that this problem could be tackled. The simplest approach would be to fit all competing models using the MCMC techniques outlined in the previous sections and compute the marginal likelihoods associated with each (Chib, 1995) in order to compute *Bayes factors*. For example, given two models, $\mathcal{M}_1$ and $\mathcal{M}_2$, and data y, it is clear that

$$\frac{P(\mathcal{M}_1|y)}{P(\mathcal{M}_2|y)} = \frac{P(\mathcal{M}_1)}{P(\mathcal{M}_2)} \times \frac{\pi(y|\mathcal{M}_1)}{\pi(y|\mathcal{M}_2)},$$

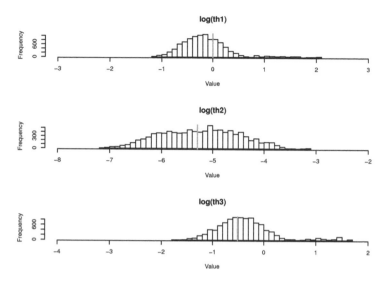

Figure 11.18 *Marginal summaries of the ABC–SMC posterior corresponding to the code in* Figure 11.17.

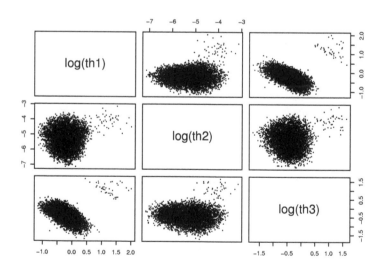

Figure 11.19 *Pairs plots for the ABC–SMC posterior corresponding to the code in Figure 11.17. There seem to be fewer sampling artefacts than for pure rejection ABC.*

and so the Bayes factor $\pi(y|\mathcal{M}_1)/\pi(y|\mathcal{M}_2)$ gives the relative strength of evidence in favour of $\mathcal{M}_1$ over $\mathcal{M}_2$.

In principle, the ABC methods outlined in the previous section provide a very simple way to compute posterior model probabilities. Given models $\mathcal{M}_1, \ldots, \mathcal{M}_n$, we can form a prior over models, $\pi(\mathcal{M})$, and a prior for the parameters of each model, $\pi(\theta|\mathcal{M})$, which together form a prior over models and parameters, $\pi(\mathcal{M})\pi(\theta|\mathcal{M})$. We can then use this prior in an ABC algorithm, sampling from it by first drawing a model and then some parameters conditional on that model. We then apply an ABC algorithm (e.g., rejection sampling), and look at the proportion of times that each model appears in the ABC posterior for an estimate of the posterior probability of each model. We can use this approach with a range of different ABC algorithms. Toni et al. (2009) explore this idea in conjunction with an ABC–SMC algorithm. This gives an easy and intuitive method for beginning to explore issues relating to model choice. However, some caution must be exercised with this approach. As ever with ABC algorithms, the choice of summary statistics is crucial. However, summary statistics that are well suited to discriminating between parameter values for a given model may not be suitable for discriminating between models. These issues are explored in Robert et al. (2011), to which the reader is referred for further details.

More sophisticated strategies might adopt reversible jump MCMC techniques (Green, 1995) in order to simultaneously infer parameters and structure. In principle, the reversible jump techniques could also be used for *ab initio* network inference, but note well the previous caveat. In particular, note that inferences are sensitive not only to the prior adopted over the set of models to be considered, but also to the prior structure adopted for the rate constants conditional on the model. Simultaneous inference for parameters and network structure is currently an active research area.

11.8 Exercises

1. Consider the immigration-death process (from Chapter 5) with immigration rate $\lambda = 1$ and death rate $\mu = 0.1$ (giving an equilibrium distribution which is $Po(10)$) and simulate some complete data from it using the Gillespie algorithm, starting from 0, up to time 30.

   (a) Use these data to form the complete-data likelihood (11.2).

   (b) Maximise the complete-data likelihood using (11.4).

   (c) How well are you able to recover the true rate constants? How much data do you need? Does this vary according to the true rate constants originally chosen?

   (d) Compute the Bayesian posterior distributions for the rate constants starting with a fairly diffuse prior (say, a $Ga(0.1, 0.1)$ distribution). Compare the mean of the posterior to the maximum likelihood estimates (MLEs).

   (e) Consider the variance of the posterior. How does this change as the amount of available data changes? How does this shed light on this issue of how many data are required for reliable parameter inference?

2. Simulate some data from the CLE approximation to the immigration-death model

considered in the previous exercise over the same time interval, using a simple Euler–Maruyama method with $\Delta t = 0.1$.

(a) Calculate the likelihood of the simulated data and maximise it numerically to find the MLEs. How well can you recover the true rate constants?

(b) Implement a simple Metropolis–Hastings algorithm to compute the posterior distribution of the rate constants. How many data are needed to reduce the posterior uncertainty to an acceptable level?

(c) Investigate the relationship between time and sampling frequency. Is it better to have 1,000 observations over a 10-second period or a 10-hour period? How does this vary over models and rate constants?

(d) Redo this exercise using data simulated exactly using the Gillespie algorithm and then discretised onto a regular grid. How much bias does the CLE approximation introduce?

3. Consider the discrete stochastic kinetic Lotka–Volterra model considered in Section 11.5.4

(a) Re-do the analysis for the initial example, using the data set LVnoise10. Is it necessary to have 100 particles in the particle filter? How few can we get away with whilst still having a chain with tolerable mixing? Is there any perceptible advantage to be had from increasing the number of particles in the filter?

(b) Implement an MCMC scheme for inference in the partially observed case, using LVpreyNoise10 in the case of unknown measurement error. What thinning is required in order to obtain a final sample exhibiting reasonable autocorrelations? How well can the parameters be identified in this case? Does it help to impose a $Ga(10, 1)$ prior for the unknown measurement error standard deviation?

(c) Run an MCMC code for the data set LVnoise30 (which has a measurement error standard deviation of 30). How do inferences compare to those obtained in (a)? Is it easier or more difficult to learn the unknown measurement standard deviation when the measurement noise is larger?

(d) The data set LVnoise10Scale10 is just LVnoise10 scaled by a factor of 10 (representing a measurement scale not in molecules). Treating the scaling constant as unknown, check if the factor is well identified by the data. Is it possible to identify both the scaling factor and the measurement error standard deviation from the data?

4. Use ABC and ABC–SMC algorithms to carry out inference for the Lotka–Volterra model using the noisy datasets LVnoise10 and LVpreyNoise10. Compare your inferences to those obtained using pMCMC. What are the pros and cons of ABC versus pMCMC approaches in the context of noisy time course data?

5. Install the pomp R package from CRAN. Work through the tutorial vignette. Use iterated filtering to compute the maximum likelihood estimates of the Lotka–Volterra parameters using LVpreyNoise10. How do they compare with the Bayesian estimates?

11.9 Further reading

Application of Bayesian inference to estimation of discrete stochastic kinetic models is examined in Boys et al. (2008). Note that this builds on previous work for the estimation of stochastic compartmental models; see Gibson and Renshaw (1998), for example. An overview of the problem of inference for multivariate diffusion processes is given in Durham and Gallant (2002), and an application to stochastic kinetic models is discussed in Golightly and Wilkinson (2005). A sequential MCMC algorithm for inference is presented in Golightly and Wilkinson (2006a), and is applied to stochastic kinetic models in Golightly and Wilkinson (2006b). An efficient global MCMC algorithm for diffusions is introduced in Golightly and Wilkinson (2008) and refined in Wilkinson and Golightly (2010). A sequential LF-MCMC scheme for stochastic kinetic models is examined in Wilkinson (2011), which also examines some of the conceptual problems associated with inference using data derived from fluorescence microscopy data. The use of the linear noise approximation for rate parameter inference is explored in Komorowski et al. (2009), and ABC methods for inference in Toni et al. (2009), Owen et al. (2015a), and Kursawe et al. (2018). For further details on particle MCMC algorithms such as the PMMH, see Andrieu et al. (2010), and for applications to stochastic kinetic models, see Andrieu et al. (2009), Golightly and Wilkinson (2011), and Owen et al. (2015b). For non-Bayesian inference using likelihood-free techniques, see Ionides et al. (2006).

CHAPTER 12

Conclusions

This book has provided an introduction to the concepts of stochastic modelling relevant for systems biology applications, and has illustrated those concepts in a systems biology context. The main area of application has been the modelling and simulation of genetic and biochemical networks (Harper et al., 2011), but it should be reasonably clear that the techniques can also be applied to a range of other modelling scenarios. Most interesting biological processes exhibit stochastic variation, and understanding the nature and effect of the randomness can often be fundamental to understanding the system.

Although Chapters 10 and 11 have touched on issues relating to inference, the primary focus of the text has been on the development of models which adequately capture system dynamics and algorithms for simulation. Similarly, much of the focus has been on intrinsic noise arising from the discreteness of molecular reaction events, rather than the many other sources of noise and heterogeneity in biological systems. The inferential approaches described have been computationally intensive Bayesian methods using MCMC and ABC. However, non-Bayesian approaches, such as iterated filtering (Ionides et al., 2006), could turn out to be useful in practice. Further, the emphasis throughout the text has been on 'toy' models, rather than realistic descriptions of real experimental systems, or on the statistical issues associated with wet-lab systems biology work. In reality, having a detailed and accurate model of the measurement process is likely to be just as important as having a good model of the biological system for satisfactory inferences to be made (Wilkinson, 2011). Many outstanding research issues remain relating to the design and analysis of statistical experiments for systems biology. We are still far from having complete and satisfactory solutions to general problems of statistical inference for networks and rate constants from routinely available experimental data. There are also difficult experimental design issues that arise in this context. For example, how do we decide on sample sizes, time points, and perturbations (chemical and genetic) to study in order to be confident of obtaining useful results? Although we can show that in principle it should be possible to separate out measurement noise from intrinsic noise and intrinsic noise from extrinsic noise, all just using only time course measurements on the system state, can we really ever hope to be able to do this in practice given the limitations of actual measurement technology? Many other questions remain. How should we build hierarchical statistical models for cell populations and multiple perturbations (chemical and genetic), and what are the most effective methods for fitting such models to experimental data? How can we combine detailed time course data on individual cells with snapshot cell population measurements made using techniques such as flow cytometry (Pfeifer et al., 2010)? Is data giving information on the population mean (such as micro-array data) useful for parametrising stochastic models? How can we best utilise new high-throughput sequencing technologies for inferring the parameters and structure of models? Do single-cell sequencing technologies give us new ways to study noise in biological systems? How can we best extend methods for parameter inference to allow model comparison and selection (Toni et al., 2009), and can we further extend such techniques to allow more general network inference?

Most existing work in the area of high-throughput data analysis builds on very simple descriptive models of the underlying processes, rather than the detailed 'mechanistic' models that have been considered in this text. This is positive in that it renders the associated design and analysis problems more tractable, but it also places limitations on what can be achieved. For example, a simple dynamic Bayesian network (DBN) fitted to some time-course micro-array data (Grzegorczyk and Husmeier, 2011) is not attempting to capture the true system dynamics, and so cannot be translated back to a detailed mechanistic model of the kind we would like to be able to simulate. Essentially, there is currently a 'gap' between the detailed modelling work that has been considered here and the applied statistical work currently being carried out by bioinformaticians. It seems likely that modelling approaches which attempt to bridge the gap between detailed stochastic kinetic descriptions and descriptive statistical models offer the best framework for addressing this problem, so models such as the CLE (Wilkinson and Golightly, 2010) and the LNA (Komorowski et al., 2009) are particularly interesting from this perspective.

Although Chapter 11 has discussed some of the issues relating to inference for stochastic kinetic models, it is not completely straightforward to use these techniques directly on large, complex models, or in conjunction with routinely available high-throughput experimental data. There is therefore a pressing need to develop techniques which allow routine experimental data to be used for 'calibrating' detailed simulation models. While this is reasonably straightforward for simple models that are fast to simulate, calibration of slow stochastic models is not easy. In particular, spatial stochastic models of cellular dynamics are likely to be important in some application areas (Andrews et al., 2010), and as we saw in Chapter 9, these are typically highly computationally intensive. Further, as systems biology modelling works up to more ambitious scales of organisation, models of cell populations, tissues, and whole organs (Timmer et al., 2010) will need to be developed, and calibrated against system level data. For deterministic models that are slow to simulate, there is a theory of Bayesian model calibration (Kennedy and O'Hagan, 2001; Santner et al., 2003) that can be applied (though it is worth noting that most existing literature in this area is concerned with models having a low-dimensional output, which is typically not the case in the systems biology context). Although it is in principle possible to extend model calibration technology for the fitting of stochastic models, it is not straightforward to do so in practice. The development of fast and approximate techniques for calibrating large and complex stochastic simulation models is currently an active research area (Henderson et al., 2009).

In conclusion, this text has covered only the essential concepts required for beginning to think seriously about systems biology from a stochastic viewpoint. The really interesting statistical problems concerned with marrying complex stochastic systems biology models to routinely available high-throughput experimental data remain largely unsolved and the subject of a great deal of ongoing research. However, the methods described in this book provide a mathematical and computational framework for thinking clearly about the problems that remain, and suggest potential avenues worthy of exploration in pursuit of their solution.

SBML Models

This appendix contains a few examples of complete SBML documents describing models discussed in the text. All of these models, and many others, can be downloaded from this book's website.

A.1 Auto-regulatory network

The model below is the SBML for the auto-regulatory network discussed in Chapter 2 and Chapter 7.

```
<?xml version="1.0" encoding="UTF-8"?>
<sbml xmlns="http://www.sbml.org/sbml/level3/version1/core" level="3" version="1
    ">
  <model id="AutoRegulatoryNetwork" name="Auto-regulatory network"
      substanceUnits="item" timeUnits="second" volumeUnits="litre" extentUnits=
      "item">
    <listOfCompartments>
      <compartment id="Cell"/>
    </listOfCompartments>
    <listOfSpecies>
      <species id="Gene" compartment="Cell" initialAmount="10"
          hasOnlySubstanceUnits="true"/>
      <species id="P2Gene" name="P2.Gene" compartment="Cell" initialAmount="0"
          hasOnlySubstanceUnits="true"/>
      <species id="Rna" compartment="Cell" initialAmount="0"
          hasOnlySubstanceUnits="true"/>
      <species id="P" compartment="Cell" initialAmount="0" hasOnlySubstanceUnits
          ="true"/>
      <species id="P2" compartment="Cell" initialAmount="0"
          hasOnlySubstanceUnits="true"/>
    </listOfSpecies>
    <listOfReactions>
      <reaction id="RepressionBinding" name="Repression binding" reversible="
          false">
        <listOfReactants>
          <speciesReference species="Gene" stoichiometry="1"/>
          <speciesReference species="P2" stoichiometry="1"/>
        </listOfReactants>
        <listOfProducts>
          <speciesReference species="P2Gene" stoichiometry="1"/>
        </listOfProducts>
        <kineticLaw>
          <math xmlns="http://www.w3.org/1998/Math/MathML">
            <apply>
              <times/>
              <ci> k1 </ci>
              <ci> Gene </ci>
              <ci> P2 </ci>
            </apply>
          </math>
          <listOfLocalParameters>
            <localParameter id="k1" value="1"/>
          </listOfLocalParameters>
```

```
      </kineticLaw>
  </reaction>
  <reaction id="ReverseRepressionBinding" name="Reverse repression binding"
        reversible="false">
    <listOfReactants>
      <speciesReference species="P2Gene" stoichiometry="1"/>
    </listOfReactants>
    <listOfProducts>
      <speciesReference species="Gene" stoichiometry="1"/>
      <speciesReference species="P2" stoichiometry="1"/>
    </listOfProducts>
    <kineticLaw>
      <math xmlns="http://www.w3.org/1998/Math/MathML">
        <apply>
          <times/>
          <ci> k1r </ci>
          <ci> P2Gene </ci>
        </apply>
      </math>
      <listOfLocalParameters>
        <localParameter id="k1r" value="10"/>
      </listOfLocalParameters>
    </kineticLaw>
  </reaction>
  <reaction id="Transcription" reversible="false">
    <listOfReactants>
      <speciesReference species="Gene" stoichiometry="1"/>
    </listOfReactants>
    <listOfProducts>
      <speciesReference species="Gene" stoichiometry="1"/>
      <speciesReference species="Rna" stoichiometry="1"/>
    </listOfProducts>
    <kineticLaw>
      <math xmlns="http://www.w3.org/1998/Math/MathML">
        <apply>
          <times/>
          <ci> k2 </ci>
          <ci> Gene </ci>
        </apply>
      </math>
      <listOfLocalParameters>
        <localParameter id="k2" value="0.01"/>
      </listOfLocalParameters>
    </kineticLaw>
  </reaction>
  <reaction id="Translation" reversible="false">
    <listOfReactants>
      <speciesReference species="Rna" stoichiometry="1"/>
    </listOfReactants>
    <listOfProducts>
      <speciesReference species="Rna" stoichiometry="1"/>
      <speciesReference species="P" stoichiometry="1"/>
    </listOfProducts>
    <kineticLaw>
      <math xmlns="http://www.w3.org/1998/Math/MathML">
        <apply>
          <times/>
          <ci> k3 </ci>
          <ci> Rna </ci>
        </apply>
      </math>
      <listOfLocalParameters>
        <localParameter id="k3" value="10"/>
      </listOfLocalParameters>
    </kineticLaw>
  </reaction>
  <reaction id="Dimerisation" reversible="false">
```

```
      <listOfReactants>
        <speciesReference species="P" stoichiometry="2"/>
      </listOfReactants>
      <listOfProducts>
        <speciesReference species="P2" stoichiometry="1"/>
      </listOfProducts>
      <kineticLaw>
        <math xmlns="http://www.w3.org/1998/Math/MathML">
          <apply>
            <times/>
            <ci> k4 </ci>
            <cn> 0.5 </cn>
            <ci> P </ci>
            <apply>
              <minus/>
              <ci> P </ci>
              <cn type="integer"> 1 </cn>
            </apply>
          </apply>
        </math>
        <listOfLocalParameters>
          <localParameter id="k4" value="1"/>
        </listOfLocalParameters>
      </kineticLaw>
    </reaction>
    <reaction id="Dissociation" reversible="false">
      <listOfReactants>
        <speciesReference species="P2" stoichiometry="1"/>
      </listOfReactants>
      <listOfProducts>
        <speciesReference species="P" stoichiometry="2"/>
      </listOfProducts>
      <kineticLaw>
        <math xmlns="http://www.w3.org/1998/Math/MathML">
          <apply>
            <times/>
            <ci> k4r </ci>
            <ci> P2 </ci>
          </apply>
        </math>
        <listOfLocalParameters>
          <localParameter id="k4r" value="1"/>
        </listOfLocalParameters>
      </kineticLaw>
    </reaction>
    <reaction id="RnaDegradation" name="RNA_Degradation" reversible="false">
      <listOfReactants>
        <speciesReference species="Rna" stoichiometry="1"/>
      </listOfReactants>
      <kineticLaw>
        <math xmlns="http://www.w3.org/1998/Math/MathML">
          <apply>
            <times/>
            <ci> k5 </ci>
            <ci> Rna </ci>
          </apply>
        </math>
        <listOfLocalParameters>
          <localParameter id="k5" value="0.1"/>
        </listOfLocalParameters>
      </kineticLaw>
    </reaction>
    <reaction id="ProteinDegradation" name="Protein_degradation" reversible="
        false">
      <listOfReactants>
        <speciesReference species="P" stoichiometry="1"/>
      </listOfReactants>
```

```
         <kineticLaw>
           <math xmlns="http://www.w3.org/1998/Math/MathML">
             <apply>
               <times/>
               <ci> k6 </ci>
               <ci> P </ci>
             </apply>
           </math>
           <listOfLocalParameters>
             <localParameter id="k6" value="0.01"/>
           </listOfLocalParameters>
         </kineticLaw>
       </reaction>
     </listOfReactions>
   </model>
 </sbml>
```

A.2 Lotka–Volterra reaction system

The model below is the SBML for the stochastic version of the Lotka–Volterra system, discussed in Chapter 6.

```
<?xml version="1.0" encoding="UTF-8"?>
<sbml xmlns="http://www.sbml.org/sbml/level3/version1/core" level="3" version="1
     ">
  <model id="LotkaVolterra" substanceUnits="item" timeUnits="second" volumeUnits
      ="litre" extentUnits="item">
   <listOfCompartments>
     <compartment id="Cell"/>
   </listOfCompartments>
   <listOfSpecies>
     <species id="Prey" compartment="Cell" initialAmount="50"
         hasOnlySubstanceUnits="true"/>
     <species id="Predator" compartment="Cell" initialAmount="100"
         hasOnlySubstanceUnits="true"/>
   </listOfSpecies>
   <listOfReactions>
     <reaction id="PreyReproduction" reversible="false">
       <listOfReactants>
         <speciesReference species="Prey" stoichiometry="1"/>
       </listOfReactants>
       <listOfProducts>
         <speciesReference species="Prey" stoichiometry="2"/>
       </listOfProducts>
       <kineticLaw>
         <math xmlns="http://www.w3.org/1998/Math/MathML">
           <apply>
             <times/>
             <ci> c1 </ci>
             <ci> Prey </ci>
           </apply>
         </math>
         <listOfLocalParameters>
           <localParameter id="c1" value="1"/>
         </listOfLocalParameters>
       </kineticLaw>
     </reaction>
     <reaction id="PredatorPreyInteraction" reversible="false">
       <listOfReactants>
         <speciesReference species="Prey" stoichiometry="1"/>
         <speciesReference species="Predator" stoichiometry="1"/>
       </listOfReactants>
       <listOfProducts>
         <speciesReference species="Predator" stoichiometry="2"/>
       </listOfProducts>
```

```
        <kineticLaw>
          <math xmlns="http://www.w3.org/1998/Math/MathML">
            <apply>
              <times/>
              <ci> c2 </ci>
              <ci> Prey </ci>
              <ci> Predator </ci>
            </apply>
          </math>
          <listOfLocalParameters>
            <localParameter id="c2" value="0.005"/>
          </listOfLocalParameters>
        </kineticLaw>
      </reaction>
      <reaction id="PredatorDeath" reversible="false">
        <listOfReactants>
          <speciesReference species="Predator" stoichiometry="1"/>
        </listOfReactants>
        <kineticLaw>
          <math xmlns="http://www.w3.org/1998/Math/MathML">
            <apply>
              <times/>
              <ci> c3 </ci>
              <ci> Predator </ci>
            </apply>
          </math>
          <listOfLocalParameters>
            <localParameter id="c3" value="0.6"/>
          </listOfLocalParameters>
        </kineticLaw>
      </reaction>
    </listOfReactions>
  </model>
</sbml>
```

A.3 Dimerisation-kinetics model

A.3.1 *Continuous deterministic version*

The model below is the SBML for the continuous deterministic version of the dimerisation kinetics model, discussed in Section 7.2.

```
<?xml version="1.0" encoding="UTF-8"?>
<sbml xmlns="http://www.sbml.org/sbml/level3/version1/core" level="3" version="1
    ">
  <model id="DimerKineticsDet" name="Dimerisation_Kinetics_(deterministic)"
      substanceUnits="mole" timeUnits="second" volumeUnits="litre" extentUnits=
      "mole">
    <listOfCompartments>
      <compartment id="Cell" size="1e-15"/>
    </listOfCompartments>
    <listOfSpecies>
      <species id="P" compartment="Cell" initialConcentration="5e-07"/>
      <species id="P2" compartment="Cell" initialConcentration="0"/>
    </listOfSpecies>
    <listOfReactions>
      <reaction id="Dimerisation" reversible="false">
        <listOfReactants>
          <speciesReference species="P" stoichiometry="2"/>
        </listOfReactants>
        <listOfProducts>
          <speciesReference species="P2" stoichiometry="1"/>
        </listOfProducts>
        <kineticLaw>
```

```
          <math xmlns="http://www.w3.org/1998/Math/MathML">
            <apply>
              <times/>
              <ci> Cell </ci>
              <ci> k1 </ci>
              <ci> P </ci>
              <ci> P </ci>
            </apply>
          </math>
          <listOfLocalParameters>
            <localParameter id="k1" value="500000"/>
          </listOfLocalParameters>
        </kineticLaw>
      </reaction>
      <reaction id="Dissociation" reversible="false">
        <listOfReactants>
          <speciesReference species="P2" stoichiometry="1"/>
        </listOfReactants>
        <listOfProducts>
          <speciesReference species="P" stoichiometry="2"/>
        </listOfProducts>
        <kineticLaw>
          <math xmlns="http://www.w3.org/1998/Math/MathML">
            <apply>
              <times/>
              <ci> Cell </ci>
              <ci> k2 </ci>
              <ci> P2 </ci>
            </apply>
          </math>
          <listOfLocalParameters>
            <localParameter id="k2" value="0.2"/>
          </listOfLocalParameters>
        </kineticLaw>
      </reaction>
    </listOfReactions>
  </model>
</sbml>
```

A.3.2 Discrete stochastic version

The model below is the SBML for the discrete stochastic version of the dimerisation
kinetics model, discussed in Section 7.2.

```
<?xml version="1.0" encoding="UTF-8"?>
<sbml xmlns="http://www.sbml.org/sbml/level3/version1/core" level="3" version="1
    ">
  <model id="DimerKineticsStoch" name="Dimerisation_Kinetics_(stochastic)"
      substanceUnits="item" timeUnits="second" volumeUnits="litre" extentUnits=
      "item">
    <listOfCompartments>
      <compartment id="Cell" size="1e-15"/>
    </listOfCompartments>
    <listOfSpecies>
      <species id="P" compartment="Cell" initialAmount="301"
          hasOnlySubstanceUnits="true"/>
      <species id="P2" compartment="Cell" initialAmount="0"
          hasOnlySubstanceUnits="true"/>
    </listOfSpecies>
    <listOfReactions>
      <reaction id="Dimerisation" reversible="false">
        <listOfReactants>
          <speciesReference species="P" stoichiometry="2"/>
        </listOfReactants>
        <listOfProducts>
```

```
            <speciesReference species="P2" stoichiometry="1"/>
          </listOfProducts>
          <kineticLaw>
            <math xmlns="http://www.w3.org/1998/Math/MathML">
              <apply>
                <divide/>
                <apply>
                  <times/>
                  <ci> c1 </ci>
                  <ci> P </ci>
                  <apply>
                    <minus/>
                    <ci> P </ci>
                    <cn type="integer"> 1 </cn>
                  </apply>
                </apply>
                <cn type="integer"> 2 </cn>
              </apply>
            </math>
            <listOfLocalParameters>
              <localParameter id="c1" value="0.00166"/>
            </listOfLocalParameters>
          </kineticLaw>
        </reaction>
        <reaction id="Dissociation" reversible="false">
          <listOfReactants>
            <speciesReference species="P2" stoichiometry="1"/>
          </listOfReactants>
          <listOfProducts>
            <speciesReference species="P" stoichiometry="2"/>
          </listOfProducts>
          <kineticLaw>
            <math xmlns="http://www.w3.org/1998/Math/MathML">
              <apply>
                <times/>
                <ci> c2 </ci>
                <ci> P2 </ci>
              </apply>
            </math>
            <listOfLocalParameters>
              <localParameter id="c2" value="0.2"/>
            </listOfLocalParameters>
          </kineticLaw>
        </reaction>
      </listOfReactions>
    </model>
</sbml>
```

APPENDIX B

Software associated with this book

There is a significant amount of software and other on-line materials supporting this text. This includes links to software and websites mentioned in the text, models in SBML-shorthand and SBML, some of which have been discussed in the text, R packages containing software implementations of ideas and algorithms discussed, tools for converting SBML-shorthand to and from SBML, and a Scala library, reimplementing the algorithms from the book in an efficient, strongly typed, compiled, functional programming language. Since software information changes rapidly, it is recommended to always check the on-line information relating to this material, and to regard the available on-line information as primary. However, for convenience, a brief introduction to some of the most important supporting software is included here.

The main website associated with the book has this address:

`https://github.com/darrenjw/smfsb`

This website is actually a GitHub repository, which can be easily downloaded or cloned. This provides a convenient way to obtain many of the on-line resources associated with the book, including all of the SBML and SBML-shorthand models, and the collection of chapter notes/links.

B.1 smfsb R package

The main software library associated with the book is the R package, smfsb. This library is available from CRAN, and hence should be possible to install on any internet-connected R installation with:

install.packages(`"smfsb"`)

This will also install a few required dependencies. The library is actually developed on R-Forge (`https://r-forge.r-project.org/projects/smfsb/`), and so the very latest version of this package can be installed with:

install.packages(`"smfsb"`,
    `repos="http://R-Forge.R-project.org"`)

should there be bug-fixes in the latest version not yet published to CRAN. See the book website for advice. Once installed, it should be possible to load the library at any time with:

library(`smfsb`)

It is documented in the usual way, and includes a vignette for getting started. This can be viewed with:

```
vignette("smfsb", package="smfsb")
```

Consult the vignette for further information relating to this package.

B.2 SBML-shorthand

SBML is the standard format for communication of biochemical network models. However, due to the verbosity of the format, this book has made extensive use of a short notation for SBML models known as SBML-shorthand. It is therefore convenient to have software tools for converting models from SBML-shorthand to SBML and back. Some Python scripts are available for this purpose, from the GitHub repo:

```
https://github.com/darrenjw/sbml-sh
```

These scripts require both Python and the Python libSBML bindings. See the libSBML website:

```
http://sbml.org/Software/libSBML
```

for further information on installing libSBML for Python, but note that on most Linux-based systems, it can be as simple as installing the package `python-libsbml` using the standard OS package manager (e.g., `apt-get` for Debian-based systems or `yum` for Fedora-based systems).

You can test that Python libSBML bindings are installed correctly by typing `import libsbml` at the Python REPL. If this returns silently, the bindings are probably correctly installed.

Once the Python libSBML bindings are correctly installed, you should just be able to copy the SBML-shorthand scripts to somewhere in your path. Then you should be able to convert the shorthand model `foo.mod` to the full SBML model `foo.xml` with an OS command like:

```
mod2sbml.py < foo.mod > foo.xml
```

See the SBML-shorthand website documentation links for further details.

B.3 smfsbSBML R package

`smfsbSBML` is a small R package for parsing an SBML file into an SPN object suitable for use with the `smfsb` R library. It consists mainly of the single function `sbml2spn`. It is not included in the main `smfsb` R package as it has a strong dependency on the `libSBML` R package, which, at the time of writing, is not in CRAN, and hence less straightforward to install. Consequently, the `smfsbSBML` package is also not on CRAN.

First install the `libSBML` R package following the latest information available from:

```
http://sbml.org/Software/libSBML
```

You can test that this package has been successfully installed by typing

library(libSBML)

at the R command prompt. If this returns silently, then the installation has probably worked.

Once you have the `libSBML` R package correctly installed, you can attempt to install the `smfsbSBML` package. Download the source package (link on the book website) and install from your OS command line with a command like:

```
R CMD INSTALL smfsbSBML_0.1.tar.gz
```

updating the version number as appropriate. You can check if the install has worked by typing

library(smfsbSBML)

at the R command prompt. If this returns silently, then the installation has probably worked. Then the command

```
spn = sbml2spn("foo.xml")
```

will parse the SBML model in the file `foo.xml` and return `spn`, an SPN object which can be passed into an `smfsb` function such as `StepGillespie` for forward simulation. Again, be sure to check the book website for the latest on this package.

B.4 scala-smfsb library

R is a nice language for analysing data and experimenting with numerical algorithms, but it is not a good language for developing efficient algorithms for scientific and statistical computing. It is therefore useful to provide examples of how to write code and algorithms in an efficient, compiled, strongly typed programming language. Scala is a convenient choice here, since it has good support for the functional programming style adopted for the R code in this text, and therefore allows a fairly direct port of the R `smfsb` package to Scala. The most commonly used version of Scala runs on the Java Virtual Machine (JVM), and therefore a JVM must be installed, and in fact, a full Java Development Kit (JDK) is preferable. On Linux and similar platforms, this can be as simple as installing a package such as `openjdk-8-jdk` or `openjdk-9-jdk` with the standard OS package manager, but on Windows it may be easier to install a JDK from Oracle:

`http://www.oracle.com/technetwork/java/javase/downloads`

Once a JDK is installed, the only other software needed is SBT, the Scala build tool. Follow the official installation instructions for your platform, available from the SBT website:

`https://www.scala-sbt.org/`

Once SBT is installed, you are ready to experiment with the `scala-smfsb` library. Although many new Scala users find it surprising, it is not necessary to do a system-wide installation of Scala in order to use the language, and there are some advantages to not doing so.

The library is developed on GitHub:

`https://github.com/darrenjw/scala-smfsb`

The GitHub website has the latest information on using the package, to which the reader is referred for further details. However, it should be noted that it is not necessary to build the library from source code, since the binaries are published onto 'The

Central repository' (https://search.maven.org/) which can be loosely considered to be the JVM equivalent of CRAN, and hence the library can be directly added to a SBT project via a simple one-line binary dependency. It should also be noted that despite the fact that Scala is a compiled, statically typed language, it can be used interactively in a very similar way to dynamic languages such as R. Further information for getting started with programming in Scala and using the library is provided at the scala-smfsb website, including a tutorial walk-through somewhat analogous to an R 'vignette'.

Bibliography

Abramowitz, M. and Stegun, I. A. (1984). *Pocketbook of Mathematical Functions*. Verlag Harri Deutsch, Frankfurt.

Alfonsi, A., Cances, E., Turinici, G., Di Ventura, B., and Huisinga, W. (2005). Adaptive simulation of hybrid stochastic and deterministic models for biochemical systems. *ESAIM: Proceedings*, 14:1–13.

Allen, L. J. S. (2011). *Stochastic Processes with Applications to Biology*. Pearson Prentice Hall, Upper Saddle River, New Jersey, second edition.

Ander, M., Beltrao, P., Di Ventura, B., Ferkinghoff-Borg, J., Foglierini, M., Kaplan, A., Lemerle, C., Tomas-Oliveira, I., and Serrano, L. (2004). SmartCell, a framework to simulate cellular processes that combines stochastic approximation with diffusion and localisation: analysis of simple networks. *IEE Systems Biology*, 1(1):129–138.

Anderson, D. F. (2007). A modified next reaction method for simulating chemical systems with time dependent propensities and delays. *The Journal of Chemical Physics*, 127(21):214107.

Anderson, D. F. and Kurtz, T. G. (2015). *Stochastic Analysis of Biochemical Systems*. Mathematical Biosciences Institute Lecture Series. Springer International Publishing.

Andrews, S. S., Addy, N. J., Brent, R., and Arkin, A. P. (2010). Detailed simulations of cell biology with Smoldyn 2.1. *PLoS Comput Biol*, 6(3):e1000705+.

Andrieu, C., Doucet, A., and Holenstein, R. (2009). Particle Markov chain Monte Carlo for efficient numerical simulation. In L'Ecuyer, P. and Owen, A. B., editors, *Monte Carlo and Quasi-Monte Carlo Methods 2008*, pages 45–60. Spinger-Verlag Berlin Heidelberg.

Andrieu, C., Doucet, A., and Holenstein, R. (2010). Particle Markov chain Monte Carlo methods (with discussion). *Journal of the Royal Statistical Society, Series B*, 72(3):269–342.

Andrieu, C. and Moulines, E. (2006). On the ergodicity properties of some adaptive MCMC algorithms. *The Annals of Applied Probability*, 16(3):1462–1505.

Andrieu, C. and Roberts, G. O. (2009). The pseudo-marginal approach for efficient Monte Carlo computations. *Annals of Statistics*, 37(2):697–725.

Arkin, A., Ross, J., and McAdams, H. H. (1998). Stochastic kinetic analysis of developmental pathway bifurcation in phage λ-infected *Escherichia coli* cells. *Genetics*, 149:1633–1648.

Ball, K., Kurtz, T. G., Popovic, L., and Rempala, G. (2006). Asymptotic analysis of multiscale approximations to reaction networks. *Annals of Applied Probability*, 16(4):1925–1961.

Beaumont, M. A. (2003). Estimation of population growth or decline in genetically monitored populations. *Genetics*, 164:1139–1160.

Beaumont, M. A., Zhang, W., and Balding, D. J. (2002). Approximate Bayesian computation in population genetics. *Genetics*, 162(4):2025–2035.

Beskos, A., Papaspiliopoulos, O., Roberts, G., and Fearnhead, P. (2006). Exact and computationally efficient likelihood-based estimation for discretely observed diffusion processes (with discussion). *Journal of the Royal Statistical Society: Series B (Statistical Methodology)*, 68(3):333–382.

Beskos, A. and Roberts, G. O. (2005). Exact simulation of diffusions. *Annals of Applied Probability*, 15(4):2422–2444.

Blossey, R., Cardelli, L., and Phillips, A. (2006). A compositional approach to the stochastic dynamics of gene networks. *Transactions on Computational Systems Biology IV*, pages 99–122.

Blum, M. G. B., Nunes, M. A., Prangle, D., and Sisson, S. A. (2013). A comparative review of dimension reduction methods in approximate Bayesian computation. *Statistical Science*, 28(2):189–208.

Bower, J. M. and Bolouri, H., editors (2000). *Computational Modeling of Genetic and Biochemical Networks*. MIT Press, Cambridge, Massachusetts.

Boys, R. J., Wilkinson, D. J., and Kirkwood, T. B. L. (2008). Bayesian inference for a discretely observed stochastic kinetic model. *Statistics and Computing*, 18(2):125–135.

Bundschuh, R., Hayot, F., and Jayaprakash, C. (2003). Fluctuations and slow variables in genetic networks. *Biophysical Journal*, 84(3):1606–15.

Burden, R. and Faires, J. (2010). *Numerical Analysis*. Cengage Learning.

Chaouiya, C. (2007). Petri net modelling of biological networks. *Briefings in Bioinformatics*, 8(4):210–9.

Chib, S. (1995). Marginal likelihood from the Gibbs output. *Journal of the American Statistical Association*, 90(432):1313–1321.

Ciocchetta, F. and Hillston, J. (2009). Bio-PEPA: A framework for the modelling and analysis of biological systems. *Theoretical Computer Science*, 410(33-34):3065–3084.

Cornish-Bowden, A. (2004). *Fundamentals of Enzyme Kinetics*. Portland Press, London, third edition.

Cotter, S. L., Zygalakis, K. C., Kevrekidis, I. G., and Erban, R. (2011). A constrained approach to multiscale stochastic simulation of chemically reacting systems. *The Journal of Chemical Physics*, 135:094102.

Cox, D. R. and Miller, H. D. (1977). *The Theory of Stochastic Processes*. Chapman and Hall, London.

Crawley, M. (2007). *The R Book*. Wiley, New York.

Cressie, N. (1993). *Statistics for Spatial Data*. Wiley, New York, second edition.

Crudu, A., Debussche, A., and Radulescu, O. (2009). Hybrid stochastic simplifications for multiscale gene networks. *BMC systems biology*, 3(1):1–25.

De Smet, R. and Marchal, K. (2010). Advantages and limitations of current network inference methods. *Nature Reviews Microbiology*, 8(10):717–729.

Del Moral, P. (2004). *Feynman-Kac formulae: genealogical and interacting particle systems with applications*. Springer series in statistics: Probability and its applications. Springer, New York.

Del Moral, P., Doucet, A., and Jasra, A. (2006). Sequential Monte Carlo samplers. *Journal of the Royal Statistical Society: Series B (Statistical Methodology)*, 68(3):411–436.

Devroye, L. (1986). *Non-uniform Random Variate Generation*. Springer Verlag, New York.

Doucet, A., de Freitas, N., and Gordon, N., editors (2001). *Sequential Monte Carlo Methods in Practice*. Springer, New York.

DuCharme, R. (1999). *XML: The Annotated Specification*. Prentice Hall PTR, Upper Saddle River, New Jersey.

Durham, G. B. and Gallant, R. A. (2002). Numerical techniques for maximum likelihood estimation of continuous time diffusion processes. *Journal of Business and Economic Statistics*, 20:279–316.

E, W., Liu, D., and Vanden-Eijnden, E. (2007). Nested stochastic simulation algorithms for chemical kinetic systems with multiple time scales. *Journal of Computational Physics*, 221(1):158–180.

Elf, J. and Ehrenberg, M. (2003). Fast evaluation of fluctuations in biochemical networks with the linear noise approximation. *Genome Research*, 13(11):2475–84.

Elf, J. and Ehrenberg, M. (2004). Spontaneous separation of bi-stable biochemical systems into spatial domains of opposite phases. *Systems Biology*, 1(2):230–236.

Elowitz, M. B., Levine, A. J., Siggia, E. D., and Swain, P. S. (2002). Stochastic gene expression in a single cell. *Science*, 297(5584):1183–1186.

Evans, T. W., Gillespie, C. S., and Wilkinson, D. J. (2008). The SBML discrete stochastic models test suite. *Bioinformatics*, 24:285–286.

Fearnhead, P. and Prangle, D. (2012). Constructing summary statistics for approximate Bayesian computation: semi-automatic approximate Bayesian computation. *Journal of the Royal Statistical Society: Series B (Statistical Methodology)*, 74(3):419–474.

Filippi, S., Barnes, C. P., Cornebise, J., and Stumpf, M. P. H. (2013). On optimality of kernels for approximate Bayesian computation using sequential Monte Carlo. *Statistical Applications in Genetics and Molecular Biology*, 12(1):87–107.

Flanagan, D. (2005). *Java in a Nutshell*. A Nutshell Handbook. O'Reilly, Sebastopol, CA.

Fuchs, C. (2013). *Inference for Diffusion Processes: With Applications in Life Sciences*. Springer Berlin Heidelberg.

Gamerman, D. and Lopes, H. (2006). *Markov chain Monte Carlo: Stochastic simulation for Bayesian inference*. Texts in statistical science. Taylor & Francis, Boca Raton, FL.

Gardiner, C. (2004). *Handbook of Stochastic Methods for Physics, Chemistry, and the Natural Sciences*. Springer series in synergetics. Springer, New York.

Gentle, J. (2003). *Random Number Generation and Monte Carlo Methods*. Statistics and computing. Springer, New York.

Gentle, J., Härdle, W., and Mori, Y. (2004). *Handbook of Computational Statistics: Concepts and Methods*. Springer, New York.

Ghosh, A., Leier, A., and Marquez-Lago, T. T. (2015). The spatial chemical Langevin equation and reaction diffusion master equations: moments and qualitative solutions. *Theoretical Biology and Medical Modelling*, 12(1):5.

Gibson, G. J. and Renshaw, E. (1998). Estimating parameters in stochastic compartmental models using Markov chain methods. *IMA Journal of Mathematics Applied in Medicine and Biology*, 15:19–40.

Gibson, M. A. and Bruck, J. (2000). Efficient exact stochastic simulation of chemical systems with many species and many channels. *Journal of Physical Chemistry A*, 104(9):1876–1889.

Gillespie, C. S. (2009). Moment-closure approximations for mass-action models. *IET Systems Biology*, 3(1):52–58.

Gillespie, D. T. (1976). A general method for numerically simulating the stochastic time evolution of coupled chemical reactions. *Journal of Computational Physics*, 22:403–434.

Gillespie, D. T. (1977). Exact stochastic simulation of coupled chemical reactions. *Journal of Physical Chemistry*, 81:2340–2361.

Gillespie, D. T. (1992a). *Markov Processes: An Introduction for Physical Scientists*. Academic Press, New York.

Gillespie, D. T. (1992b). A rigorous derivation of the chemical master equation. *Physica A*, 188:404–425.

Gillespie, D. T. (2000). The chemical Langevin equation. *Journal of Chemical Physics*, 113(1):297–306.

Gillespie, D. T. (2001). Approximate accelerated stochastic simulation of chemically reacting systems. *Journal of Chemical Physics*, 115(4):1716–1732.

Gillespie, D. T. and Petzold, L. R. (2003). Improved leap-size selection for accelerated stochastic simulation. *Journal of Chemical Physics*, 119(16):8229–8234.

Golightly, A. and Wilkinson, D. J. (2005). Bayesian inference for stochastic kinetic models using a diffusion approximation. *Biometrics*, 61(3):781–788.

Golightly, A. and Wilkinson, D. J. (2006a). Bayesian sequential inference for nonlinear multivariate diffusions. *Statistics and Computing*, 16:323–338.

Golightly, A. and Wilkinson, D. J. (2006b). Bayesian sequential inference for stochastic kinetic biochemical network models. *Journal of Computational Biology*, 13(3):838–851.

Golightly, A. and Wilkinson, D. J. (2008). Bayesian inference for nonlinear multivariate diffusion models observed with error. *Computational Statistics and Data Analysis*, 52(3):1674–1693.

Golightly, A. and Wilkinson, D. J. (2011). Bayesian parameter inference for stochastic biochemical network models using particle MCMC. *Interface Focus*, 1(6):807–820.

Golightly, A. and Wilkinson, D. J. (2015). Bayesian inference for Markov jump processes with informative observations. *Statistical Applications in Genetics and Molecular Biology*, 14(2):168–188.

Golub, G. H. and Van Loan, C. F. (1996). *Matrix Computations*. Johns Hopkins University Press, Baltimore, Maryland, third edition.

Goss, P. J. E. and Peccoud, J. (1998). Quantitative modeling of stochastic systems in molecular biology by using stochastic Petri nets. *Proceedings of the National Academy of Science USA*, 95:6750–6755.

Green, P. J. (1995). Reversible jump Markov chain Monte Carlo computation and Bayesian model determination. *Biometrika*, 82:711–732.

Grzegorczyk, M. and Husmeier, D. (2011). Improvements in the reconstruction of time-varying gene regulatory networks: dynamic programming and regularization by information sharing among genes. *Bioinformatics*, 27(5):693–699.

Haas, P. (2002). *Stochastic Petri nets: Modelling, Stability, Simulation*. Springer series in operations research. Springer, New York.

Hardy, S. and Robillard, P. N. (2004). Modeling and simulation of molecular biology systems using Petri nets: modeling goal of various approaches. *Journal of Bioinformatics and Computational Biology*, 2(4):619–637.

Harper, C. V., Finkenstädt, B., Woodcock, D. J., Friedrichsen, S., Semprini, S., Ashall, L., Spiller, D. G., Mullins, J. J., Rand, D. A., Davis, J. R. E., and White, M. R. H. (2011). Dynamic analysis of stochastic transcription cycles. *PLoS Biol*, 9(4):e1000607+.

Haseltine, E. L. and Rawlings, J. B. (2002). Approximate simulation of coupled fast and slow reactions for stochastic chemical kinetics. *Journal of Chemical Physics*, 117(15):6959–6969.

Hastings, W. K. (1970). Monte Carlo sampling methods using Markov chains and their applications. *Biometrika*, 57:97–109.

Hattne, J., Fange, D., and Elf, J. (2005). Stochastic reaction-diffusion simulation with MesoRD. *Bioinformatics*, 21(12):2923–2924.

Henderson, D. A., Boys, R. J., Krishnan, K. J., Lawless, C., and Wilkinson, D. J. (2009). Bayesian emulation and calibration of a stochastic computer model of mitochondrial DNA deletions in substantia nigra neurons. *Journal of the American Statistical Association*, 104(485):76–87.

Higham, D., Intep, S., Mao, X., and Szpruch, L. (2011). Hybrid simulation of autoregulation within transcription and translation. *BIT Numerical Mathematics*, pages 1–20.

Hlavacek, W. S., Faeder, J. R., Blinov, M. L., Posner, R. G., Hucka, M., and Fontana, W. (2006). Rules for modeling Signal-Transduction systems. *Sci. STKE*, 2006(344):re6+.

Hoops, S., Sahle, S., Gauges, R., Lee, C., Pahle, J., Simus, N., Singhal, M., Xu, L., Mendes, P., and Kummer, U. (2006). COPASI — a complex pathway simulator. *Bioinformatics*, 22(24):3067–3074.

Hucka, M., Finney, A., Sauro, H. M., Bolouri, H., Doyle, J. C., Kitano, H., Arkin, A. P., Bornstein, B. J., Bray, D., Cornish-Bowden, A., Cuellar, A. A., Dronov, S., Gilles, E. D., Ginkel, M., Gor, V., Goryanin, I. I., Hedley, W. J., Hodgman, T. C., Hofmeyr, J.-H., Hunter, P. J., Juty, N. S., Kasberger, J. L., Kremling, A., Kummer, U., Novere, N. L., Loew, L. M., Lucio, D., Mendes, P., Minch, E., Mjolsness, E. D., Nakayama, Y., Nelson, M. R., Nielsen, P. F., Sakurada, T., Schaff, J. C., Shapiro, B. E., Shimizu, T. S., Spence, H. D., Stelling, J., Takahashi, K., Tomita, M., Wagner, J., and Wang, J. (2003). The systems biology markup language (SBML): a medium for representation and exchange of biochemical network models. *Bioinformatics*, 19(4):524–531.

Iacus, S. M. (2008). *Simulation and inference for stochastic differential equations: with R examples*. Springer series in statistics. Springer, New York.

Ionides, E. L., Breto, C., and King, A. A. (2006). Inference for nonlinear dynamical systems. *Proceedings of the National Academy of Sciences*, 103:18438–18443.

Isaacson, S. A. (2008). Relationship between the reaction–diffusion master equation and particle tracking models. *Journal of Physics A: Mathematical and Theoretical*, 41(6):065003.

Isaacson, S. A. and Isaacson, D. (2009). Reaction–diffusion master equation, diffusion–limited reactions, and singular potentials. *Phys. Rev. E*, 80:066106.

Jahnke, T. and Huisinga, W. (2007). Solving the chemical master equation for monomolecular reaction systems analytically. *Journal of Mathematical Biology*, 54:1–26. 10.1007/s00285-006-0034-x.

Johnson, N. L. and Kotz, S. (1969). *Distributions in Statistics*, volume 1. Wiley, New York.

Kendall, D. G. (1950). An artificial realization of a simple "birth-and-death" process. *Journal of the Royal Statistical Society. Series B (Methodological)*, 12(1).

Kennedy, M. C. and O'Hagan, A. (2001). Bayesian calibration of computer models (with discussion). *Journal of the Royal Statistical Society, Series B*, 63:425–464.

Kiehl, T. R., Mattheyses, R. M., and Simmons, M. K. (2004). Hybrid simulation of cellular behavior. *Bioinformatics*, 20(3):316–322.

King, A. A., Ionides, E. L., and Bretó, C. M. (2008). *pomp: Statistical inference for partially observed Markov processes.*

Kingman, J. F. C. (1993). *Poisson processes.* Oxford science publications. Clarendon Press.

Kitano, H., editor (2001). *Foundations of Systems Biology.* MIT Press, Cambridge, Massachusetts.

Klipp, E., Herwig, R., Kowald, A., Wierling, C., and Lehrach, H. (2005). *Systems Biology in Practice: Concepts, Implementation and Application.* Wiley-VCH, New York.

Kloeden, P. E. and Platen, E. (1992). *Numerical Solution of Stochastic Differential Equations.* Springer, New York.

Koch, I., Reisig, W., and Schreiber, F. (2010). *Modeling in Systems Biology: The Petri Net Approach.* Computational Biology. Springer, New York.

Komorowski, M., Finkenstädt, B., Harper, C. V., and Rand, D. A. (2009). Bayesian inference of biochemical kinetic parameters using the linear noise approximation. *BMC Bioinformatics*, 10:343.

Kursawe, J., Baker, R. E., and Fletcher, A. G. (2018). Approximate Bayesian computation reveals the importance of repeated measurements for parameterising cell-based models of growing tissues. *Journal of Theoretical Biology*, 443:66–81.

Kurtz, T. G. (1972). The relationship between stochastic and deterministic models for chemical reactions. *The Journal of Chemical Physics*, 57(7):2976–2978.

Latchman, D. (2002). *Gene Regulation: A Eukaryotic Perspective.* Garland Science, fourth edition.

Le Novère, N., Bornstein, B., Broicher, A., Courtot, M., Donizelli, M., Dharuri, H., Li, L., Sauro, H., Schilstra, M., Shapiro, B., Snoep, J. L., and Hucka, M. (2006). BioModels database: a free, centralized database of curated, published, quantitative kinetic models of biochemical and cellular systems. *Nucleic Acids Research*, 34(suppl 1):D689–D691.

Le Novère, N. and Shimizu, T. S. (2001). STOCHSIM: modelling of stochastic biomolecular processes. *Bioinformatics*, 17(6):575–576.

Lee, M. W., Vassiliadis, V. S., and Park, J. M. (2009). Individual-based and stochastic modeling of cell population dynamics considering substrate dependency. *Biotechnology and Bioengineering*, 103(5):891–9.

Leimkuhler, B. and Matthews, C. (2013). Rational construction of stochastic numerical methods for molecular sampling. *Applied Mathematics Research eXpress*, 2013(1):34–56.

Lemerle, C., Di Ventura, B., and Serrano, L. (2005). Space as the final frontier in stochastic simulations of biological systems. *FEBS Letters*, 579(8):1789–1794.

Lewis, P. A. W. and Shedler, G. S. (1979). Simulation of non-homogeneous Poisson processes by thinning. *Naval Research Logistics Quarterly*, 26:403–413.

Lord, G. J., Powell, C. E., and Shardlow, T. (2014). *An Introduction to Computational Stochastic PDEs*. Cambridge University Press.

Lotka, A. J. (1925). *Elements of Physical Biology*. Williams and Wilkins, Baltimore.

Marjoram, P., Molitor, J., Plagnol, V., and Tavare, S. (2003). Markov chain Monte Carlo without likelihoods. *Proc. Natl. Acad. Sci. U.S.A.*, 100(26):15324–15328.

Matsumoto, M. and Nishimura, T. (1998). Mersenne twister: a 623-dimensionally equidistributed uniform pseudo-random number generator. *ACM Trans. Model. Comput. Simul.*, 8(1):3–30.

McAdams, H. H. and Arkin, A. (1997). Stochastic mechanisms in gene expression. *Proceedings of the National Acadamy of Science USA*, 94:814–819.

McAdams, H. H. and Arkin, A. (1999). It's a noisy business: genetic regulation at the nanomolecular scale. *Trends in Genetics*, 15:65–69.

McQuarrie, D. A. (1967). Stochastic approach to chemical kinetics. *Journal of Applied Probability*, 4:413–478.

Metropolis, N., Rosenbluth, A. W., Rosenbluth, M. N., Teller, A. H., and Teller, E. (1953). Equations of state calculations by fast computing machines. *Journal of Chemical Physics*, 21:1087–1092.

Miller, I. and Miller, M. (2004). *John E. Freund's Mathematical Statistics with Applications*. Pearson Prentice Hall, Upper Saddle River, New Jersey, seventh edition.

Milner, P., Gillespie, C. S., and Wilkinson, D. J. (2011). Moment closure approximations for stochastic kinetic models with rational rate laws. *Mathematical Biosciences*, 231(2):99–104.

Morgan, B. J. T. (1984). *Elements of Simulation*. Chapman & Hall/CRC Press, London.

Mugler, A., Walczak, A., and Wiggins, C. (2009). Spectral solutions to stochastic models of gene expression with bursts and regulation. *Physical Review E*, 80(4):1–19.

Murata, T. (1989). Petri nets: properties, analysis and applications. *Proceedings of the IEEE*, 77(4):541–580.

Murphy, K. (2012). *Machine Learning: A Probabilistic Perspective*. MIT Press.

Norris, J. R. (1998). *Markov chains*. Cambridge series on statistical and probabilistic mathematics. Cambridge University Press, Cambridge.

Odersky, M., Spoon, L., and Venners, B. (2008). *Programming in Scala*. Artima.

O'Hagan, A. and Forster, J. J. (2004). *Bayesian Inference*, volume 2B of *Kendall's Advanced Theory of Statistics*. Arnold, London.

Øksendal, B. (2003). *Stochastic Differential Equations: An Introduction with Applications*. Springer-Verlag, Heidelberg, sixth edition.

Owen, J., Wilkinson, D. J., and Gillespie, C. S. (2015a). Likelihood free inference for Markov processes: a comparison. *Statistical Applications in Genetics and Molecular Biology*, 14(2):189–209.

Owen, J., Wilkinson, D. J., and Gillespie, C. S. (2015b). Scalable inference for Markov processes with intractable likelihoods. *Statistics and Computing*, 25(1):145156.

Pahala Gedara, J., Gupta, P., Li, B., Madsen, C., Oyebamiji, O., Gonzalez-Cabaleiro, R., Rushton, S., Bridgens, B., Swailes, D., Allen, B., McGough, A. S., Zuliani, P., Ofiteru, I. D., Wilkinson, D. J., Chen, J., and Curtis, T. P. (2017). A mechanistic individual-based model of microbial communities. *PLoS ONE*, 12(8):e0181965.

Paszek, P., Ryan, S., Ashall, L., Sillitoe, K., Harper, C. V., Spiller, D. G., Rand, D. A., and White, M. R. H. (2010). Population robustness arising from cellular heterogeneity. *Proceedings of the National Acadamy of Sciences*, (107):11644–11649.

Perkins, T. and Swain, P. (2009). Strategies for cellular decision-making. *Mol Syst Biol*, 5:326.

Pfeifer, A., Kaschek, D., Bachmann, J., Klingmuller, U., and Timmer, J. (2010). Model-based extension of high-throughput to high-content data. *BMC Systems Biology*, 4(1).

Phillips, A. and Cardelli, L. (2007). *Efficient, Correct Simulation of Biological Processes in the Stochastic Pi-calculus*, volume 4695 of *Lecture Notes in Computer Science*, chapter 13, pages 184–199. Springer Berlin / Heidelberg.

Pinney, J. W., Westhead, D. R., and McConkey, G. A. (2003). Petri net representations in systems biology. *Biochemical Society Transactions*, 31(6):1513–1515.

Pitt, M. K., dos Santos Silva, R., Giordani, P., and Kohn, R. (2012). On some properties of Markov chain Monte Carlo simulation methods based on the particle filter. *Journal of Econometrics*, 171(2):134 – 151.

Proctor, C. J., Söti, C., Boys, R. J., Gillespie, C. S., Shanley, D. P., Wilkinson, D. J., and Kirkwood, T. B. L. (2005). Modelling the action of chaperones and their role in ageing. *Mechanisms of Ageing and Development*, 126(1):119–131.

Puchalka, J. and Kierzek, A. M. (2004). Bridging the gap between stochastic and deterministic regimes in the kinetic simulations of the biochemical reaction networks. *Biophysical Journal*, 86:1357–1372.

R Development Core Team (2005). *R: A Language and Environment for Statistical Computing*. R Foundation for Statistical Computing, Vienna, Austria. ISBN 3-900051-07-0.

Rao, C. V. and Arkin, A. P. (2003). Stochastic chemical kinetics and the quasi-steady-state assumption: application to the Gillespie algorithm. *Journal of Chemical Physics*, 118:4999–5010.

Reisig, W. (1985). *Petri Nets: An Introduction*. Monographs on Theoretical Computer Science. Springer, New York.

Rice, J. (2007). *Mathematical Statistics and Data Analysis*. Duxbury advanced series. Thomson/Brooks/Cole, Belmont, CA, third edition.

Ripley, B. D. (1987). *Stochastic Simulation*. Wiley, New York.

Robert, C. P., Cornuet, J.-M., Marin, J.-M., and Pillai, N. S. (2011). Lack of confidence in approximate Bayesian computation model choice. *Proceedings of the National Academy of Sciences*, 108(37):15112–15117.

Roberts, G. O. and Stramer, O. (2001). On inference for non-linear diffusion models using Metropolis-Hastings algorithms. *Biometrika*, 88(3):603–621.

Ross, S. M. (1996). *Stochastic Processes*. Wiley, New York.

Ross, S. M. (2009). *Introduction to Probability Models*. Academic Press, New York.

Salis, H. and Kaznessis, Y. (2005a). Accurate hybrid stochastic simulation of a system of coupled chemical or biochemical reactions. *Journal of Chemical Physics*, 122:054103.

Salis, H. and Kaznessis, Y. (2005b). Numerical simulation of stochastic gene circuits. *Computers and Chemical Engineering*, 29:577–588.

Samant, A., Ogunnaike, B. A., and Vlachos, D. G. (2007). A hybrid multiscale Monte Carlo algorithm (HyMSMC) to cope with disparity in time scales and species populations in intracellular networks. *BMC bioinformatics*, 8:175+.

Santillán, M., Mackey, M. C., and Zeron, E. S. (2007). Origin of bistability in the lac Operon. *Biophysical Journal*, 92(11):3830–3842.

Santner, T. J., Williams, B. J., and Notz, W. I. (2003). *The Design and Analysis of Computer Experiments*. Springer, New York.

Satir, G. and Brown, D. (1995). *C++: The Core Language*. O'Reilly Series. O'Reilly & Assoc., Inc., Sebastopol, CA.

Schnoerr, D., Grima, R., and Sanguinetti, G. (2016). Cox process representation and inference for stochastic reaction–diffusion processes. *Nature Communications*, 7:11729.

Schnoerr, D., Sanguinetti, G., and Grima, R. (2015). Comparison of different moment-closure approximations for stochastic chemical kinetics. *The Journal of Chemical Physics*, 143(18):185101.

Singh, A. and Hespanha, J. P. (2010). Stochastic hybrid systems for studying biochemical processes. *Physical and Engineering Sciences*, 368(1930):4995–5011.

Sisson, S. A., Fan, Y., and Tanaka, M. M. (2007). Sequential Monte Carlo without likelihoods. *Proc. Natl. Acad. Sci. U.S.A.*, 104(6):1760–1765.

Stramer, O. and Bognar, M. (2011). Bayesian inference for irreducible diffusion processes using the pseudo-marginal approach. *Bayesian Analysis*, 6(2):231–258.

Stryer, L. (1988). *Biochemistry*. Freeman, New York, third edition.

Swain, P. S., Elowitz, M. B., and Siggia, E. D. (2002). Intrinsic and extrinsic contributions to stochasticity in gene expression. *Proceedings of the National Acadamy of Sciences*, 99(20):12795–12800.

Szpruch, L. and Higham, D. (2010). Comparing hitting time behavior of Markov jump processes and their diffusion approximations. *Multiscale Modeling & Simulation*, 8(2):605.

Timmer, J., Henney, A., Moore, A., and Klingmuller, U. (2010). Systems biology of mammalian cells: A report from the Freiburg conference. *Bioessays*, 32(12):1099–1104.

Toni, T., Welch, D., Strelkowa, N., Ipsen, A., and Stumpf, M. P. H. (2009). Approximate Bayesian computation scheme for parameter inference and model selection in dynamical systems. *J. R. Soc. Interface*, 6(31):187–202.

Ullah, M. and Wolkenhauer, O. (2009). Investigating the two-moment characterisation of subcellular biochemical networks. *Journal of Theoretical Biology*, 260(3):340–352.

Van Kampen, N. G. (1992). *Stochastic Processes in Physics and Chemistry*. North-Holland.

Volterra, V. (1926). Fluctuations in the abundance of a species considered mathematically. *Nature*, 118:558–560.

W3C (2000). W3c math home.

Walczak, A., Mugler, A., and Wiggins, C. (2009). A stochastic spectral analysis of transcriptional regulatory cascades. *Proceedings of the National Academy of Sciences*, 106(16):6529.

Wallace, E. W. J., Gillespie, D. T., Sanft, K. R., and Petzold, L. R. (2012). Linear noise approximation is valid over limited times for any chemical system that is sufficiently large. *IET Systems Biology*, 6:102–115(13).

Wilkinson, D. J. (2005). Parallel Bayesian computation. In Kontoghiorghes, E. J., editor, *Handbook of Parallel Computing and Statistics*, pages 481–512. Marcel Dekker/CRC Press, New York.

Wilkinson, D. J. (2006). *Stochastic Modelling for Systems Biology*. Chapman & Hall/CRC Press, Boca Raton, Florida.

Wilkinson, D. J. (2009). Stochastic modelling for quantitative description of heterogeneous biological systems. *Nature Reviews Genetics*, 10:122–133. 10.1038/nrg2509.

Wilkinson, D. J. (2011). Parameter inference for stochastic kinetic models of bacterial gene regulation: a Bayesian approach to systems biology (with discussion). In Bernardo, J. M., editor, *Bayesian Statistics 9*, pages 679–706. Oxford Science Publications.

Wilkinson, D. J. and Golightly, A. (2010). Markov chain Monte Carlo algorithms for SDE parameter estimation. In Lawrence, N. D., editor, *Learning and Inference for Computational Systems Biology*, pages 253–275. MIT Press.

Wilkinson, D. J. and Yeung, S. K. H. (2002). Conditional simulation from highly structured Gaussian systems, with application to blocking-MCMC for the Bayesian analysis of very large linear models. *Statistics and Computing*, 12:287–300.

Wilkinson, D. J. and Yeung, S. K. H. (2004). A sparse matrix approach to Bayesian computation in large linear models. *Computational Statistics and Data Analysis*, 44:493–516.

Wilkinson, R. (2013). Approximate Bayesian computation (ABC) gives exact results under the assumption of model error. *Statistical Applications in Genetics and Molecular Biology*, 12(2):129–141.

Index

ABC, *see* approximate Bayesian
  computation
analysis of simulation output, 208
approximate Bayesian computation, 299,
  335
AR(1), 134
associative law, 54
auto-regulation, 16
  stochastic, 214

Bayes factor, 344
Bayes' theorem, 62
Bayesian
  inference, 273
  model calibration, 352
  network (dynamic), 352
binomial distribution, 73
Bio-PEPA, 37
bipartite graph, 24
birth–death process, 5, 234
bootstrap particle filter, 320
Box–Muller method, 107
Brownian motion, 154
  time-change, 162

C programming language, 332
C++ programming language, 332
CDF, *see* cumulative distribution function
cell population modelling, 218
central limit theorem, 94
Chapman–Kolmogorov equations, 127, 139
Chebyshev's inequality, 83
chemical kinetics, 173
chemical Langevin equation, 231
  inference, 314
chemical master equation, 196, 233
chemical reactions, 8
CIR process, *see* Cox–Ingersoll–Ross
  process

CLE, *see* chemical Langevin equation
CLT, *see* central limit theorem
coefficient of variation, 98
commutative law, 54
compartments, 40
complement, 54
complete data, 304
  likelihood, 305
concentration, 173
conservation law, 33
continuous probability models, 79
counting process, 147
coupled reactions, 23
Cox–Ingersoll–Ross process, 165
cumulative distribution function, 65
  continuous, 80

degradation, 15
DeMorgan's laws, 55
density function, 79
deSolve, 180
detailed balance, 130, 135
diffusion, 251
  approximations, 155
  bridge, 317
  inference, 314
  Kolmogorov backward equation, 159
  Kolmogorov forward equation, 160
  process, 154
dimensionality reduction, 211
dimer, 9
dimerisation kinetics
  deterministic, 177
  stochastic, 205
directed graph (digraph), 24
discrete probability models, 64
disjoint, 54
disjoint union, 55
dispersion index, 98
distribution